Mathematical Modeling, Simulation, Visualization and e-Learning

Dialla Konaté
Editor

Mathematical Modeling, Simulation, Visualization and e-Learning

Proceedings of an International Workshop held at Rockefeller Foundation's Bellagio Conference Center, Milan, Italy, 2006

 Springer

Dialla Konaté
Department of Mathematics
Virginia Tech
McBryde Hall #460
Blacksburg, VA 24060
USA
dkonate@math.vt.edu

ISBN: 978-3-540-74338-5   e-ISBN: 978-3-540-74339-2

Library of Congress Control Number: 2007938161

Mathematics Subject Classification Numbers (2000): 34, 35, 41, 42, 47, 49, 65, 68, 70, 74, 76, 92, 97

© 2008 Springer-Verlag Berlin Heidelberg

Cover design: WMX Design GmbH, Heidelberg

Printed on acid-free paper

9 8 7 6 5 4 3 2 1

springer.com

Preface

The **Bellagio International Conference on Mathematical Modeling, Simulation, Visualization and e-Learning** was held from November 20 to November 26, 2006, with primary financial support from the Rockefeller Foundation. It also benefited from the financial and material support of Virginia Tech and Winston-Salem State University.

Taken together mathematical modeling, simulation, and visualization have become a major tool in scientific investigation. They are also a growing interdisciplinary subject in collegiate curricula, and are expected to attract more students from underserved communities to science and technology because of their universality and because they provide a visual aspect to what is taught and learned, even in environments where physical implementation and experimentation may not be possible.

The *Bellagio Initiative*, which was the underlying idea behind the Bellagio International Conference, consists in creating a forum where excellent mathematicians from underserved countries and communities and the world's leading scientists can meet, work together and build networks in order to share scientific knowledge. This requires that all contributions presented in such a forum have a high scientific level and a self-contained style. The objective of this forum is that skilled scientists from all groups and countries be incorporated within the mainstream of scientific discovery.

It is our goal that scientists from all around the world be welcomed to aid in investigating the great challenges our world is facing. These challenges include, among others, the spread of infectious diseases, the management of water resources, and efforts to reduce air and water pollution. Mathematical modeling and simulation is among the paths which may lead to a better understanding of complex phenomena like those known but not fully understood in biological and chemical processes.

In this volume the reader will find excellent contributions from some very well known and other less known scientists. We have made a conscious decision not to focus on one specific area but to open the forum to a wider set of problems. This provides an opportunity for more scientists to come in and be part of the network. An important condition we have imposed on all the contributions to this volume is that each presents a collection of open questions which can fuel undergraduate or graduate research activities, even in smaller or more isolated scientific communities.

This volume is divided into four main sections devoted to:

- numerical methods and problem solving;
- modeling and control of phenomena;
- simulation and visualization of processes and phenomena;
- e-learning.

The dispersion of the different contributions between these four sections is purely arbitrary. The strong link between these four sections is the power they can, together, provide to scientific investigation.

We are very grateful to Springer-Verlag, which has agreed to partner with the *Bellagio Initiative* to publish the current book and be part of this important effort. This book is sold at a special discounted price which does not include any remuneration for authors or institutions. This supports our objective, which is to allow scientists from all countries to have better access to scientific knowledge and actively contribute to scientific discovery. This is why I am honored and proud to introduce to you the present volume which I hope, as the first of a series of publications to come, belongs on scientists, applied mathematicians, and engineers' bookshelves.

Editing a book written by many authors on different continents is a difficult task. So in closing, I would like to thank everyone who has supported me while I was coordinating the work of publishing this volume. I would also like to apologize to my wife and my children for the time I should have spent and missed spending with them.

Blacksburg, August 2007   Dialla Konaté

Names of Officers to the Conference

Professor Dialla Konate: Director and P.I (dkonate@vt.edu or konated@wssu.edu)

Organizing Committee

Professors William Greenberg (Virginia Tech, USA), Dialla Konate (Winston-Salem State University, USA); Oluwole Makinde (University of Limpopo, South Africa); Ousseynou Nakoulima (University of Antilles & Guiana, France & the Caribbean); Eitan Tadmor (University of Maryland, College Park); Oleg I. Yordanov (Bulgarian Academy of Science, Budapest, Bulgaria).

Team Leaders and Invited Speakers

Omrane Abdennebi (University of Antilles & Guiana, France & the Caribbean); John Georgiadis (University of Illinois at Urbana Champaign, USA), David Kinderlehrer (Carnegie-Mellon University, USA); Reinhard Laubenbacher (Virginia Tech, USA), Tashakkori Rahman (Appalachian State University, USA); Roger Temam (University of Indiana, USA); Michael Renardy (Virginia Tech, USA).

Contents

Part IV e-Learning

Part I

Numerical Methods and Problem Solving

Quasi-Analytical Computation of Energy Levels and Wave Functions in a Class of Chaotic Cavities with Inserted Objects

F. Seydou, O.M. Ramahi, and T. Seppänen

Summary. A simple multipole expansion method for analytically calculating the energy levels and the corresponding wave functions in a class of chaotic cavities is presented in this work. We will present results for the case when objects, which might be perfect electric conductors and/or dielectrics, are located inside the cavity. This example is demonstrative of typical experiments used in chaotic cavities to study the probabilistic eigenvalue distribution when objects are inserted into the cavity.

1 Introduction

In the relatively new field of quantum chaos [11], billiards have been of great interest. The simplest billiard prototype is defined as a perfectly conducting enclosure filled with a homogeneous medium. It is known that integrable systems (which have the same number of constants of motion as their dimension), such as billiards with regular shape, are non-chaotic, whereas non-integrable systems (with fewer constants of motion than their dimensionality), such as generic billiards, are chaotic [1, 18]. Examples of regular billiards include rectangular, circular, and elliptical geometries. Generic billiards are one of the simplest examples of conservative dynamical systems with chaotic classical trajectories. Examples of chaotic billiards include the stadium billiard, the D-shape truncated circle, the Sinai billiard and the bowtie shape billiard. There are also examples of geometries that are neither chaotic nor integrable. For more details about all these geometries we refer to [7] and the references therein.

In general, billiard systems of many kinds have proven to be fruitful in modeling systems in the field of quantum physics and electronics. Indeed, quantum billiards, coupled with advances in crystal growth and lithographic techniques have made it possible to produce very small and clean devices, known as nanodevices [3], which at sufficiently low temperatures, could be regarded as an experimental realization of a quantum billiard.

Since some nanodevices are governed by single-particle physics, it can be described by solving the time-independent Schrödinger equation, subject to some properly chosen boundary conditions, the simplest example of which is the problem of a particle in an infinite potential well. This amounts to solving an eigenvalue problem. These eigenfunctions (the wave functions) are important for understanding the behavior of mesoscopic structures, and will be crucial for the design of the nanoscale electronic devices. In addition, a great deal of interesting physics can be explored by means of an understanding of the behavior of the wave functions (stationary states) in irregular shaped devices, including wave localization and wave function fluctuations.

Since the Schrödinger equation for a free particle assumes the form of the well-known Helmholtz equation the problem of determining the stationary states of the particle in the infinite well amounts to the calculation of the Helmholtz eigenvalues and eigenfunctions for Dirichlet boundary conditions along the boundary of the well. This analogy between the Schrödinger and Helmholtz equations also allows to compare the obtained numerical results with the experimentally determined eigenmodes of a vibrating membrane, or the resonant modes of the oscillating electromagnetic field in a resonant cavity, of the same shape as the billiard.

The degree of difficulty in solving the Helmholtz eigenvalue problem depends on the actual shape of the infinite well (the quantum billiard system). When the shape of the billiard is highly regular, such as square, rectangular, or circular, then the problem can be solved by means of separation of variables [18]. The problem of determining the stationary states of a quantum billiard, with arbitrary shape, cannot be solved analytically. Instead, a tedious and costly numerical calculation is typically expected.

The matrix diagonalization method [12, 13] is a typical method for finding the stationary states. But, this strategy is inherently limited, and cannot be used for the purpose of finding high-lying eigenstates [6]. Moreover, the method requires a heavy numerical task due to 2D grid calculations.

One of the most popular strategies for numerical solution of the quantum billiard problem is the plane-wave decomposition method (PWDM), [14, 24]. The scaling method based on the PWDM [20–22] has also been efficiently used for convex billiards. While the PWDM technique is found to be extremely efficient in practice, for non-convex billiards, in general, it is not efficient [10].

The main useful numerical strategies that have been suggested in the literature in order to find the eigenstates of the Helmholtz equation, for arbitrary billiards, are based on boundary integral approach and often referred to as boundary integral method (BIM) [23]. Several versions of the BIM exist depending on the choice of basis functions [6, 19]. While the BIM equation is exact, its convergence is, in general, very slow for non-smooth boundaries. Moreover a meshing of the boundary is required and singularities have to be dealt with appropriately.

Most of the techniques considered in the literature are based on billiards filled with homogeneous medium whereas much less is reported when objects are included in the billiard, which is demonstrative of typical experiments used in chaotic cavities to study the probabilistic eigenvalue distribution [26].

The aim of this paper is to present a quasi-analytical method for calculating the stationary states in a large and important class of quantum billiards described by closed cavities (containing non-intersecting microcavities) formed by cylinders that are tangent to each other. A generic description of such cavities is shown in Fig. 1, where we show a cavity described by five circular cylinders of varying radii. Notice that the cylinders in Fig. 1 are touching but *not* intersecting. A familiar chaotic cavity that falls under this category is the so-called bowtie cavity, shown in Fig. 2, which is formed by four circular cylinders of equal or varying radii.

The paper is organized as follows. We start with the problem formulation. We then describe the method in the subsequent sections. Finally we present the numerical results and summarize our findings.

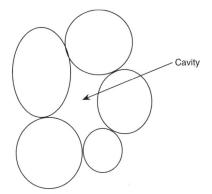

Fig. 1. A cavity made of five touching circles

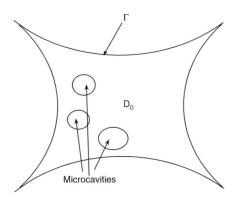

Fig. 2. Geometry of the bowtie shaped cavity containing microcavities

2 Theory

2.1 Formulation of the Problem

We consider a two-dimensional cavity D_0 with boundary Γ and a constant refractive index $n_0 = \sqrt{\epsilon_0 \mu_0}$, where μ_0 is the permittivity and ϵ_0 is the permeability. We denote the wave number by $k_0 = \omega n_0$, where ω is the frequency. Furthermore we assume that the cavity contains M parallel microcavities D_j, which might be perfect electric conductors (PEC) and/or dielectrics of constant wave numbers k_j, $j = 1, \ldots, M$. See Fig. 2.

The solution of Maxwell's equations for the vortices of the electromagnetic field is given in [4], and leads to an equation for the electric (magnetic) field that is similar to the conventional Schrödinger equation. The vector character of the fields implies, however, that one has to distinguish two possible polarization directions with differing boundary conditions. The situation where the electric (magnetic) field is parallel to the cylinder (z) axis is called TM (TE) polarization, with the magnetic (electric) field being thus transverse. The total field can be characterized by a single scalar function u, which represents either the E_z (H_z) component for the case of TM (TE) polarization.

We then have $M + 1$ homogeneous regions D_j, and in each region the function u is written as follows:

$$u = \xi_j u^0 + u_j,$$

where u^0 either represents the field generated by sources or should be assumed zero if eigenfrequencies are being sought, ξ_j is equal to unity if the source is located in D_j and zero otherwise. Assuming a TM polarization with perfectly conducting walls and a non-magnetic medium, $e^{ik_z z}$ dependence leads to the Helmholtz equation

$$\left(\nabla^2 + k^2 + k_z^2 \right) u = 0$$

with a vanishing u on Γ. On the boundaries of D_j, $j = 1, \ldots, M$, we have a continuity of u and its normal derivative for the case of dielectrics and a homogeneous Dirichlet boundary condition, i.e., $u = 0$ for the case of PEC. Here k_z is the propagation constant along the z-direction and k represents k_j in D_j, $j = 0, 1, \ldots, M$. In the following we will set k_z to zero corresponding to a wave in the $x - y$ plane.

The quantum analogue of the Helmholtz equation is given by time-independent Schrödinger equation for two-dimensional systems [8], i.e.,

$$\left(\nabla^2 + 2m \frac{\epsilon_n - V}{h^2} \right) \Psi_n = 0,$$

which describes the nth excited eigenfunction Ψ_n of a particle of mass m in a potential V, where ϵ_n is the energy of the nth excited state, and h is Planck's constant divided by 2π. The probability density $|\Psi_n|^2$ in the Schrödinger equation for the quantum mechanical problem is analogous to $|u|^2$ in the Helmholtz equation for the electromagnetic problem, as long as the $2m(\epsilon_n - V)/h^2$ and k^2 terms in these equations are constant. The solutions to the Helmholtz equation with perfectly conducting walls are equivalent to the solutions of the two-dimensional Schrödinger equation with hard wall boundaries ($\Psi_n = 0$ at the boundary Γ) of the same geometry.

Now, assume the cavity of interest is has a shape such as the geometry given in Fig. 1 (with inclusions of microcavities) or Fig. 2, and denoted by D_0, we would like to find:

1. The values of $|u_0|$ inside the cavity;
2. Non-trivial eigenvalues k_n and eigenfunctions $u_{0,n}$ that satisfy the two-dimensional Helmholtz equation with homogeneous Dirichlet boundary condition, as formulated above.

2.2 Cavities Formed by Touching Multiple Cylinders

The quasi-analytical approach (QAA) method introduced here, for calculating the energy spectrum of billiards, is simple and allow us determine both the energy levels and the corresponding wave functions for quantum billiards formed by touching circular cylinders.

To tackle the problem, we take as center points O_j, and circumscribe them by circles Γ_j (describing the cross-sections of the cylinders) of radii $R_j, j = 1, 2, \ldots, J$, where J denotes the number of circles, and separated by a variable a, as shown in Fig. 3. Next, we assume a scattering of electromagnetic waves by J infinitely long circular cylinders centered at O_j with radius R_j. Then, in the outer region D_a, we have the Helmholtz equation

$$\left(\nabla^2 + k^2\right) u_a = 0,$$

where u_a satisfies the Dirichlet boundary condition on each of the boundaries. When we take the limit as a goes to zero we obtain a cavity formed by touching

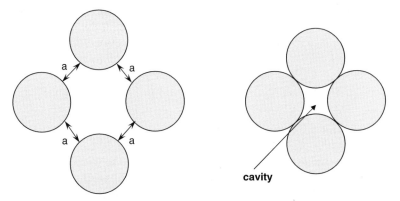

Fig. 3. Four circles separated by a variable a. The bowtie shape cavity is obtained when we let a tend to zero

circles (see Fig. 3). Our goal is to find the solution in the cavity, denoted by D_0, as an approximation of the solution in D_a when a goes to zero. The electromagnetic problem we have to solve is thus to find the solution u_a of the Helmholtz equation in D_a with Dirichlet boundary conditions on the circles. Then, when a goes to zero, we find the solution in the cavity.

Assuming that u_a satisfies the Sommerfeld radiation condition, and using the Green's representation theorem [5], we have

$$u_a(\mathbf{x}) = -\int_{\cup_{j=1}^J \Gamma_j} \left(\Phi(\mathbf{x}, \mathbf{y}) \frac{\partial u_a(\mathbf{y})}{\partial \nu} - u_a(\mathbf{y}) \frac{\partial \Phi(\mathbf{x}, \mathbf{y})}{\partial \nu} \right) ds(\mathbf{y}),$$

with

$$\Phi(\mathbf{x}, \mathbf{y}) = \frac{i}{4} H_0^{(1)}(k|\mathbf{x} - \mathbf{y}|),$$

where ν is the unit outward normal and $H_0^{(1)}$ is the Hankel function of the first kind and order zero.

We will use the addition theorem for Hankel function (cf. [4], p. 591)

$$H_m^{(1)}(k|\mathbf{x} - \mathbf{y}|)e^{im\theta''(\mathbf{x})} = \sum_{n=-\infty}^{\infty} e^{in\theta(\mathbf{x})} J_n^B[kr(\mathbf{x})] H_{n-m}^{(1)}[kr(\mathbf{y})]e^{i(m-n)\theta(\mathbf{y})},$$

when $r(\mathbf{y}) \geq r(\mathbf{x})$, and

$$H_m^{(1)}(k|\mathbf{x} - \mathbf{y}|)e^{im\theta''(\mathbf{x})} = \sum_{n=-\infty}^{\infty} e^{in\theta(\mathbf{x})} H_n^{(1)}[kr(\mathbf{x})] J_{n-m}^B[kr(\mathbf{y})]e^{i(m-n)\theta(\mathbf{y})},$$

when $r(\mathbf{y}) < r(\mathbf{x})$, where $r(\mathbf{x})$ and $\theta(\mathbf{x})$ are the polar coordinates of $\mathbf{x}$, θ'' is the polar angle of $R_{xy} = \mathbf{x} - \mathbf{y}$, $H_m^{(1)}$ is the Hankel function of the first kind and order m, and J_p^B is the Bessel function of order p. Hence, for a point $\mathbf{x} \in D_a$, $u_a(\mathbf{x})$ can be written in the form

$$u_a(\mathbf{x}) = \sum_{j=1}^J \sum_{n=-\infty}^{\infty} b_n^j H_n^{(1)}[kr_j(\mathbf{x})]e^{in\theta_j(\mathbf{x})}$$

with

$$b_n^j = \begin{array}{l} -\int_{\Gamma_j} \frac{\partial}{\partial \nu(\mathbf{y})} \{ J_n^B[kr_j(\mathbf{y})]e^{-in\theta_j(\mathbf{y})} \} u_a(\mathbf{y}) \, ds(\mathbf{y}) \\ -\int_{\Gamma_j} J_n^B[kr_j(\mathbf{y})]e^{-in\theta_j(\mathbf{y})} \frac{\partial}{\partial \nu(\mathbf{y})} u_a(\mathbf{y}), \end{array}$$

where $r_j(\mathbf{x})$ and $\theta_j(\mathbf{x})$ are the polar coordinates of a point $\mathbf{x} \in D_a$ in the coordinate system $X_j O_j Y_j$, and the constants b_n^j, $j = 1, \ldots, J$ and $-\infty \leq n \leq \infty$, are the unknown coefficients to be found by the boundary conditions.

When M circular objects D_j of radii r_j, $j = J+1, \ldots, J+M$, are inserted inside the cavity formed by the touching circles it is clear that in the sum above we have to replace J by $J + M$. Let us define k in D_j by k_j whereas in the rest of the cavity k is taken to be k_0.

Using again the addition theorem above and the boundary conditions we obtain an infinite system of equations. This system must be truncated, at a number N, to obtain a finite system of equations in the form

$$\mathbf{Sb} = \mathbf{0}.$$

The vector $\mathbf{b}$ contains the variables b_n^j, $j = 1, \ldots, J, J + 1, \ldots, J + M$ and $-N \leq n \leq N$, and the matrix $\mathbf{S}$ is given by

$$\mathbf{S} = \begin{pmatrix} \mathbf{s}_1 & \mathbf{T}_{1,2} & \cdots & \mathbf{T}_{1,J} & \mathbf{T}_{1,J+1} & \cdots & \mathbf{T}_{1,J+M} \\ \mathbf{T}_{2,1} & \mathbf{s}_2 & \cdots & \mathbf{T}_{2,J} & \mathbf{T}_{2,J+1} & \cdots & \mathbf{T}_{2,J+M} \\ \cdots & \cdots & \cdots & \cdots & \cdots & \cdots & \cdots \\ \mathbf{T}_{J,1} & \mathbf{T}_{J,2} & \cdots & \mathbf{s}_J & \mathbf{T}_{J,J+1} & \cdots & \mathbf{T}_{J,J+M} \\ \cdots & \cdots & \cdots & \cdots & \cdots & \cdots & \cdots \\ \mathbf{T}_{J+M,1} & \mathbf{T}_{J+M,2} & \cdots & \mathbf{T}_{J+M,J} & \mathbf{T}_{J+M,J+1} & \cdots & \mathbf{s}_{J+M} \end{pmatrix}.$$

In the above matrix, for $j, l = 1, \ldots, J + M$ and $-N \leq n, m \leq N$, $\mathbf{T}_{l,j}$ is a square matrix of the (m, n)th element $T_{l,j,m,n}$ given by

$$T_{l,j,m,n} = e^{i(n-m)\theta_l^j} H_{m-n}^{(1)}(k_0 r_l^j),$$

with r_l^j being the distance between O_l and O_j, and θ_l^j is the angle between $O_l O_j$ and the x-axis. On the other hand $\mathbf{s}_j$ is the column matrix of the mth element $s_{j,m}$ whose expression depends on the type of object inserted into the cavity. In particular, let us assume that we have K_1 PECs and K_2 dielectrics, where $K_1 + K_2 \leq M$, Then for the PEC case $s_{j,m}$ is given by

$$s_{j,m} = \frac{H_m^{(1)}(k_0 r_j)}{J_m(k_0 r_j)}, \quad j = 1, \ldots, J + K_1 \text{ and } -N \leq m \leq N,$$

whereas for dielectric insertions the values of $s_{j,m}$ differ from the PEC case for $j > J$ in which case it is given by

$$s_{j,m} = -\frac{k_0 J_m'(k_0 r_j) J_m(k_j r_j) - k_j J_m'(k_j r_j) J_m(k_0 r_j)}{k_0 H_m^{(1)'}(k_0 r_j) J_m(k_j r_j) - k_j J_m'(k_j r_j) H_m^{(1)}(k_0 r_j)}.$$

By solving the system of equations, for $j, l = 1, \ldots, J + M$ and $-N \leq n, m \leq N$, we obtain an approximation of the coefficients b_n^j. Then the field u_a can be recovered from the above sum.

This method is a semi-analytical technique. In some sense it is analytical since it is based on solutions in the form of infinite series. At the same time the method is numerical, since it requires inversion of a matrix for determining coefficients in the series.

To validate our method, we develop a numerical solution using the BIM [5]. Using the BIM, we seek solutions of the Helmholtz equation in the cavity as a combination of layer potentials and Green's identity. Then a Nyström discretization of the integral with singularity extraction [5] is used. More details of this method will be published elsewhere in a near future.

2.3 Eigenvalues

Eigenvalues play a critical role in the analysis of chaotic cavities. Determining whether a cavity is chaotic or not is made possible by analyzing the statistical distribution of the eigenvalues. Determining the eigenvalues reduces to finding the minima of either the tension, the smallest singular value of $\mathbf{S}$, or the determinant of $\mathbf{S}$, $det(\mathbf{S})$ [18]. In this paper we consider only the case of $det(\mathbf{S})$. Strictly speaking, to find the eigenvalues in a cavity, we need to look for wave numbers k such that $det(\mathbf{S}) = \mathbf{0}$. We note that the function $\hat{D}(k) = det(\mathbf{S})$ is complex-valued and, therefore, its real roots k_n (the sought eigenvalues) must be simultaneously zeros both real and imaginary parts. On the other hand we expect the numerical solutions of the equation $\hat{D} = 0$ to be complex with a small imaginary part.

To find the eigenvalues, i.e., the roots of $\hat{D} = 0$ we may use an iterative computational technique. We may use a Newton Raphson iterative solver. The eigenvalues can be computed using the following iteration:

1. Pick an initial value $k_n^{(0)}$
2. Iterate

$$k_n^{(m+1)} = k_n^{(m)} - \frac{1}{tr\left(\mathbf{S}^{-1}(k_n^{(m)})\mathbf{S}'(k_n^{(m)})\right)} \quad m = 0, 1, 2, \ldots,$$

where tr is the trace of a matrix and the derivative is taken with respect to $k_n^{(m)}$.

Once the eigenfrequency is determined, we may get the eigenfunctions $u_{a,n}$ by the inverse power method, i.e., suppose $k_n^{(m_0)}$ is a good approximation, but $\mathbf{S}(k_n^{(m_0)})$ is still invertible we iterate by

$$\mathbf{S}(k_n^{(m_0)})u_{a,n}^{(m+1)} = u_{a,n}^{(m)} \quad m = 0, 1, 2, \ldots.$$

In our numerical results below we look for the eigenvalues over a segment of wave number. Theoretically, if the wave number is an eigenvalue, then the minima of $|\hat{D}|$ should be zero. Eigenvalues are seen as peaks in the figures where we plot $|\hat{D}|$ against the wave number.

3 The Numerical Results of the Classical Bowtie Cavity Containing Circular Objects

The bowtie cavity (Fig. 2) is well known in the chaos literature [8]. It is realized by forming four circular cylinders of equal radius r_0 (generalized bowtie cavities can also be formed by cylinders with different radii).

In the numerical results presented here, we first excite the cavity by a complex-source-point (CSP) given by

$$\Phi_s(\mathbf{x}, \mathbf{x}_s) = H_0^{(1)}(k|\mathbf{x} - \mathbf{x}_s|)$$

located at a point $\mathbf{x}_s$ inside the cavity. The CSP is useful in the spectra study of cavities [2]. We then compare the QAA and the BIM methods.

Taking $a = 10^{-6}$, for a given point source located in the cavity above the x-axis we plot $|u_0|$ against the x-axis inside the cavity using the BIM and the QAA methods, for different wave numbers k_0 when $k_1 = 2k_0$ and $k_2 = k_0/2$. The results are reported in Figs. 4 and 5 for the cases when the bowtie cavity contains two PECs and two dielectrics, respectively. We see a very good match for the two methods.

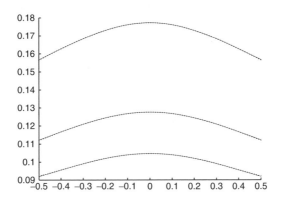

Fig. 4. Given a source inside the cavity placed at $(0, 0.6)$ and two circular PECs with radius 0.1 centered at $(0, 0.3)$ and $(0.1, -0.2)$ we plot the field $|u_0|$ on the line $y = 0$ using the BIM (*solid line*) and the QAA (*dots*) techniques for different wave numbers $k_0 = 2$ (*first from the top*), $k_0 = 4$ (*second from the top*) and $k_0 = 4$ (*bottom*). In all the figures $k_1 = 2k_0$ and $k_2 = k_0/2$. An excellent match is observed

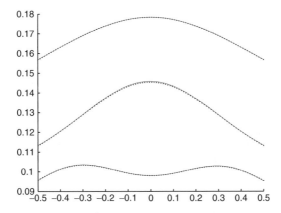

Fig. 5. Given a source inside the cavity placed at $(0, 0.6)$, one dielectric and one PECs with radius 0.1 centered at $(0, 0.3)$ and $(0.1, -0.2)$ we plot the field $|u_0|$ on the line $y = 0$ using the BIM (*solid line*) and the QAA (*dots*) techniques for different wave numbers $k_0 = 2$ (*first from the top*), $k_0 = 4$ (*second from the top*) and $k_0 = 4$ (*bottom*). In all the figures $k_1 = 2k_0$ and $k_2 = k_0/2$. An excellent match is observed

To see the location of the eigenvalues, in Figs. 6–10, we plot $log(|\hat{D}|)$ against the frequency using the QAA and the BIM methods. First we see that the determinant is very small in both cases. On the other hand, although the determinant does not have the same values for the two methods we see that the minima have the same locations. We would therefore expect to have very similar eigenvalues for the two approaches.

Let us consider the case when one circular object, that might be either a PEC or a dielectric with refractive index n is located inside the cavity. We shall look for two cases: when the object is located at the center of the cavity and when it is off the center at the point $(0, 0.6)$.

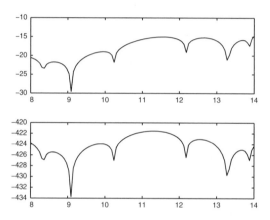

Fig. 6. Here the logarithm of the absolute value of the determinant is plotted against the wave number k_0 for the case when a perfect electric conducting circular cylinder is placed at the center of the bowtie shaped cavity, using the QAA (*top*) and the BIM (*bottom*) techniques

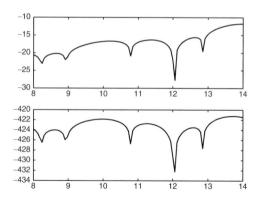

Fig. 7. Here the logarithm of the absolute value of the determinant is plotted against the wave number k_0 for the case when a perfect electric conducting circular cylinder is placed off the center of the bowtie shaped cavity, using the QAA (*top*) and the BIM (*bottom*) techniques

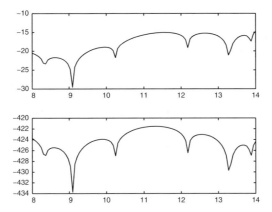

Fig. 8. Here the logarithm of the absolute value of the determinant is plotted against the wave number k_0 for the case when a dielectric circular cylinder, with refractive index $n = 4$, is placed at the center of the bowtie shaped cavity, using the QAA (*top*) and the BIM (*bottom*) techniques

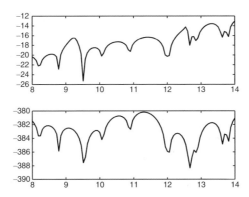

Fig. 9. Here the logarithm of the absolute value of the determinant is plotted against the wave number k_0 for the case when a dielectric circular cylinder, with refractive index $n = 4$, is placed off the center of the bowtie shaped cavity, using the QAA (*top*) and the BIM (*bottom*) techniques

First, for the case of PEC (Figs. 6 and 7) we observe that the number of eigenvalues is reduced when the object is moved from the center compared to when it is placed at the center. Turning to the case when the object inside the cavity is a dielectric (Figs. 8 and 9) we assume that the refractive index is taken as $n = 4$. We see that there are more eigenvalues when the dielectric cylinder is moved off the center compared to when it is placed at the center. Finally for the case of two dielectrics the number of eigenvalues is different than previous cases (Fig. 10).

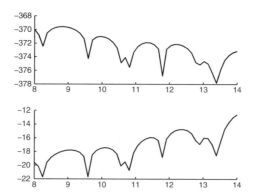

Fig. 10. Here the logarithm of the absolute value of the determinant is plotted against the wave number k_0 for the case when two dielectric circular cylinders, with refractive indices $n_1 = 2k_0$ and $n_2 = k_0/2$, are placed off the center of in the bowtie shaped cavity, using the QAA (*top*) and the BIM (*bottom*) techniques

4 Conclusion

In this paper we have implemented a quasi-analytical method to compute stationary states and wave functions for an important class of chaotic cavities containing dielectrics and/or PECs. The method is based on the use of multipole expansion for solving the Helmholtz equation. Our results, for a bowtie shaped cavity, were compared to a BIM approach and very similar results were obtained. The main advantage of our method is the closed form nature of the matrix in the resulted linear system, and its derivatives can be computed exactly. Our results show that the location and the number of the eigenvalues depend greatly on the location, the number and the type of objects inserted into the cavity.

References

1. V. Berry, Regularity and chaos in classical mechanics, illustrated by three deformations of circular "billiard", Eur. J. Phys. 2, 91–102 (1981).
2. S.V. Boriskina, P. Sewell, and T.M. Benson, Accurate simulation of two-dimensional optical microcavities with uniquely solvable boundary integral equations and trigonometric Galerkin discretization, J. Opt. Soc. Am. A 21 (2004).
3. F.A. Buot, Mesoscopic physics and nanoelectronics: nanoscience and nanotechnology, Phys. Rep. 234, 73–174 (1993).
4. W.C. Chew, Waves and Fields in Inhomogeneous Media (Van Nostrand Reinhold, New York, 1990).
5. D. Colton and R. Kress, Integral Equation Methods in Scattering Theory (Wiley, New York, 1983).

6. D. Cohen, N. Lepore, and E.J. Heller, Consolidating boundary methods for finding the eigenstates of billiards, J. Phys. A: Math. Gen. 37, 2139–2161 (2004). PII: S0305–4470(04)69126-8.

7. V. Galdi, I.M. Pinto, and L.B. Felsen, Wave propagation in ray-chaotic enclosures: paradigms, oddities and examples, IEEE Antennas Propag. Mag. 47, 62–81 (2005).

8. A. Gokirmak, D.-H. Wu, J.S.A. Bridgewater, and S.M. Anlage, Scanned perturbation technique for imaging electromagnetic standing wave patterns of microwave cavities, Rev. Sci. Instrum. 69, No. 9 (1998).

9. G. Gouesbet, S. Meunier-Guttin-Cluzel, and G. Grehan, Opt. Commun. 201, 223 (2002).

10. B. Gutkin, Can billiard eigenstates be approximated by superpositions of plane waves? J. Phys. A: Math. Gen. 36, 8603–8622 (2003).

11. M.C. Gutzwiller, Chaos in Classical and Quantum Mechanics (Springer, Berlin, Heidelberg, New York, 1990).

12. D.B. Haidvogel and T.A. Zang, The accurate solution of Poisson's equation by expansion in Chebyshev polynomials, J. Comput. Phys. 30, 167–180 (1979).

13. P. Haldenwang, G. Labrosse, S. Abboudi, and M. Deville, Chebyshev 3d spectral and 2d pseudospectral solvers for the Helmholtz equation, *J. Comput. Phys.* 5, pp. 115–128, (1984).

14. E.J. Heller, in: M.-J. Giannoni, A Voros, and J. Zinn-Justin (eds.), Chaos and Quantum Systems (Amsterdam, Elsevier, 1991) p. 548

15. S. Hemmady, X. Zheng, E. Ott, T. Antonsen, and S. Anlage, Universal impedance fluctuations in wave chaotic systems, Phys. Rev. Lett. 94, 014102 (2005).

16. S. Hemmady, X. Zheng, T. Antonsen, E. Ott, and S. Anlage, Universal statistics of the scattering coefficient of chaotic microwave cavities, Phys. Rev. E71, 056215 (2005).

17. M. Hentschel and K. Richter, Quantum chaos in optical systems: the annular billiard, Phys. Rev. E66, 056207 (2002).

18. D.L. Kaufman, I. Kosztin, and K. Schulten, Expansion method for stationary states of quantum billiards, Am. J. Phys. 67 (1999).

19. I. Kosztin and K. Schulten, Boundary integral method for stationary states of two-dimensional quantum systems, Int. J. Modern Phys. C8, 293 (1997).

20. Vergini and M. Saraceno, Phys. Rev. E52, 2204 (1995).

21. Vergini, Ph.D. Thesis, Universidad de Buenos Aires, 1995.

22. Barnett, Ph.D. Thesis, Harvard University, 2000.

23. J. Wiersig, Boundary element method for resonances in dielectric microcavities, J. Opt. A: Pure Appl. 5360 PII: S1464–4258(03) 39803–4 Opt. 5 (2003).

24. C. Zhang, J. Liu, M.G. Raizen, and Q. Niu, Quantum chaos of Bogoliubov waves for a Bose–Einstein condensate in stadium billiards, Phys. Rev. Lett. 93, 074101 (2004).

25. X. Zheng, S. Hemmady, T. Antonsen, S. Anlage, and E. Ott, Characterization of fluctuations of impedance and scattering matrices in wave chaotic scattering, Phys. Rev. E73, 046208 (2006).

26. http://www.ireap.umd.edu/MURI-2001/Review_8June02/ 02_Anlage.pdf# search=%22chaos %20dielectric%20inside%22

Existence Results and Open Problems in the Kinetic Theory of Dense Gases

W. Greenberg

Summary. Although the Boltzmann equation is the earliest and best known of the classic equations in kinetic theory, its weakness in modeling non-dilute gases has long been recognized. Some 85 years ago, Enskog proposed modifications for dense gases which generated more accurate transport coefficients than the Boltzmann equation. However, the Enskog equation does not model intermolecular potentials. We wish to outline some recent advances in improving the Enskog equation, and to highlight a number of problems which remain open.

1 Introduction

Although there is an extensive literature on discrete velocity Boltzmann equations [1], and a smaller literature on discrete velocity Enskog equations [2], the study of kinetic equations on spatially discrete domains is extremely limited. Here we would like to present models of the Enskog equation on a three-dimensional spatial lattice both with the full velocity dependent Enskog collision operator and also with a discrete velocity set. The first type of model is referred to as a semi-discrete model, since only the spatial variable is discretized, while the second is referred to as a fully discrete model. Moreover, we will extend the behavior of the Enskog collision operator (which describes only hard sphere collisions) to include next-nearest neighbor interactions, thereby modeling a square well interaction potential, or, more generally, a piecewise linear local interaction potential.

The Boltzmann equation, first posed in 1876, is the best known equation in the kinetic theory of gases [3]. However, this equation, which describes molecules as point particles and yields transport equations only of an ideal gas, is an accurate portrayal of a dilute gas. In order to have a more accurate description of moderately dense gases, Enskog in 1921 proposed the equation subsequently bearing his name [4]. The Enkog equation, revised in the 1960s to represent exact hydrodynamics, takes into account the non-zero diameter of real molecules, and has turned out to be an accurate description of dense gases up to 10% of close packing. Because the Enskog equation models only hard sphere collisions without intermolecular potential, Greenberg et al. have considered an Enkog type collision operator with square well, and, more generally, local piecewise constant, potential [5, 6]. Although discrete velociy models of the Boltzmann equation have an extensive literature going back more than 40 years, the spatially discrete Boltzmann equation was introduced more recently by Greenberg and coworkers [7, 8]. In these models the spatial variable was replaced by a finite periodic lattice.

In this article we will present a lattice version of the Enskog equation, studied in a Banach space of absolutely integrable functions of the velocity variables, i.e., only the spatial variable will be discretized. We will discuss both the analog of a hard sphere collision model and an Enskog model with local (next-nearest neighbor) interaction.

Throughout, we will point to some open problems in the kinetic theory of dense gases.

2 Kinetic Equations and Streaming Operator

For perspective, let us write the Enskog equation in a three-dimensional spatial domain:

$$\left[\frac{\partial}{\partial t} + \vec{v} \cdot \nabla_{\vec{r}}\right] f(\vec{r}, \vec{v}, t) = C_E(f(\vec{r}, \vec{v}, t), f(\vec{r}, \vec{v}, t)) \tag{1}$$

for the distribution function $f : R^3_{\vec{r}} \times R^3_{\vec{v}} \times R_+ \to R_+$ representing the differential density of particles at position $\vec{r}$ at time t with velocity $\vec{v}$. Here, $C_E(f, f)$ is the Enskog collision operator

$$C_E(f, f)(\vec{r}, \vec{v}, t) = \iint_{R^2 \times S^2_+} [Y(\vec{r}, \vec{r} - a\vec{\epsilon}) f(\vec{r}, \vec{v}', t) f(\vec{r} - a\vec{\epsilon}, \vec{v}'_1, t) \tag{2}$$

$$- Y(\vec{r}, \vec{r} + a\vec{\epsilon}) f(\vec{r}, \vec{v}, t) f(\vec{r} + a\vec{\epsilon}, \vec{v}_1, t)] < \vec{\epsilon}, \vec{v} - \vec{v}_1 > d\vec{\epsilon} \, d\vec{v}_1,$$

$$\vec{v}' = \vec{v} - \vec{\epsilon} < \vec{\epsilon}, \vec{v} - \vec{v}_1 >, \qquad \vec{v}'_1 = \vec{v}_1 + \vec{\epsilon} < \vec{\epsilon}, \vec{v} - \vec{v}_1 > \tag{3}$$

for a gas of molecules of diameter a, with $\vec{\epsilon}$-integration over $\{\vec{\epsilon} \in R^3 : ||\vec{\epsilon}|| = 1, \vec{\epsilon} \cdot (\vec{v} - \vec{v}_1) > 0\}$.

In this section we consider both semi-discrete and fully discrete Enskog equations on a lattice. The free streaming operator, corresponding to the continuum operator $\vec{v} \cdot \nabla_{\vec{r}}$, is constructed in the same fashion for both types of models.

The semi-discrete equation for the distribution function $f_i(\vec{v}, t)$ is written:

$$\frac{\partial f_i}{\partial t}(\vec{v}, t) + (Af)_i(\vec{v}, t) = J(f, f)_i. \tag{4}$$

The index i is the spatial index denoting the ith lattice point in the periodic three-dimensional cubic lattice Λ^3, and $\vec{v}$ is the (dimensionless) velocity vector.

The operators A and J will be defined below. We seek solutions in the Banach space $C([0,T], \mathcal{X})$ where $\mathcal{X} = L^1(\Lambda^3 \times S)$ with norm $||f|| = \sum_i ||f_i(\vec{v})||_1$. The sum is over N^3 lattice sites, $S = R^3$ in the case of the semi-discrete model and S is a finite set in the case of the fully discrete model. We denote with $\mathcal{T}_+$ the cone of positive functions in $\mathcal{X}$, and by $\mathcal{G}(\mathcal{T}_+)$ the cone of measurable functions $f(\cdot) : R_+ \to \mathcal{T}_+$.

The operator A is the finite difference approximation to the gradient term. To give A specifically, let π be an identification between the lattice Λ^3 and Z^3. Then A is an $N^3 \times N^3$ matrix:

$$A_{ij} = \sum_{\hat{u}} (\vec{v} \cdot \hat{u}) \Delta_{ij}^{\hat{u}}(\vec{v}), \tag{5}$$

where

$$\Delta_{ij}^{\hat{u}}(\vec{v}) = \delta_{ij} - \delta_{i, \pi(\pi^{-1}(j) + \hat{u})}, \quad \vec{v} \cdot \hat{u} > 0, \quad \Delta^{\hat{u}}(-\vec{v}) = \Delta^{\hat{u}}(\vec{v})^* \tag{6}$$

and the sum is over the three orthogonal coordinate vectors $\hat{u}$. We have, for convenience, taken the lattice spacing to be of unit length. The periodic boundary conditions are imposed by viewing the lattice as a three-dimensional torus, and thus $\pi^{-1}(j) + \hat{u} \in \Lambda^3$ for every j.

A representation of A may be written as follows. If $v_x, v_y, v_z \geq 0$ and the $n \times n$ matrix E is defined by

$$E_{ij} = \begin{cases} \delta_{n,j}, & i = 1 \\ \delta_{i,j+1}, & i > 1 \end{cases} \tag{7}$$

then

$$A = (v_x + v_y + v_z)I \otimes I \otimes I - v_x(E \otimes I \otimes I) - v_y(I \otimes E \otimes I) - v_z(i \otimes I \otimes E). \tag{8}$$

Note $E^n = I$ and that the representation of A if $v_i < 0$ can be obtained from (6).

Let us consider first the semi-discrete Enskog collision operator:

$$J(f,f)_i(\vec{v},t) = G_0(f,f)_i(\vec{v},t) - f_i(\vec{v},t)L_0(f)_i(\vec{v},t) = \tag{9}$$

$$= \sum_{\hat{e} \in \Gamma} \int_{R^3} d\vec{v}_1 \left[Y_{i,i-\epsilon} f_i(\vec{v}',t) f_{i-\epsilon}(\vec{v}_1',t) - Y_{i,i+\epsilon} f_i(\vec{v},t) f_{i+\epsilon}(\vec{v}_1,t) \right] \cdot$$

$$\cdot < \hat{e}, \vec{v} - \vec{v}_1 > \theta(\hat{e} \cdot (\vec{v} - \vec{v}_1)),$$

where θ is the Heaviside function and $\vec{v}'$, $\vec{v}_1'$ are given in (2). The geometric factor Y is a functional of f, $Y_{i,j} = Y(n_i(t), n_j(t))$, where $n_i(t) = \int_{R^3} d\vec{v} f_i(\vec{v},t)$, and is assumed positive, jointly continuous, bounded and symmetric in its arguments. The set $\Gamma \subset S^2$ is the set of unit vectors in R^3 pointing in the direction of nearest neighbors, taken periodically, e.g., the unit coordinate vectors in a rectangular lattice. Indices such as $Y_{i,i+\epsilon}$ are written in shorthand for $Y_{i, \pi(\pi^{-1}(i) + \hat{e})}$.

Equations (4)–(9) are the discrete version of the (revised) Enskog equation (1a), which models hard sphere collisions. The square well Enskog equation, derived by Davis et al. [9] and Greenberg et al. [5], models, in the continuum case, an intermolecular potential of the form

$$\phi(||\vec{r}_1 - \vec{r}_2||) = \begin{cases} \infty, & 0 < ||\vec{r}_1 - \vec{r}_2|| < a \\ -q, & a < ||\vec{r}_1 - \vec{r}_2|| < R \\ 0, & ||\vec{r}_1 - \vec{r}_2|| \geq R \end{cases} \tag{10}$$

for a single square well of depth q and width R, and a sequence of such wells for a piecewise constant local potential. The resultant kinetic equation has a collision term containing precisely the Enskog collision operator, on account of the hard sphere collision, and, in the case of a single square well, three very similar collision terms representing the molecule at $||\vec{r}_1 - \vec{r}_2|| = R$ (i) entering the well, (ii) exiting the well, and (iii) reflecting off the well if energy is not sufficient for an escape (or a penetration for a repulsive well). The last cannot take place, of course, for an attractive well unless an intermediate collision has occurred while the particle is in the well. In the case that the well consists of m piecewise constant steps, the collision operator will contain $3m+1$ Enskog-like collision terms.

As we are interested in lattice models, we will defer writing out the continuum equation for square well potentials, recommending the reader to the quoted literature, and restrict ourselves to writing the lattice equation. In the case of a single well, which translates into a strictly next-nearest neighbor interaction, the semi-discrete lattice collision operator is:

$$J(f,f)_i(\vec{v},t) = \sum_{\hat{\epsilon}\in\Gamma_0} \int_{R^3} d\vec{v}_1 \left[Y_{i,i+\epsilon} f_i(\vec{v}',t) f_{i+\epsilon}(\vec{v}_1',t) - Y_{i,i-\epsilon} f_i(\vec{v},t) f_{i-\epsilon}(\vec{v}_1,t) \right] \cdot$$

$$\cdot \hat{\epsilon}\cdot(\vec{v}-\vec{v}_1)\theta(\hat{\epsilon}\cdot(\vec{v}-\vec{v}_1)) + \tag{11}$$

$$+ \gamma \sum_{\hat{\epsilon}\in\Gamma_1} \int_{R^3} d\vec{v}_1 \left[Y_{i,i+\epsilon} f_i(\vec{v}'',t) f_{i+\epsilon}(\vec{v}_1',t) - Y_{i,i-\epsilon} f_i(\vec{v},t) f_{i-\epsilon}(\vec{v}_1,t) \right]\cdot$$

$$\cdot \hat{\epsilon}\cdot(\vec{v}-\vec{v}_1)\theta(\hat{\epsilon}\cdot(\vec{v}-\vec{v}_1)) +$$

$$+ \gamma \sum_{\hat{\epsilon}\in\Gamma_1} \int_{R^3} d\vec{v}_1 \left[Y_{i,i-\epsilon} f_i(\vec{v}''',t) f_{i-\epsilon}(\vec{v}_1''',t) - Y_{i,i+\epsilon} f_i(\vec{v},t) f_{i+\epsilon}(\vec{v}_1,t) \right]\cdot$$

$$\cdot \hat{\epsilon}\cdot(\vec{v}-\vec{v}_1)\theta(\hat{\epsilon}\cdot(\vec{v}-\vec{v}_1) - \sqrt{4q}) +$$

$$+ \gamma \sum_{\hat{\epsilon}\in\Gamma_1} \int_{R^3} d\vec{v}_1 \left[Y_{i,i-\epsilon} f_i(\vec{v}',t) f_{i-\epsilon}(\vec{v}_1',t) - Y_{i,i+\epsilon} f_i(\vec{v},t) f_{i+\epsilon}(\vec{v}_1,t) \right]\cdot$$

$$\cdot \theta(\sqrt{4q} - \hat{\epsilon}\cdot(\vec{v}-\vec{v}_1))\hat{\epsilon}\cdot(\vec{v}-\vec{v}_1)\theta(\hat{\epsilon}\cdot(\vec{v}-\vec{v}_1)) =$$

$$= \sum_{k=0}^{3} [G_k(f,f)_i - f_i L_k(f)_i] = G(f,f)_i - f_i L(f)_i,$$

where Γ_0 is related to the set of nearest neighbor vectors, Γ_1 to the set of next-nearest neighbor vectors, and $\gamma > 0$. Here, the double and triple primed velocities are derived by conservation of momentum and energy, just as were the velocity transformations in (1b). For example,

$$\vec{v}'' = \vec{v} - \frac{1}{2}\vec{\epsilon}\{<\vec{\epsilon}, \vec{v} - \vec{v}_1> -[<\vec{\epsilon}, \vec{v} - \vec{v}_1>^2 +4q]^{\frac{1}{2}}\}, \qquad (12)$$

$$\vec{v}_1'' = \vec{v}_1 + \frac{1}{2}\vec{\epsilon}\{<\vec{\epsilon}, \vec{v} - \vec{v}_1> -[<\vec{\epsilon}, \vec{v} - \vec{v}_1>^2 +4q]^{\frac{1}{2}}\}, \qquad (13)$$

with a similar transformation for $\vec{v}'''$, $\vec{v}_1'''$ [3]. It is not difficult to show the conservation (of mass) property

$$\int_{R^3} d\vec{v}\,\{G(f,f)_i - f_i L(f)_i\} = 0, \quad f \in \mathcal{X}. \qquad (14)$$

In the case of next^i-nearest neighbor interactions for $i = 1, \ldots, m$, there will be $3m$ additional collision terms with corresponding transformations of (primed) outgoing velocities, obtained by the conservation laws and taking into account the energy levels q_i. In this case, there will be summations over next^i-nearest neighbor vectors Γ_i and $J(f,f) = \sum_{k=0}^{3m}[G_k(f,f) - fL_k(f)]$. As the functional analysis to be considered in what follows carries over in a transparent way to these additional collision terms, we will, for convenience, pose the lemmas for the case of nearest neighbor interaction only, i.e., the Enskog lattice collision operator (9), commenting only on any points for which the lattice model with interaction potential might differ.

Next, let us consider (4) for the fully discrete lattice model. In this case, the variable $\vec{v}$ belongs to a set S of finite cardinality. We shall label such velocities $\vec{v}_j$ for $j = 1, \ldots, card(S)$. If the probability of the collision of two particles of velocities $\vec{v}_i$ and $\vec{v}_j$ producing velocities $\vec{v}_k$ and $\vec{v}_l$ at transfer angle (vector) $\hat{\epsilon}$ is given by P_{ij}^{kl}, then for a nearest neighbor interaction model P must satisfy:

$$0 \le P_{ij}^{kl}(\hat{\epsilon}) \le 1, \qquad (15)$$

$$P_{ij}^{kl}(\hat{\epsilon}) \ne 0 \text{ if and only if } \begin{cases} <\vec{v}_i - \vec{v}_j, \hat{\epsilon}> \,\ge 0 \\ <\vec{v}_k - \vec{v}_l, \hat{\epsilon}> \,\le 0 \\ \vec{v}_i + \vec{v}_j = \vec{v}_k + \vec{v}_l \\ v_i^2 + v_j^2 = v_k^2 + v_l^2 \end{cases} \qquad (16)$$

$$\sum_{kl} P_{ij}^{kl}(\hat{\epsilon}) = \sum_{ij} P_{ij}^{kl}(\hat{\epsilon}) = 1, \qquad (17)$$

$$P_{ij}^{kl}(\hat{\epsilon}) = P_{ji}^{lk}(-\hat{\epsilon}). \qquad (18)$$

The distribution function f now should be written as $f_{m,i}$, where the first index refers is the spatial variable and the second is the velocity variable. Then the fully discrete kinetic equation is written

$$\frac{\partial f_{m,i}}{\partial t}(t) + (Af)_{m,i}(t) = J(f(t), f(t))_{m,i}, \tag{19}$$

where

$$J(f,f)_{m,i} = G(f,f)_{m,i} - f_{m,i}L(f)_{m,i} = \tag{20}$$

$$\sum_{j,k,l} \sum_{\hat{\epsilon}} Y^+_{m,m-\epsilon} P^{ij}_{kl}(\hat{\epsilon})m, k f_{m-\epsilon,l} < \hat{\epsilon}, \vec{v}_l - \vec{v}_k > -Y_{m,m+\epsilon} P^{kl}_{ij}(\hat{\epsilon}) f_{m,i} f_{m+\epsilon,j}.$$

$$< \hat{\epsilon}, \vec{v}_i - \vec{v}_j >$$

The sum over $\hat{\epsilon}$ is a sum over nearest neighbors as discussed in the semi-discrete model. The streaming operator A is defined again by (5), however with the velocities $\vec{v}$ confined to the set S. The geometric factor Y_{m_1,m_2} is a function of $\sum_i f_{m_1,i}$, $\sum_i f_{m_2,i}$ and is assumed positive and symmetric in its arguments.

An example of such a nearest neighbor model for a cubic lattice is a generalization of the Carleman model for the Boltzmann equation [10]. In this case, the discrete velocity set is

$$S = \{\hat{x}, -\hat{x}, \hat{y}, -\hat{y}, \hat{z}, -\hat{z}\}, \tag{21}$$

where the velocities in S are vectors parallel to the coordinate axes, normalized for the sake of convenience. A more complicated mode, related to the Broadwell model [10], is

$$S = \tag{22}$$
$$\{\pm\hat{x}, \pm\hat{y}, \pm\hat{z}, \pm(\hat{x} - \hat{y}), \pm(\hat{x} - \hat{z}), \pm(\hat{y} - \hat{z}), \pm(\hat{x} + \hat{y}), \pm(\hat{x} + \hat{z}), \pm(\hat{y} + \hat{z}), 0\},$$

The analog of the square well or local piecewise linear interaction in the continuum model will here be a next^i-nearest neighbor interaction model, as for the semi-discrete model. However, in this case the transition matrix P^{kl}_{ij} and the velocity set under consideration must be modified. In particular, for a finite set of velocities S generating a nearest neighbor model, the relationships such as (13) describing outgoing velocities generated by the well or piecewise linear potential will create an enhanced set, still called S, and the last case in (16) will require energy conservation with respect to the local potential. (Note momentum conservation is maintained.) The collision operator $J(f, f)$ will, of course, require a sum over the enhanced velocity set. In what follows, we will refer to the nearest neighbor model, which corresponds to strictly hard sphere collisions. The results, in general, properly altered to deal with bounded, rather than conserved, energy, do carry over to (next)^i-nearest neighbor interaction models; we will indicate when validity is only for the simpler model.

For the Boltzmann equation there is an extensive and rich literature on discrete velocity models. For the Enskog equation, only a minimal number of discrete velocity models exist [13–15]. This is a wide open field of research.

In the next three sections, we will confine ourselves to the semi-discrete model, although the results in Sects. 3 and 4 almost entirely carry over to the fully discrete model, with the appropriate change in notation to the finite velocity set. In Sect. 6 we will return to the fully discrete model.

3 Semi-Group and Iterations

We discuss here the semi-discrete equation (4). Throughout this and the following section, we will consider only the velocity cutoff model of the semi-discrete equation, which includes in the collision kernel the additional factor $\theta(p - ||\vec{v} - \vec{v}_1||)$ for some fixed $p > 0$. Then G and L are bounded functionals: $||G(f, f)|| \le k_1||f||^2$ and $||L(f)|| \le k_2||f||$ for constants k_1, k_2 depending on p.

Throughout, we will suppress the position variable (index) when the meaning remains clear.

It is easy to see that A generates a c_0-semi-group $U_A(t)$ and $A + L(f)$ a two-parameter evolution operator T_f, i.e., $T_f(t, s)\xi_0$ is a solution of the homogeneous equation

$$\frac{dg}{dt} + Ag + L(f)g = 0, \qquad g(s) = \xi_0. \tag{23}$$

The most relevant properties these operators can be summarized in the following lemma.

Lemma 3.1. *(a)* $U_A(t)$ *and* $T_f(t_2, t_1)$ *are invariant on the cone of positive functions* $\mathcal{T}_+ \subset \mathcal{X}$ *for* t *and* $t_2 - t_1$ *positive, and* $f \in \mathcal{G}(\mathcal{T}_+)$.
(b) $U(t)$ *is a contraction semi-group and continues analytically to a bounded holomorphic semi-group* $U(z)$.
(c) $T_f(t_2, t_1)$ *is a contraction mapping on* $\mathcal{X}$ *for* $t_2 - t_1$ *positive and* $f \in \mathcal{G}(\mathcal{T}_+)$.
(d) *For all* $f \in \mathcal{X}$,

$$\sum_{i=1}^{N^3} (U(t)f)_i = \sum_{i=1}^{N^3} f_i. \tag{24}$$

Proof. First one shows that every element of e^{-tA} is a power series in $t|\vec{v} \cdot \hat{u}|$ with positive coefficient to lowest order. Hence, for $t|\vec{v} \cdot \hat{u}|$ sufficiently small, $(e^{-tA}f)_i \ge 0$ for $f \in \mathcal{T}_+ \cap M_N$, where M_N denotes the union of subspaces of functions in $\mathcal{T}_+$ with support in the hypercube about the origin with sides of length $2N$. Then, by exponential addition, this extends to arbitrary t.

On M_N, $\mathcal{A}(t) = -A - L(f(t))$ is a bounded operator, and $T_f(s, t)$ is given explicitly by

$$T_f(t, s) = \text{s.}\lim_{m \to \infty} \exp \int_{t_{m-1}}^{t_m} \mathcal{A}(t')dt' \ldots \exp \int_{t_0}^{t_1} \mathcal{A}(t')dt' \tag{25}$$

with the limit taken over n-partitions $t = t_m > t_{m-1} > \cdots > t_1 > t_0 = s$. Using the Lie product formula and the uniform boundedness of each of the exponentials, one can represent $T_f(t, s)$ as the double limit

$$T_f(t,s) = \text{s. } \lim_{m \to \infty} \lim_{n \to \infty} \left\{ \left[U_A \left(\frac{t_m - t_{m-1}}{n} \right) \exp \frac{- \int_{t_{m-1}}^{t_m} L(f(s))ds}{n} \right]^n \right.$$

$$\left. \times \cdots \times \left[U_A \left(\frac{t_1 - t_0}{n} \right) \exp \frac{- \int_{t_0}^{t_1} L(f(s))ds}{n} \right]^n \right\}. \qquad (26)$$

But L is diagonal and positive on $\mathcal{T}_+$, and therefore so are the exponentials in (26). Thus $T_f(t,s)\mathcal{T}_+ \subset \mathcal{T}_+$, completing the proof of (a).

To prove (b), one employs the important identity

$$\sum_{i=1}^{N^3} (A^m)_{ij} = 0, \qquad \forall m \in Z_+ \qquad (27)$$

and the expansion

$$e^{s(-I+E)} = \frac{1}{n} \sum_{\alpha=0}^{n-1} \sum_{i=1}^{n} e^{s(w_i - 1)} w_{-\alpha i} E_\alpha. \qquad (28)$$

where $E_0 = I$, $E_1 = E$, $E_\alpha = E_{\alpha-1} E$ and w_α are the primitive nth roots of unity. From the first, one may see that $U_A(t)$ is isometric on $\mathcal{T}_+$, and therefore, by invariance, on $\mathcal{X}$. Writing $U_A(s) = U_x(s) \otimes U_y(s) \otimes U_z(s)$, to show $U_A(s)$ is a bounded holomorphic semi-group, it is sufficient to show that $U_x(s)$ and $sAU_x(s)$ are bounded uniformly in a sector $\mathcal{S}_\theta \subset C$,

$$\mathcal{S}_\theta = \{z \in C | |\arg z| < \theta < \frac{\pi}{2}\}. \qquad (29)$$

The uniform boundedness of $U_A(t)$ follows from the expansion (28) and some straightforward complex analysis. The uniform boundedness of $sAU(s)$ is an immediate consequence of the boundedness of $g(\xi) = \xi e^{-\xi}$, $Re\ \xi \geq 0$. This proves (b).

Since L is positive, $\exp\{-\int ds\, L(f(s))\}$ is contractive, and by the representation (26), part (c) is proved, and (d) is an easy consequence of (27) and part (a). $\square$

Let us consider the integral equation

$$f(t) = U_A(t)f_0 + \int_0^t ds\, U_A(t-s)\{G(f(s), f(s)) - f(s)L(f(s))\} \qquad (30)$$

as well as the integral equation

$$f(t) = T_f(t,0)f_0 + \int_0^t ds\, T_f(t,s)G(f(s), f(s)). \qquad (31)$$

Note that for next-nearest neighbor interactions, the integral in (31) will be

$$\int_0^t ds\, T_f(t,s) \sum_{k=0}^{3m} G_0(f(s), f(s)). \tag{32}$$

Equation (30) is convenient for iterating a solution of the kinetic equation, and (31) is useful for establishing positivity.

We wish to solve both equations by iteration. Define

$$f^{(0)}(\vec{v}, t) = f_0(\vec{v}) \tag{33}$$

$$f^{(n)}(\vec{v}, t) = U_A(t) f_0(\vec{v}) + \int_0^t ds\, U_A(t-s) J(f^{(n-1)}(s), f^{(n-1)}(s)) \tag{34}$$

and

$$g^{(0)}(\vec{v}, t) = f_0(\vec{v}) \tag{35}$$

$$g^{(n+1)}(\vec{v}, t) = T_{g^{(n)}}(t,0) f_0(\vec{v}) + \int_0^t ds\, T_{g^{(n)}}(t,s) G(g^{(n)}(s), g^{(n)}(s)) \tag{36}$$

Lemma 3.2. *The iterative schemes converge in $\mathcal{X}$ to solutions $f(t)$ of (30) and $g(t)$ of (31) for t sufficiently small, and $g(t) \in \mathcal{T}_+$. The solutions are continuous functions of the initial datum f_0.*

Proof. As a result of the boundedness of the collision operator, for t sufficiently small, $||f^{(n)}|| \le M$, independent of t and n, and

$$||f^{(n+1)} - f^{(n)}|| \le t||J(f^{(n)}, f^{(n)}) - J(f^{(n-1)}, f^{(n-1)})||$$

$$\le 2tM||J||||f^{(n)} - f^{(n-1)}||. \tag{37}$$

Consequently, the iterative scheme (34) converges to a solution $f(t)$ of (30) for

$$t < min\{\frac{1}{4||J||||f_0||}, \frac{1}{2||J||M}\} \tag{38}$$

and $f(t)$ is a continuous function of f_0.

To see the sequence $\{g^{(n)}(t)\}$ is Cauchy, define

$$g^{(n+\frac{1}{2})}(\vec{v}, t) = T_{g^{(n-1)}}(t,0) f_0(\vec{v}) + \int_0^t ds\, T_{g^{(n-1)}}(t,s) G(g^{(n)}(s), g^{(n)}(s)) \tag{39}$$

and write S_s^t for $\sup_{0 \le s \le t}$. Then we have easily from Lemma 3.1,

$$||g^{(n+\frac{1}{2})}(t) - g^{(n)}(t)|| \le 2t||G||M_1 S_s^t ||g^{(n)}(s) - g^{(n-1)}(s)|| \tag{40}$$

and

$$||g^{(n+1)}(t) - g^{(n+\frac{1}{2})}(t)|| \leq ||T_{g^{(n)}}(t,0) - T_{g^{(n-1)}}(t,0)||||f_0||+$$

$$+tS_s^t||T_{g^{(n)}}(t,s) - T_{g^{(n-1)}}(t,s)||||G||||g^{(n)}(s)||^2. \tag{41}$$

for t sufficiently small. Define $\chi(t) = (T_{g^{(n)}}(t,s) - T_{g^{(n-1)}}(t,s))\xi_0$ for fixed s. Then χ is the solution of the coupled system

$$\frac{dg}{dt} + [A + L(g^{(n)}(t))]g = 0, \quad g(s) = \xi_0 \tag{42}$$

$$\frac{d\chi}{dt} + [A + L(g^{(n-1)}(t))]\chi = L(g^{(n-1)}(t) - g^{(n)}(t))g(t), \qquad \chi(s) = 0. \tag{43}$$

Then from

$$\chi(t) = \int_0^t ds\, T_{g^{(n)}}(t,s)L(g^{(n-1)}(s) - g^{(n)}(s))g(s) \tag{44}$$

we have

$$||g^{(n+1)}(t) - g^{(n+\frac{1}{2})}(t)|| \leq \tag{45}$$

$$\leq ||L||(t||f_0|| + t^2||G||M_1^2)S_s^t||g^{(n)}(s) - g^{(n-1)}(s)||$$

Collecting these results, it is sufficient to assume $0 \leq t \leq T_0$ for $T_0^{-1} = 8||G||M_1 + 8M_1 + 5||L||||f_0||$ to obtain

$$||g^{(n+\frac{1}{2})}(t) - g^{(n)}(t)|| \leq \frac{1}{4}S_s^t||g^{(n)}(s) - g^{(n-1)}(s)|| \tag{46}$$

$$||g^{(n+1)}(t) - g^{(n+\frac{1}{2})}(t)|| \leq \frac{1}{4}S_s^t||g^{(n)}(s) - g^{(n-1)}(s)||. \tag{47}$$

From these estimates it is evident that the sequence $\{g^{(n)}\}$ is Cauchy. The remainder of the lemma is clear. $\quad\square$

4 Global Solutions

We continue our discussion of the semi-discrete equation (4) with velocity cutoff. We will extend the mild local solutions of Sect. 3 to global-in-time solutions, and then prove they are classical solutions.

Lemma 4.1. *Let g_1, g_2 be solutions of (30),(31), respectively, satisfying $g_1(0) = g_2(0) = f_0$. Then $g_1 = g_2$.*

Proof. One can compute from (31)

$$g_2(t+s) = T_{g_2}(t+s,t)\left[T_{g_2}(t,0)f_0 + \int_0^t dt'\, T_{g_2}(t,t')G(g_2(t'))\right] + \qquad (48)$$

$$+ \int_t^{t+s} dt'\, T_{g_2}(t+s,t')G(g_2(t')) =$$

$$= T_{g_2}(t+s,t)[g_2(t) + sG(g_2(t)) + O(s)].$$

Define $\eta(s) = [T_{g_2}(t+s,t) - U(s)]\xi_0$. From

$$\frac{\partial \eta}{\partial s} = -[A + L(g_2(t+s))]\eta(s) - L(g_2(t+s))U_A(s)\xi_0 \qquad (49)$$

and $\eta(0) = 0$, we have

$$\eta(s) = -\int_0^s dt'\, T_{g_2}(t+s,t+t')L(g_2(t+t'))U_A(t')\xi_0. \qquad (50)$$

Combining this with (48), we may write

$$g_2(t+s) = U_A(s)[g_2(t) + sG(g_2)) - sL(g_2(t))g_2(t)] + O(s). \qquad (51)$$

On the other hand,

$$g_1(t+s) = U_A(s)U_A(t)f_0 + \int_0^t dt'\, U_A(s)U(t-t')J(g_1(t'),g_1(t')) +$$

$$+ \int_t^{t+s} U_A(t+s-t')J(g_1(t'),g_1(t')) \qquad (52)$$

$$= U_A(s)[g_1(t) + sJ(g_1(t),g_1(t))] + O(s)$$

Writing $\alpha(t) = ||g_2(t) - g_1(t)||$, we have

$$\alpha(t+s) - \alpha(t) \le s||J||(||g_2(t)|| + ||g_1(t)||)||g_2(t) - g_1(t)|| \qquad (53)$$

or

$$\mathcal{D}^+\alpha(t) \le 2||J||\alpha(t)\sup_{i,0\le t\le T_0} ||g_i(t)||. \qquad (54)$$

The Gronwall Lemma completes the proof. □

We wish to show that the solution of (30) is differentiable in t, and thus a solution of (4). Since $U_A(t)$ is a holomorphic semi-group, it is sufficient, by Kato's theorem, to show that $J(f(t), f(t))$ is Hölder continuous [11, pp. 487–491]. This will follow from the Hölder continuity of $f(t)$.

Lemma 4.2. *For $f_0 \in \mathcal{T}_+$, the functions $\{f^{(n)}(t)\}$ given by the iterative scheme (34) are differentiable on some interval $[0, T_0]$, are uniformly Lipschitz, and the derivatives $\{f^{(n)\prime}(t)\}$ are uniformly bounded in t and n.*

Proof. From the estimate

$$f^{(n)}(t+h) - f^{(n)}(t) = \tag{55}$$

$$= (U_A(h) - I)\left\{U_A(t)f_0 + \int_0^t dt'\, U_A(t-t')[J(f^{(n-1)}(t'), f^{(n-1)}(t'))\right.$$

$$-J(f^{(n-1)}(t), f^{(n-1)}(t))] + \int_0^t dt'\, U_A(t-t')J(f^{(n-1)}(t), f^{(n-1)}(t))\Big\} +$$

$$+hU_A(h)J(f^{(n-1)}(t), f^{(n-1)}(t)) + O(h)$$

the right derivative is

$$\mathcal{D}^+ f^{(n)}(t) = AU_A(t)f_0 + A[\int_0^t U_A(t-t')J(f^{(n-1)}(t'), f^{(n-1)}(t'))] +$$

$$+U_A(t)J(f^{(n-1)}(t), f^{(n-1)}(t)). \tag{56}$$

Then a bound on $\mathcal{D}^+ f^{(n)}(t)$ is obtained by the estimate $||AU_A(t)|| \le K/t$ for some constant K (by analyticity). In particular, for $n \ge 2$,

$$||\mathcal{D}^+ f^{(n)}(t)|| \le ||Af_0|| + tK||J||2m\frac{||f^{(n-1)}(t') - f^{(n-1)}(t)||}{t' - t} + ||J||m^2 \tag{57}$$

and a uniform bound is obtained inductively by estimating Lipschitz constants K_n for each $f^{(n)}$. Indeed, $\mathcal{D}^+ f^{(1)}(t) = U_A(t)Af_0 + U_A(t)J(f_0, f_0)$ and $f^{(1)}$ is Lipschitz with constant $K_1 = ||Af_0|| + ||J||||f_0||^2$ and for $n \ge 2$ from the estimate above we have

$$||f^{(n)}(t) - f^{(n)}(s)|| \le (||Af_0|| + ||J||m^2 + 2tKm||J||K_{n-1})|t - s|, \tag{58}$$

so that

$$K_n = \alpha + t\beta K_{n-1} \tag{59}$$

with $\alpha = ||Af_0|| + ||J||m^2$ and $\beta = 2Km||J||$, which is uniformly bounded for $t < 1/(2Km||J||)$. This completes the proof. $\square$

Theorem 4.3. *Suppose $f_0 \in \mathcal{T}_+$. Then there exists a unique positive solution $f(t)$ of the integral equation (30), or equivalently (31), for all $t \ge 0$, and $f(t)$ is a continuously differentiable solution of the Enskog lattice equation (4) for the velocity cutoff model. Further, $f(t) \in \mathcal{T}_+$ and f depends continuously upon the initial datum f_0.*

Proof. It remains only to note that $||f(t)|| = ||f_0||$ for t sufficiently small. For, integrating (30) over $\vec{v}$ and summing over i, recalling $U_A(t)$ is an isometry on $\mathcal{T}_+$, we have (for t sufficiently small)

$$||f|| = ||f_0|| + \sum_{\hat{u}\in\Gamma}\int_{R^3} d\vec{v}\int_0^t ds\sum_{i=1}^{N^3}[U_A(s)J(f(s), f(s))]_i. \tag{60}$$

Using Lemma 3.1, this becomes

$$||f|| = ||f_0|| + \int_0^t ds \sum_{i=1}^{N^3} \sum_{\hat{e} \in \Gamma} \int_{R^3} d\vec{v} \, [J(f(s), f(s))]_i. \tag{61}$$

But 1 is a collision invariant. Hence the integral term vanishes, and $||f|| = ||f_0||$. Now, the procedure can be repeated, and the theorem follows. $\square$

Note that the theorem is valid both for the lattice system with (cutoff) Enskog collision operator and with next^m-nearest neighbor interaction.

5 Removal of the Cutoff

Finally, let us consider the semi-discrete lattice model (4) with next^i-nearest neighbor interaction (without cutoff). We continue to suppress the spatial index i when possible. We assume given an initial distribution on the lattice $f_0(\vec{v})_i \in \mathcal{I}_+$ with finite mass, energy and entropy:

$$\sum_i \int_{R^3} d\vec{v} \, f_0(\vec{v})_i \{1 + v^2 + |\log f_0(\vec{v})_i|\} < \infty. \tag{62}$$

By the results of Sect. 4, for each positive integer p, the cutoff lattice equation (4) has a classical solution $f^{(p)}(\vec{v}, t)_i$ satisfying $f^{(p)}(\vec{v}, 0)_i = f_0(\vec{v})_i$. Fix a time interval $[0, T]$. Then, for these solutions, the equality

$$\sum_i \int_{R^3} d\vec{v} \, f^{(p)}(\vec{v}, t)_i = \sum_i \int d\vec{v} \, f^{(p)}(\vec{v}, 0)_i \tag{63}$$

and estimate

$$\sum_i \int_{R^3} d\vec{v} \, v^2 f^{(p)}(\vec{v}, t)_i \le k_1 \sum_i \int_{R^3} d\vec{v} v^2 f^{(p)}(\vec{v}, 0)_i + k_2 \tag{64}$$

for $t \in [0, T]$ and constants k_1, k_2 depending on T are a result of the symmetry of the collision kernel [6].

Lemma 5.1. *Let* $H(f) = \sum_i \int_{R^3} d\vec{v} \, f(\vec{v})_i \log f(\vec{v})_i$. *Then* $H(f^{(p)}(\vec{v}, t)) \le H(f^{(p)}(\vec{v}, 0)) + k_3$, *where* k_3 *is a function of* T *and* $f_0(\vec{v})$.

Proof. Since $U_A(t)_{ij} \ge 0$ and $\sum_i U_A(t)_{ij} = 1$ by the proof of Lemma 3.1, then for fixed velocity $\vec{v}$, $U_A(t)$ is the transition matrix for a discrete Markov system. Since any space-independent distribution is a fixed point of $U_A(t)$, standard arguments [11] prove that $H(U_A(t) f_0(\vec{v})_i)$ is nonincreasing. Now the lemma follows from estimates in [6]. $\square$

Theorem 5.2. *Suppose* $\sum_i \int_{R^3} d\vec{v} \, f_0(\vec{v})_i \{1 + v^2 + |\log f_0(\vec{v})_i|\} < \infty$, $f^{(p)}(\vec{v}, t)_i$ *is a solution of the lattice equation with cutoff* p, *and* $f^{(p)}(\vec{v}, 0)_i = f_0(\vec{v})_i$. *Then* $\{f^{(p)}\}$ *contains a subsequence which converges weakly in* $\mathcal{X}$. *The limit function* $f(\vec{v}, t)_i$ *is continuous in* t *and satisfies the integral equation (30) with unbounded collision kernel.*

Proof. The Dunford–Pettis property of L^1 and the mass, energy, entropy bounds previously demonstrated prove the existence of a subsequence (as $p \to \infty$) converging weakly in $\mathcal{X}$ to a function $f(t)$ for a denumerable dense set of t. Extension to all t follows from the equicontinuity of the family $\{f^{(p)}\}$. Indeed, let $\mathcal{X}_v = \{f \in \mathcal{X} : (1 + v^2)^{\frac{1}{2}} f((\vec{v}) \in \mathcal{X}\}$. Since $f^{(p)}$ is a solution of (4), $||f^{(p)\prime}(t)|| \le ||Af^{(p)}(t)|| + K||f^{(p)}||_v^2$, with K independent of p. Further, $||Af^{(p)}|| \le 6||f^{(p)}||_v$. Then using $||f^{(p)}||_v \le ||f_0||_v$, equicontinuity of the sequence follows.

Since $\mathcal{X}_v$ also satisfies the Dunford–Pettis property, and $J : \mathcal{X}_v \times \mathcal{X}_v \to \mathcal{X}$ is weakly continuous, $J(f^{(p)}(t))$ converges weakly to $J(f(t))$ pointwise in t. Then, using the integral equation for $f^{(p)}(t)$, the dominated convergence theorem, and the continuity of J, one can see that the limit function $f(t)$ satisfies (30). □

Again, the result is equally valid for the Enskog lattice equation and for the lattice equation with next^i-nearest neighbor interaction. The cost of treating the equation without velocity cutoff is the weakness of the solution and the loss of a uniqueness proof.

Another continuous model extending the Enskog equation for which existence theorems are known is the Vlasov–Enskog system [16]. This system models a weak long-range Coulomb intermolecular potential. There are no results in the literature concerning any semi-discrete models of a Vlasov–Enskog system.

6 Discrete Velocity Model

Let us return to the fully discrete lattice model for both nearest neighbor and (next)^i-nearest neighbor interactions. We consider the kinetic equation (19) with initial condition

$$f_{m,i}(0) = (f_0)_{m,i}. \tag{65}$$

As in Sect. 3, the streaming operator A generates a (matrix-valued) semigroup $U_A(t)$ on $\mathcal{X} = L^1(\Lambda^3 \times S)$ and an evolution operator T_ψ which satisfy the properties of Lemma 3.1.

Let $\psi_{m,i}$ be any function on $\Lambda^3 \times S$ and $I = \sum_{m,i} \psi_{m,i} J(f,f)_{m,i}$. Then a sequence of rearrangements, utilizing the properties of the transition matrix P_{ij}^{kl} as given in (16), leads to

$$I = \sum_{ijkl} \sum_m \sum_{\hat{e}} [\psi_{m,k} + \psi_{m+\epsilon,l} - \psi_{m,i} - \psi_{m+\epsilon,j}] Y_{m,m+\epsilon}.$$

$$P_{ij}^{kl}(\hat{e}) < \hat{e}, \vec{v}_i - \vec{v}_j > \tag{66}$$

Since for $\psi_{m,i} = 1$, $\vec{v}_i$, v_i^2 one has $I = 0$, any solutions of the Enskog lattice equation with nearest neighbor interaction only will satisfy conservation of mass, momentum and energy. In the case of next^i-nearest neighbor interactions, conservation of mass and momentum can still be demonstrated. In particular, for both nearest neighbor and (next)^i-nearest neighbor interaction, 1 is a collision invariant, and the collision operator is, of course bounded: $||J(f, f)|| \leq ||J|| \, ||f||^2$.

Now consider the integral equation (30) and iterative scheme (34). Then

$$||f^{(n)}|| \leq ||U(t)f_0|| + t||U(t - s)J(f^{(n-1)}, f^{(n-1)})|| \tag{67}$$

$$\leq ||f_0|| + t||J|| \, ||f^{(n-1)}||^2. \tag{68}$$

Consequently, for $t||f_0|| < 1/(4||J||)$, $||f^{(n)}|| < 2||f_0||$ and the iterative scheme is Cauchy. Thus for t in the indicated interval, the limit function f satisfies (30). Summing over position and using Lemma 3.1(d),

$$\sum_m f_{m,i}(t) = \sum_m (f_0)_{m,i} + \int_0^t ds \sum_m J(f(s), f(s))_{m,i}. \tag{69}$$

Then, summing over velocity,

$$||f(t)|| = ||f_0|| + \int_0^t ds \sum_i \sum_m J(f(s), f(s))_{m,i}. \tag{70}$$

but the integrand vanishes by (14) for any of the lattice models. Thus

$$||f(t)|| = ||f_0|| \tag{71}$$

and the local solution extends to a global solution. The differentiability of $f(t)$ is immediate and positivity may be demonstrated as in Sect. 3. We have

Theorem 6.1. *For any initial condition $f_0 \in \mathcal{X}$, $f_0 \geq 0$, the initial value problem (19)–(65) for the fully discrete kinetic equation has a unique solution $f \in C([0, \infty), \mathcal{X})$, and $f(t) \geq 0$. The result is valid for both nearest neighbor and (next)^i-nearest neighbor interactions.*

Here is a rich source of open problems. Fully discrete models are systems of first order ordinary differential equations. This presents the possibility of numerical computation using elementary codes. Almost nothing is known in this venue. How do the discrete models compare? What are the effects of the square well potential? Does a deep well lead to clustering? What are the possibilities of non-periodic boundary conditions? What would a fully discrete Vlasov–Enskog system look like? The author encourages further speculation and research.

References

1. R. Gatignol, *Theorie cinetique de gaz a repartition discrete de vitesses*, Lecture Notes in Physics 36, Springer-Verlag, New York, 1975.
2. G.Toscani, *On the discrete velocity models of the Boltzmann equation in several dimensions*, Ann. Matem. Pura Appl. **138**, 279–308 (1984).
3. C. Cercignani, *Theory and Application of the Boltzmann Equation*, Elsevier, New York, 1978.
4. N. Bellomo, M. Lachowicz, J. Polewczak and G. Toscani, *Mathematical Topics in Nonlinear Kinetic Theory II: the Enskog Equation*, World Scientific, London, 1991.
5. W. Greenberg, P. Lei and R.S. Liu, *Stability theory for the kinetic equations of a moderately dense gas*, in Rarefied Gas Dynamics: Theory and Simulations (Progress in Astronautics and Aeronautics, vol. 159), B.D. Shizgal and D.P. Weaver, eds., American Institute of Aeronautics and Astronomy, Washington, DC, 1994, pp. 599–607.
6. W. Greenberg and A. Yao, *Kinetic equations for gases with piecewise constant intermolecular potentials*, Transport Theor. Stat. Phys. **27**, 137–150 (1997).
7. C. Cercignani, W. Greenberg and P.F. Zweifel, *Global solutions of the Boltzmann equations on a lattice*, J. Stat. Phys. **20**, 449–462 (1979).
8. W. Greenberg, J. Voigt and P.F. Zweifel, *Discretized Boltzmann equation: lattice limit and non-Maxwellian gases*, J. Stat. Phys. **21**, 649–657 (1979).
9. H.T. Davis, S.A. Rice and J.V. Sengers, *On the kinetic theory of dense fluids. IX. the fluid of rigid sphere well attraction*, J. Chem. Phys. **35**, 2210 (1961).
10. T. Platkowski and R. Illner, *Discrete velocity models of the Boltzmann equation: a survey on the mathematical aspects of the theory*, SIAM Rev. **30**, 239–253 (1988).
11. T. Kato, *Perturbation Theory for Linear Operators*, Springer-Verlag, New York, 1966.
12. M. Moreau, *Note on the entropy production in a discrete Markov system*, J. Math. Phys. **19**, 2494–2498 (1978).
13. G. Borgioli, G. Lauro and R. Monaco, *On the discrete velocity models of the Enskog equation*, in Nonlinear Kinetic Theory an Mathematical Aspects of Hyperbolic Systems, V. Boffi, F. Bampi, G. Toscani, eds., Advances in Mathematics for Applied Sciences 9, Singapore, World Scientific, 1992, pp. 38–47.
14. G. Borgioli, V. Gerasimenko and G. Lauro, *Derivation of a discrete Enskog equation from the dynamics of particles*, Rend. Sem. Mat. Univ. Polit. Torino **56**, 59–70 (1998).
15. G. Borgioli, V. Gerasimenko and G. Lauro, *Many particle dynamical systems formulation for the discrete Enskog gas*, Transport Theor. Stat. Phys. **25**, 588–592 (1996).
16. W. Greenberg and P. Lei, *A Vlasov-Enskog equation with thermal background in gas dynamics*, in Differential and Difference Equations and Applications, R.P. Agarwal and K. Perera, eds., Hindawi Publ., New York, 2006, pp. 423–432.

The High Performance Asymptotic Method in Numerical Simulation

D. Konaté

Summary. Numerical simulation of complex phenomena involving large or multiples scales requires the use of methods that are highly precise and fast. Classical numerical methods consist in replacing a given problem, thanks to discretization, with a chain of algebraic equations to be solved.

In the current paper we come up with the idea to rather introduce, some "local" equations of the same type as the given one. Then using the strategy of asymptotic expansion as in singular perturbation, (cf. [1, 2, 3]) we replace each local equation with a chain of simpler equations to be solved. The solution is obtained in the form of a local analytical solution. We use previous known analytical solutions (cf. [4, 5]).

1 The Riemann Integration and the Hp-Asymptotic Method

We start with the classical problem of computing the Riemann integral of a regular function f. The interest here is to consider it over a large interval $]0, T[$. Consider a regular function f a real constant T (which can be large) and a problem of finding an unknown function u such that

$$u(x) = \int_0^x f(t)dt + \alpha; \quad x \in \Omega =]0, T[. \tag{1}$$

Of course (1) is equivalent to the following problem of finding a function u such that

$$\begin{cases} u'(x) = f(x); & x \in \Omega =]0, T[\\ u(0) = \alpha_o \end{cases} \tag{2}$$

Our starting point is the following question: " *Is it possible to compute u at any arbitrary and prescribed order, say h^q; for $q \in \mathbf{N}$; $h \in \mathbf{R}$; $0 < h < 1$ at - an acceptable - numerical cost?* " The answer is given in the following theorem

Theorem 1.1 *Consider a regular given function f defined over an interval $\Omega = [0, T]$. For any real constant h; $0 < h < 1$ and any subinterval $\Omega_i =]x_i, x_i + h[\subset \Omega$ there exists at least a list of $q+1$ homogeneous functions*

$$v_0(x), v_1(x), \cdots, v_i(x), \cdots, v_q(x)$$

such that

$$\bar{u}_i(x) = \sum_{j=0}^{q} \frac{(x-x_i)^j}{j!} v_{(j)}(x) + \alpha_i$$

is for any prescribed and arbitrary q a h^q order approximation to

$$u(x) = \int_{x_i}^{x} f(t)dt + \alpha_i; \quad x \in \Omega_i.$$

More precisely, we have

$$|u(x) - \bar{u}_i(x)| \leq C.h^{q+2} \tag{3}$$

for $x \in \Omega_i$ and C is a constant independent of h.

Proof of Theorem 1.1. In first place, we discretize the interval of reference $\Omega =]0, T[$ to get some N sub-intervals $\Omega_i =]x_i, x_{i+1}[$ such that $x_0 = 0$; $x_{i+1} = x_i + h$; $h = T/N$; $x_N = T$. Then we consider the resulting N "local" problems such that

$$\begin{cases} u'(x) = f(x); \quad x \in \Omega_i =]x_i, x_{i+1}[\\ u(x_i) = \alpha_i \\ 1 \leq i \leq N. \end{cases} \tag{4}$$

To solve each local problem, to function f we substitute its q-th order Taylor expansion $\bar{f}$ which is

$$\bar{f}(x) = \sum_{j=0}^{q} \frac{(x-x_i)^j}{j!} f^{(j)}(x_i) \tag{5}$$

and we look for an approximation to u over Ω_i say $\bar{u}_i$ such as

$$\bar{u}_i = \alpha_i + \sum_{j=0}^{q} \frac{(x-x_i)^j}{j!} v_j(x) \tag{6}$$

where the $q+1$ coefficient functions v_j are to be identified. We take $\bar{u}_i$ and $\bar{f}$ into (4) and equate the coefficients of the $q+1$ monomials $\frac{(x-x_i)^j}{j!}$; $0 \leq j \leq q$ to get

$$\begin{cases} v'_k(x) + v_{k+1}(x) = f^{(k)}(x_i) \\ 0 \leq k \leq q-1 \\ v'_q(x) = f^{(q)}(x_i). \end{cases} \tag{7}$$

We supplement (7) with

$$v_k(x_i) = 0; \quad 0 \le k \le q. \tag{8}$$

Equality (8) provides the boundary condition to the differential problem (4) which is compatible with $\bar{u}_i(x_i) = \alpha_i$. Putting together (7) and (8) allows a backward determination of the $q+1$ unknown functions v_j which are easily computed to be

$$v_j(x) = \sum_{k=1}^{q-j+1} (-1)^{k+1} \frac{(x - x_i)^k}{k!} f^{(j+k-1)}(x_i); \quad 0 \le j \le q. \tag{9}$$

For $x \in \Omega_i$, we use the definition of u and $\bar{u}_i$, and we recall that $f(x) - \bar{f}(x) = (x - x_i)^{q+1} g(x)$ where g is a regular function to reach

$$u(x) - \bar{u}_i(x) = \int_0^x f(t)dt - \int_0^x \bar{f}(t)dt = \int_0^x (t - x_i)^{q+1} g(t)$$

$$|u(x) - \bar{u}_i(x)| \le \bar{g} \frac{(x - x_i)^{q+2}}{(q+2)!} \le \bar{g} \frac{h^{q+2}}{(q+2)!}.$$

Within this last chain of inequations, set $C = \frac{\bar{g}}{(q+2)!}$ where $\bar{g} = \max_{x \in \Omega_1} |g(x)|$ to obtain inequality (3) and complete the proof of Theorem (1.1) □

2 The Hp-Asymptotic Method on First Order Differential Equations

For some given and regular functions a; b; f; with $a(x) \ne 0$ over the interval of study, we consider the problem of finding a numerical solution, on a large interval $\Omega=]0, T[$ to

$$\begin{cases} a(x)u'(x) + b(x)u(x) = f(x) & x \in \Omega \\ u(0) = \alpha_o \end{cases} \tag{10}$$

We use the discretization of the domain Ω obtained in the previous section. In each sub-interval $\Omega_i =]x_i, x_{i+1}[$ we consider the problem of finding a solution to

$$\begin{cases} a(x)u'(x) + b(x)u(x) = f(x) & x \in \Omega_i \\ u(0) = \alpha_i \end{cases} \tag{11}$$

Using the Taylor expansion for functions a, b, and f, we set

$$\bar{a}(x) = \sum_{j=0}^q \frac{(x - x_i)^j}{j!} a^{(j)}(x_i) + \frac{(x - x_i)^{q+1}}{(q+1)!} \tilde{a}(x)$$

$$\bar{b}(x) = \sum_{j=0}^{q} \frac{(x-x_i)^j}{j!} b^{(j)}(x_i) + \frac{(x-x_i)^{q+1}}{(q+1)!} \tilde{b}(x)$$

$$\overline{f}(x) = \sum_{j=0}^{q} \frac{(x-x_i)^j}{j!} f^{(j)}(x_i) + \frac{(x-x_i)^{q+1}}{(q+1)!} \tilde{f}(x)$$

where $\tilde{a}$, $\tilde{b}$, $\tilde{f}$ are some regular functions.

Consider the vector space $\mathcal{V}$ generated by the base

$$\mathcal{B} = \{v_0(x), v_1(x), \cdots, v_i(x), \cdots, v_q(x)\}$$

where $v_j(x) = \frac{(x-x_i)^j}{j!}$. We state

Theorem 2.1 *Consider some regular and given functions a, b, defined over an interval $\Omega =]0, T[$. For any real constant h; $0 < h < 1$ and any subinterval $\Omega_i =]x_i, x_i + h[\subset \Omega$, for any prescribed and arbitrary q, there exists a unique element $\overline{u}_i \in \mathcal{V}$ such that $\overline{u}_i$ is a $(q+2)$-th order approximation to the solution of problem (10). More precisely, we have*

$$|u(x) - \overline{u}_i(x)| \le C.h^{q+2} \tag{12}$$

for $x \in \Omega_i$ and C is a constant independent of h.

Proof of Theorem 2.1. Assume

$$\overline{u}_i(x) = \sum_{j=0}^{q} \frac{(x-x_i)^j}{j!} v_j(x).$$

We have to prove that the coefficients functions v_j exist, are unique and that inequality (12) holds true.

We determine, by construction, the coefficient functions v_j such that $\overline{u}_i$ is solution to

$$\begin{cases} \bar{a}(x)\overline{u}_i(x) + \bar{b}(x)\overline{u}_i(x) = \overline{f}(x) & x \in \Omega_i \\ \overline{u}_i(x_i) = \alpha_i \end{cases} \tag{13}$$

Using the precedent Taylor expansions, we get

$$a(x)\overline{u}_i'(x) = \sum_{\substack{k+j=0 \\ k \le q; \ j \le q}}^{2q-1} \frac{(x-x_i)^{k+j}}{k!j!} a^{(k)}(x_i)\left(v_{j+1} + v_j'\right) +$$

$$\sum_{k=0}^{q} \frac{(x-x_i)^{k+q}}{k!q!} a^{(k)}(x_i) + v_q'$$

and

$$b(x)\overline{u}_i(x) = \sum_{\substack{k+j=0 \\ k\leq q;\ j\leq q}}^{2q} \frac{(x-x_i)^{k+j}}{k!j!} b^{(k)}(x_i)v_j.$$

Restricting the computing to the q-th order terms, we drop off the terms of power in $(x-x_i)$ greater than q to get

$$a(x)\overline{u}_i'(x) = \sum_{\substack{k+j=0 \\ k\leq q;\ j\leq q-1}}^{q} \frac{(x-x_i)^{k+j}}{k!j!} a^{(k)}(x_i)\left(v_{j+1} + v_j'\right) +$$

$$+ \frac{(x-x_i)^q}{q!} a(x_i)v_q'. \tag{14}$$

and

$$b(x)\overline{u}_i(x) = \sum_{\substack{k+j=0 \\ k\leq q;\ j\leq q-1}}^{q} \frac{(x-x_i)^{k+j}}{k!j!} b^{(k)}(x_i)v_j + \frac{(x-x_i)^q}{q!} b(x_i)v_q. \tag{15}$$

The identification of alike powers in $(x-x_i)$ leads to the following system of first order differential equations with constant coefficients

$$\begin{cases} \sum_{\substack{k+j=t;\ 0\leq t\leq q-1 \\ k\leq q;\ j\leq q-1}} \left[\frac{(k+j)!}{k!j!} a^{(k)}(x_i)\left(v_{j+1}+v_j'\right) + \right. \\ \qquad\qquad \left. + \frac{(k+j)!}{k!j!} b^{(k)}(x_i)v_j\right] = f^{(t)}(x_i) \\[2ex] \sum_{\substack{k+j=q \\ k\leq q;\ j\leq q-1}} \left[\frac{(k+j)!}{k!j!} a^{(k)}(x_i)\left(v_{j+1}+v_j'\right) + \right. \\ \qquad\qquad \left. + \frac{(k+j)!}{k!j!} b^{(k)}(x_i)v_j\right] + \\[1ex] \qquad + a(x_i)v_q' + b(x_i)v_q = f^{(q)}(x_i). \end{cases} \tag{16}$$

The System (16) can be written as

$$A.V' + B.V = F \tag{17}$$

where A and B are two $(q+1) \times (q+1)$ constant matrices and V and V' are two $q+1$ dimensional vectors such that $V(i) = v_i(x); V'(i) = v_i'(x); 1 \leq i \leq q+1$.

Supplemented with some initial conditions given at $x = x_i$, the system (17) has a unique solution, say V. From (11) and (13), we get that $w = u - \bar{u}_i$ is solution to

$$\begin{cases} a(x)w'(x) + b(x)w(x) = \\ \\ \frac{(x-x_i)^{q+1}}{(q+1)!} \left[\tilde{a}(x)\bar{u}_i'(x) + \tilde{b}\bar{u}_i(x) + \tilde{f} \right] \quad x \in \Omega_i \\ \\ w(x_i) = 0. \end{cases} \tag{18}$$

Set $g(x) = -\frac{(x-x_i)^{q+1}}{(q+1)!} \left[\tilde{a}(x)\bar{u}_i'(x) + \tilde{b}\bar{u}_i(x) - \tilde{f} \right]$. Since the coefficient functions in (13) are regular, $\bar{u}_i$ and $\bar{u}_i'$ are continuous. Set $M_0 \in \mathbf{R}, M_0 > 0$ to be such that

$$\text{Max} \left[\underset{0 \le i \le q}{\text{Max}} \left(\underset{x \in \Omega_i}{\text{Max}} |\bar{u}_i(x)| \right), \underset{0 \le i \le q}{\text{Max}} \left(\underset{x \in \Omega_i}{\text{Max}} |\bar{u}_i'(x)| \right) \right] \le M_0$$

The solution w has the following representation form

$$w(x) = \exp\left(-\int_{x_i}^x \frac{b(y)}{a(y)} dy \right) \times \int_{x_i}^x \left[\frac{g(y)}{a(y)} \exp\left(\int_{x_i}^x \frac{b(y)}{a(y)} dy \right) \right] dy. \tag{19}$$

Since $a(x) \ne 0$; $x \in \Omega$ and a is continuous over Ω, there exist a constant τ and a constant M_1 such that $0 < \tau \le |a(x)| \le M_1; x \in \Omega_i$. Same way, since b is continuous over Ω then there exists a constant M_2 such that $|a(x)| \le M_1; x \in \Omega_i$. Of course the constants τ, M_1, M_2 are independent of Ω_i. Taking the estimates M_0, τ, M_1, M_2 into the representation (19), setting $h = x_{i+1} - x_i$, we get to

$$|w(x)| \le Ch^{q+2}$$

where C is a constant independent of h. This ends the proof of Theorem (2.1). $\square$

2.1 A Linear Example

To make clear the numerical method described above, we consider the problem of solving

$$3y' + 2y = 2\cos(2x); \quad x \in]0, T[; y(0) = 0. \tag{20}$$

In every subinterval $]x_i, x_{i+1}[$, $x_{i+1} = x_i + h$ we will be looking for a 3th order approximation in the constant h. Assume we know $v_i(x_i) = y(x_i)$ and $v_i'(x_i) = y'(x_i)$. We set

$$v_i(x) = v0(x) + (x - x_i)v1(x) \tag{21}$$

$f(x) = 2\cos(2x)$ and substitute v_i for y in the equations of problem (20). To get the unknown coefficient functions $v0$ and $v1$ we have to solve the following system of first order differential equations

$$3v0' + 2v0 + 3v1 = f(x_i)$$
$$3v1' + 2v1 = f'(x_i). \tag{22}$$

Set D to be the differential operator with regard to the variable x. Then the operational determinant Δ to system (22) is

$$\begin{vmatrix} L1 & L3 \\ L3 & L4 \end{vmatrix}$$

where
$$L1 = 3D + 2, \quad L2 = 0, \quad L3 = 3, \quad L4 = 3D + 2.$$

The coefficient functions are then solutions to

$$(3D + 2)^2 v0 = 2f(x_i) - 3f'(x_i)$$
$$(3D + 2)^2 v1 = 2f'(x_i) \tag{23}$$

They have the following general forms

$$v0(x) = c01 \exp(-2x/3) + c02\, x \exp(-2x/3) + (1/4)\left[2f(x_i) - 3f'(x_i)\right]$$
$$v1(x) = c11 \exp(-2x/3) + c12\, x \exp(-2x/3) + (1/2)f'(x_i)$$

As in singular perturbation, we select the constant coefficients to make the functions $v0$ and $v1$ are "stable" as possible. We set $c02 = c12 = 0$ to get

$$v0(x) = c01 \exp(-2x/3) + (1/4)\left[2f(x_i) - 3f'(x_i)\right]$$
$$v1(x) = c11 \exp(-2x/3) + (1/2)f'(x_i). \tag{24}$$

The determination of $v0$ and $v1$ is achieved in determining the remaining constant coefficients from the conditions

$$v0(x_i) = v_i(x_i)$$
$$v0'(x_i) + v1(x_i) = v_i'(x_i)$$
$$v_i(x) = v0(x) + (x - x_i)v1(x) \tag{25}$$
$$v_i'(x) = v0'(x) + v1(x) + (x - x_i)v1'(x); \quad x \in]x_i, x_{i+1}[$$
$$y(x_{i+1}) = v_i(x_{i+1})$$

The following picture shows the outcome of a simulation with a meshsize $h = 1/1000$. over an interval $I =]0, 5[$.

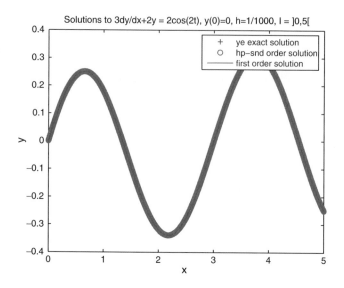

2.2 A Nonlinear Example

Consider the following nonlinear first order differential problem

$$y'(x) = y^2(x), \quad x \in]2, T[; \quad y(2) = -1 \tag{26}$$

We discretize the interval $I =]2, T[$ along with a mesh size h. We consider the sub-interval $I_i =]x_i, x_{i+1}[$. We set

$$v_i(x) = v0(x) + (x - x_i)v1(x). \tag{27}$$

Assume we know $v_i(x_i) = y(x_i)$ and $v_i'(x_i) = v_i^2(x_i)$. From the following identity which is valid within I_i

$$v0' + v1 + (x - x_i)v1' = y^2(x_i) + 2(x - x_i)y^3(x_i) \tag{28}$$

we draw the following system which will lead to the determination of the coefficient functions $v0$ and $v1$:

$$v0' + v1 = y^2(x_i)$$
$$v1' = 2y^3(x_i). \tag{29}$$

The operational determinant to system (29) is

$$\begin{vmatrix} L1 & L3 \\ L3 & L4 \end{vmatrix}$$

where

$$L1 = D, \quad L2 = 0, \quad L3 = 1, \quad L4 = D.$$

The unknown coefficient functions are solution to

$$D^2 v0 = -2y^3(x_i)$$
$$D^2 v0 = 0.$$

We select among the solutions $v0 = -y^3(x_i)x^2 + b0x + c0$ and $v1 = b1x + c1$ the most stable solutions which are constant solutions. The following algorithm guides along the entire interval I to determine the hp-asymptotics approximation to y:

$$
\begin{aligned}
c0 &= v_i(x_i) \\
c1 &= v_i^2(x_i) \\
v0(x_i) &= c0 \\
v1(x_i) &= c1 \\
v_i(x) &= v0(x) + (x - x_i)v1(x), \quad x \in]x_i, x_{i+1}[\\
y(x_{i+1}) &= v_i(x_{i+1}).
\end{aligned}
\tag{30}
$$

Some outputs from this computation are shown in the pictures below.

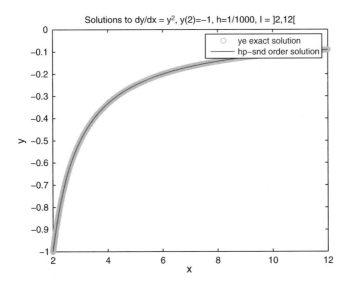

Solutions to dy/dx = y², y(2)=−1, h=1/1000, I =]2,12[

3 The Hp-Asymptotic Method on Higher Order Ordinary Equations

Consider a nth order ordinary differential equation with regular coefficients functions defined over a large interval $\Omega =]0, T[$

$$
\begin{cases}
\sum_{k=0}^{n} a_k(x)u^{(k)}(x) = f(x); & x \in \Omega \\[2mm]
u^{(k)}(0) = \alpha_k \\[2mm]
0 \leq k \leq n-1.
\end{cases}
\tag{31}
$$

Set

$$
\overline{u}(x) = \sum_{j=0}^{q} \frac{(x-x_i)^j}{j!} u_j(x)
$$

$$
\overline{a}_k(x) = \sum_{j=0}^{q} \frac{(x-x_i)^j}{j!} a_j^{(j)}(x_i); \quad 0 \leq k \leq n
$$

$$
\overline{f}(x) = \sum_{j=0}^{q} \frac{(x-x_i)^j}{j!} f^{(j)}(x_i).
$$

We consider, for $0 \leq k \leq n$:

$$
\overline{a}_k(x)\overline{u}^{(k)}(x) = \left[\sum_{\substack{m+j=0 \\ m \leq q; \ j \leq q}}^{q} \frac{(x-x_i)^j}{j!} a_k^{(j)} \right] \times
$$

$$
\left[\sum_{\substack{p=0 \\ k \leq m}}^{q} \binom{k}{p} \frac{(x-x_i)^{m-k}}{(m-k)!} v_m^{(k-p)}(x) + \right.
$$

$$
\left. \sum_{r=0}^{q} \left(\sum_{\substack{p=0 \\ k \geq j+1}}^{k} \binom{k}{p} v_r^{(k-p)}(x) \right) \right] = \overline{f}(x)
\tag{32}
$$

Over Ω_i, we will supplement (32) with some imposed conditions at the mesh point x_i. These conditions are

$$
u^k(x_i) = \overline{u}^k(x_i) = \alpha_{k,i}; \quad 0 \leq k \leq n-1.
$$

Our concern will be to determine the approximation function $\overline{u}$ via its coefficient functions u_j.

In fact since we are looking for analytical solutions which are resistant to numerical oscillations, we can take the mesh size h as small as possible and then content ourself with a qth approximation solution with $q = 1$. So within the current section from now on and unless otherwise stated we assume $q = 1$.

Over $\Omega_i =]x_i, x_{i+1}[$ we set

$$\bar{a}_k(x) = a_k(x_i) + (x - x_i)a'_k(x_i); \quad 0 \le k \le n$$

$$\bar{u}(x) = u_0(x) + (x - x_i)u'_1(x)$$
$$\bar{f}(x) = f(x_i) + (x - x_i)f'(x_i)$$

Then we have

$$\bar{u}^{(k)}(x) = u_0^{(k)}(x) + ku_1^{(k-1)}(x) + (x - x_i)u_1^{(k)}(x); \quad k \in \mathbf{N}$$

and

$$\bar{a}_k\bar{u}^{(k)} = a_k(x_i)\left[u_0^{(k)}(x) + ku_1^{(k-1)}(x)\right] +$$

$$(x - x_i)\left[a_k(x_i)u_1^{(k)}(x) + a'_k(x_i)\left(u_0^{(k)}(x) + ku_1^{(k-1)}(x)\right)\right].$$

Equation (32) supplemented with the conditions set at $x = x_i$ becomes

$$\begin{cases} \sum_{k=0}^{n} \bar{a}_k\bar{u}^{(k)} = \bar{f}(x); \quad x \in \Omega_i \\ \bar{u}^k(x_i) = \alpha_{k,i}; \quad o \le k \le n-1 \end{cases} \tag{33}$$

To determine the coefficients functions u_j we equate power terms in $x - x_i$ in the left hand side to alike terms in the right hand to get

$$\begin{cases} a_0(x_i)u_0(x) + \sum_{k=1}^{n} a_k(x_i)\left[u_0^{(k)}(x) + ku_1^{(k-1)}(x)\right] = f(x_i) \\ \\ a_0(x_i)u_1(x) + a'_0(x_i)u_0(x) + \\ \sum_{k=1}^{n}\left[a_k(x_i)u_1^{(k)}(x) + a'_k(x_i)\times \right. \\ \left. \left(u_0^{(k)}(x) + ku_1^{(k-1)}(x)\right)\right] = f'(x_i). \end{cases} \tag{34}$$

We claim

Theorem 3.1 $\bar{u}$ *and all its* $(n-1)th$ *first derivatives converge strongly to* u *and its correspondent derivatives. More precisely, we have for* $x \in \Omega_i$

$$|(u - \bar{u})^{(i)}(x)| \le C\frac{1}{3!}h^3 exp(h\|A\|_m); \quad 0 \le i \le n-1 \tag{35}$$

where $h = x_{i+1} - x_i$, $\|.\|_m$ *stands for a matrix norm and* A *is a constant matrix to be identified.*

Proof of Theorem 3.1. Set $w = u - \overline{u}$ and set X to be a nth dimensional vector such that $X(1) = X_1 = w$ and $X(i) = X_i = w^{i-1}$; $2 \le i \le n$. We notice in first place that

$$a_k(x) - \overline{a}_k(x) = \frac{(x - x_i)^2}{2!} \tilde{a}_k(x); \quad 0 \le k \le n$$

$$f(x) - \overline{f}(x) = \frac{(x - x_i)^2}{2!} \tilde{f}(x)$$

where $\tilde{a}$ and $\tilde{f}$ are two regular functions. We have

$$\sum_k a_k u^{(k)} - \sum_k \overline{a}_k \overline{u}^{(k)} = \frac{(x - x_i)^2}{2!} \tilde{f}(x); \tag{36}$$

where we may further compute the left hand side term to get

$$\sum_k a_k u^{(k)} - \sum_k \overline{a}_k \overline{u}^{(k)} = \sum_k a_k w^{(k)} + \frac{(x - x_i)^2}{2!} \sum_k \tilde{a}_k(x) \overline{u}^{(k)}(x). \tag{37}$$

From (36) and (37) we get

$$\sum_k a_k w^{(k)} = \frac{(x - x_i)^2}{2!} \left(\tilde{f}(x) - \tilde{a}_k(x) \overline{u}^{(k)}(x) \right); \quad 0 \le k \le n \tag{38}$$

Converting (38) into a system of n first order differential equations completed with the initial conditions we get

$$\begin{cases} \frac{d}{dt} X(x) = AX(x) + \frac{(x - x_i)^2}{2!} F(x); & x \in \Omega_i \\ X(x_i) = 0 \end{cases} \tag{39}$$

where

$$F(x) = \begin{pmatrix} 0 \\ 0 \\ \vdots \\ \tilde{f}(x) - \tilde{a}_k(x) \overline{u}^{(k)}(x) \end{pmatrix}$$

and

$$A = \begin{pmatrix} 0 & 1 & 0 & 0 & \cdots & 0 \\ 0 & 0 & 1 & 0 & \cdots & 0 \\ \vdots & \cdots & \cdots & \ddots & \cdots & 0 \\ \vdots & \cdots & \ddots & \ddots & \ddots & 0 \\ \vdots & \cdots & \cdots & \ddots & \ddots & 1 \\ -\frac{a_1}{a_k} & -\frac{a_2}{a_k} & -\frac{a_3}{a_k} & -\frac{a_4}{a_k} & \cdots & -\frac{a_{n-1}}{a_k} \end{pmatrix}.$$

Problem (39) is a Cauchy problem in finite dimension which solution is given by

$$X(x) = \int_{x_i}^{x} \exp\left(A(x-t)\right) F(t)dt. \tag{40}$$

We have, using the definition of the matrix exponential

$$\|\exp\left(A(x-t)\right)\|_m \leq \sum_{j=1}^{+\infty} \frac{h^j \|A\|_m^j}{j!} = \exp\left(t\|A\|_m\right). \tag{41}$$

Set $C = \underset{x \in \Omega_1}{\text{Max}} |F(x)|$ to get

$$|X(x)| \leq Ch^3 \frac{1}{3!} \exp\left(t\|A\|_m\right).$$

This ends the proof of Theorem (3.1). $\square$

3.1 An Example: A Second Order Equation

$$\begin{cases} y'' + 2y' + y = -2\sin(x) \\ y(0) = 5, \quad y'(0) = 1 \end{cases}. \tag{42}$$

The exact solution to system (42) is

$$y_e(x) = 4\exp(-x) + 5x\exp(-x) + \cos(x).$$

In a subinterval $I =]x_i, x_{i+1}[$, we set

$$v_i(x) = v0(x) + (x - x_i)v1(x). \tag{43}$$

We apply the same reasoning as previously. We select stable hp-asymptotic coefficients which in each subinterval are

$$v0(x) = a0\exp(-x) + f(x_i) - 2f'(x_i)$$

and

$$v1(x) = a0\exp(-x) + f'(x_i).$$

Then if $v_i(x_i) = y(x_i)$ and $v_i'(x_i) = y'(x_i)$ are known, the algorithm for the computation is defined by

$$\begin{aligned} a0 &= \exp(x_i)[v_i(x_i) - f(x_i) + 2f'(x_i)] \\ b0 &= \exp(x_i)[v_i'(x_i) - f'(x_i)] \\ v0 &= a0\exp(-x) - f(x_i) + 2f'(x_i) \\ v1(x) &= b0\exp(-x) + f'(x_i) \\ v_i(x_{i+1}) &= v0(x_{i+1}) + hv1(x_{i+1}) \\ v_i'(x_{i+1}) &= v0'(x_{i+1}) + v1(x_{i+1}) + hv1'(x_{i+1}). \end{aligned} \tag{44}$$

The outcome of a simulation is shown below.

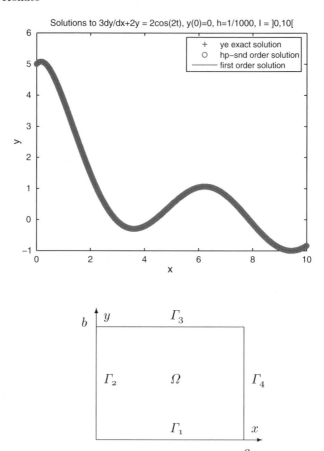

4 The Hp-Asymptotic Method on Elliptic Equations

We consider a large rectangular domain $\Omega =]0, a[\times]0, b[$ where the constants a and b are considered very large. We call Γ the boundary of Ω. We set $\Gamma = \Gamma_1 \cup \Gamma_2 \cup \Gamma_3 \cup \Gamma_4$. Consider the elliptic problem in two dimensions

$$
\begin{aligned}
\Delta u(x, y) &= f(x, y); \quad (x, y) \in \Omega \\
u(x, 0) &= g_1(x) \quad x \in \Gamma_1 \\
u(0, y) &= g_2(y) \quad y \in \Gamma_2 \\
u(x, b) &= g_3(x) \quad x \in \Gamma_3 \\
u(a, y) &= g_4(y) \quad y \in \Gamma_4
\end{aligned}
\tag{45}
$$

We assume that f is regular and the functions g_i on the boundary are regular and compatible that is

$$
g_1(0) = g_2(0); \; g_2(b) = g_3(0); \; g_3(a) = g_4(b); \; g_4(0) = g_1(a)
\tag{46}
$$

We decompose u to find some analytical approximations to it. We set

$$u = u_0 + u_1 + u_2 + u_4 \tag{47}$$

where

$$\begin{aligned} \Delta u_0(x,y) &= f(x,y); \quad (x,y) \in \Omega \\ u_0 &= 0 \quad \text{on} \quad \Gamma; \end{aligned} \tag{48}$$

$$\begin{aligned} \Delta u_1(x,y) &= 0; \quad (x,y) \in \Omega \\ u_1(x,0) &= g_1(x) \quad x \in \Gamma_1 \\ u_1 &= 0 \quad \text{on} \quad \Gamma_2 \cup \Gamma_3 \cup \Gamma_4 \end{aligned} \tag{49}$$

$$\begin{aligned} \Delta u_2(x,y) &= 0; \quad (x,y) \in \Omega \\ u_2(0,y) &= g_2(y) \quad y \in \Gamma_2 \\ u_2 &= 0 \quad \text{on} \quad \Gamma_1 \cup \Gamma_3 \cup \Gamma_4 \end{aligned} \tag{50}$$

$$\begin{aligned} \Delta u_3(x,y) &= 0; \quad (x,y) \in \Omega \\ u_3(x,b) &= g_3(x) \quad x \in \Gamma_3 \\ u_3 &= 0 \quad \text{on} \quad \Gamma_1 \cup \Gamma_2 \cup \Gamma_4 \end{aligned} \tag{51}$$

$$\begin{aligned} \Delta u_4(x,y) &= 0; \quad (x,y) \in \Omega \\ u_4(a,y) &= g_4(y) \quad y \in \Gamma_4 \\ u_4 &= 0 \quad \text{on} \quad \Gamma_1 \cup \Gamma_2 \cup \Gamma_3 \end{aligned} \tag{52}$$

To solve every system above, one has to compute the separated solutions to the attached Sturm-Liouville problems and then apply the classical Fourier method. We get:

$$\begin{aligned} u_1(x,y) &= \sum_{n=1}^{+\infty} \alpha_n \sin\left(\frac{n\pi x}{a}\right) \sinh\left(\frac{n\pi(b-y)}{a}\right) \\ u_2(x,y) &= \sum_{n=1}^{+\infty} \beta_n \sin\left(\frac{n\pi y}{b}\right) \sinh\left(\frac{n\pi(a-x)}{b}\right) \\ u_3(x,y) &= \sum_{n=1}^{+\infty} \gamma_n \sin\left(\frac{n\pi x}{a}\right) \sinh\left(\frac{n\pi y}{a}\right) \\ u_4(x,y) &= \sum_{n=1}^{+\infty} \delta_n \sin\left(\frac{n\pi y}{b}\right) \sinh\left(\frac{n\pi x}{b}\right) \end{aligned} \tag{53}$$

where the coefficients are such that

$$
\begin{cases}
\alpha_n = a_n/A_n; & \gamma_n = c_n/A_n; \\
\beta_n = b_n/B_n; & \delta_n = d_n/B_n; \\
A_n = \sinh\left(\frac{n\pi b}{a}\right); & B_n = \sinh\left(\frac{n\pi a}{b}\right),
\end{cases}
\tag{54}
$$

and

$$
a_n = \frac{2}{a}\int_0^a g_1(x)\sin\frac{n\pi x}{a}\,dx
$$

$$
b_n = \frac{2}{b}\int_0^b g_2(y)\sin\frac{n\pi y}{b}\,dy
$$

$$
c_n = \frac{2}{a}\int_0^a g_3(x)\sin\frac{n\pi x}{a}\,dx
\tag{55}
$$

$$
d_n = \frac{2}{b}\int_0^b g_4(y)\sin\frac{n\pi y}{b}\,dy
$$

$$
n \geq 1.
$$

We look for a particular solution to system (48). It is really about the computing of u_0 that the hp-asymptotic method is used. We discretize the domain Ω. Set

$$
\omega_{i,j} = \{(x,y); x \in]x_i, x_{i+1}[;\ y \in]y_j, y_{j+1}[\}.
$$

Over the cell $\omega_{i,j}$, we have

$$
f(x,y) = f(x_i, y_j) + (x - x_i)\frac{\partial f}{\partial x}(x_i, y_j) + (y - y_j)\frac{\partial f}{\partial y}(x_i, y_j).
\tag{56}
$$

We assume that for $x \in]x_i, x_{i+1}[$ and $y \in]y_j, y_{j+1}[$, u_0 is of the form

$$
u_0(x,y) = w_0(x,y) + (x - x_i)w_1(y) + (y - y_j)w_2(x).
\tag{57}
$$

From equality (57) we draw

$$
\Delta u_0 = \Delta w_0 + (y - y_j)\frac{d^2 w_2}{dx^2}(x) + (x - x_i)\frac{d^2 w_1}{dy^2}(y).
\tag{58}
$$

Put together this equality above and system (48), and identify alike terms in the factors $(x - x_i)$ and $(y - y_j)$ to obtain the following equations depending on the coefficient functions w_0, w_1, and w_3, to be determined,

$$
\Delta w_0(x,y) = f(x_i, y_j)
$$

$$
\frac{d^2 w_1}{dy^2}(y) = \frac{\partial f}{\partial x}(x_i, y_j)
$$

$$
\frac{d^2 w_2}{dx^2}(x) = \frac{\partial f}{\partial y}(x_i, y_j)
\tag{59}
$$

$$
x \in]x_i, x_{i+1}[;\quad y \in]y_j, y_{j+1}[.
$$

Together system (48) and identity (57) allow us to determine the boundary conditions attached to the equations above. Those boundary conditions are found to be

$$\begin{cases} w_0(x,0) = 0, \ w_0(0,y) = 0, \ w_0(x,b) = 0, \ w_0(a,y) = 0, \\ w_1(0) = 0, \ w_1(b) = 0, \\ w_2(0) = 0, \ w_1(a) = 0. \end{cases} \tag{60}$$

The functions w_1 and w_2 are easy to compute and

$$w_1(y) = \frac{1}{2}\frac{\partial f}{\partial x}(x_i, y_j) \left[y\left(y - b\right) \right],$$

$$w_2(x) = \frac{1}{2}\frac{\partial f}{\partial y}(x_i, y_j) \left[x\left(x - a\right) \right]. \tag{61}$$

To solve the system (60), first we turn it into an homogeneous problem. We set

$$w_4(x,y) = w_0(x,y) - w_3(x,y) \tag{62}$$

where

$$w_3(x,y) = \frac{1}{2}f(x_i, y_j) \left[y\left(y - b\right) \right] \tag{63}$$

is the particular solution to

$$\frac{d^2 w_3}{dy^2}(x,y) = f(x_i, y_j)$$

obtained under the conditions

$$w_3(0) = 0, \quad w_3(b) = 0.$$

We obtain then that w_4 satisfies the Laplace equation

$$\Delta w_4(x,y) = 0; \tag{64}$$

which separated solutions via its Sturn Liouville problems have the form

$$\sin\left(\frac{n\pi y}{b}\right)\left[a_n \sinh\left(\frac{n\pi x}{b}\right) + b_n \cosh\left(\frac{n\pi x}{b}\right) \right].$$

We use the following boundary conditions to determined the coefficients a_n and b_n

$$w_4(x,0) = 0, \ w_4(0,y) = -w_3(0,y), \ w_4(x,b) = 0, \ w_4(a,y) = -w_3(a,y)$$

We get

$$w_4(x,y) = w_{4,2}(x,y) + w_{4,4}(x,y)$$

$$w_{4,2}(x,y) = \sum_{n=1}^{+\infty} \beta_{n,w} \sin\left(\frac{n\pi y}{b}\right) \sinh\left(\frac{n\pi(a - x)}{b}\right) \tag{65}$$

$$w_{4,4}(x,y) = \sum_{n=1}^{+\infty} \delta_{n,w} \sin\left(\frac{n\pi y}{b}\right) \sinh\left(\frac{n\pi x}{b}\right)$$

50   D. Konaté

where

$$\begin{cases} \beta_{n,w} = b_{n,w}/B_{n,w}; \quad \delta_{n,w} = d_{n,w}/B_{n,w}; \\ B_{n,w} = \sinh\left(\frac{n\pi a}{b}\right), \end{cases} \quad (66)$$

and

$$b_{n,w} = -\frac{2}{b}\int_0^b w_3(y)\sin\frac{n\pi y}{b}\,dy$$
$$d_{n,w} = b_{n,w}, \quad n \geq 1.$$

Further computation brings

$$\int_0^b w_3(y)\sin\frac{n\pi y}{b}\,dy = f(x_i,y_j)\left(\frac{b}{n\pi}\right)^3[(-1)^n - 1] \quad (67)$$

which leads to

$$b_{n,w} = -2f(x_i,y_j)\frac{b^2}{(n\pi)^3}[(-1)^n - 1],$$
$$\beta_{n,w} = b_{n,w}/B_{n,w}, \quad (68)$$
$$B_{n,w} = \sinh\left(\frac{n\pi a}{b}\right)$$

or

$$b_{n,w} = 0, \quad n \text{ even}$$
$$b_{n,w} = 4f(x_i,y_j)\frac{b^2}{(n\pi)^3}, \quad n \text{ odd}$$
$$\beta_{n,w} = b_{n,w}/B_{n,w}, \quad (69)$$
$$B_{n,w} = \sinh\left(\frac{n\pi a}{b}\right)$$

References

1. W. Eckhaus, (1973) *Matched asymptotic expansions and singular perturbations* North Holland - Amsterdam.
2. D. Konaté, (2006) *Asymptotic analysis of two coupled diffusion-reaction problems*, in Ravi P. Agarwal and Kanishka Perera (eds.) Proceedings of the Conference on Differential & Difference Equations and Applications, Hindawi Publishing Corporation, pp. 575–592.
3. R. O'Malley Jr., (1991) *Singular Perturbation for Ordinary Differential Equations*, Applied Mathematical Sciences, 89, Berlin Heidelberg New York, Springer.
4. M. Pinsky, (1984) *Introduction to partial differential equations with applications*, Newyork, McGraw-Hill.
5. W. Wasow, (1965) *Asymptotic expansions for ordinary differential equations*, Newyork, Interscience publication.

Modelling the Thermal Operation in a Catalytic Converter of an Automobile's Exhaust

O.D. Makinde

Summary. Catalytic converter in an automobile's exhaust system is made up of a finely divided platinum–iridium catalyst (i.e. forming a porous matrix) and provides a platform for exothermic chemical reaction where unburned hydrocarbons completely combust. In this paper, the steady-state solutions of a strongly exothermic reaction of a viscous combustible fluid (fuel) in a catalytic converter-modelled as a cylindrical pipe filled with a saturated porous medium under Arrhenius kinetics, neglecting reactant consumption, are presented. The Brinkman flow model is employed. Having known the velocity distribution, the nonlinear energy equation is solved using a perturbation technique together with a special type of Hermite–Padé approximants and the important properties of the temperature field including bifurcations and thermal criticality are discussed.

Keywords: Catalytic converter; Reactive viscous flow; Bifurcation study; Hermite–Padé approximants

1 Introduction

It is indeed important nowadays to have vehicles that will help you reach your destinations; however, vehicles are among those that contribute much to pollution. Good thing, innovations are fast to its guard; technology has introduced techniques to lessen or better yet to prevent vehicles from emitting harmful elements that can contribute to pollution. Aside from the many techniques invented, laws and regulations regarding such problems were introduced. Such laws stated standards on emission controls. Catalytic converter is one innovation that had helped a lot in emission problems.

Before catalytic converters, there are lots of other techniques to lessen the effects of vehicle emissions to the air. Standards have been passed and approved for the redesigning of engines and engine systems; auto manufacturers and engine makers have made ways to convert elements from engine's combustions into less harmful elements [2, 6, 10]. Nowadays, performance quality catalytic converters are all you need and every vehicle produced are required to have at least one catalytic converter. These performance parts are important as they help clean the gases produced by the vehicles as it burns fuel when operating. They help control harmful gas emissions from your vehicles like hydrocarbon, carbon monoxide and nitrogen oxide [9]. Usually, catalytic converters are being installed between the exhaust manifold and the muffler; it uses chemical that act as catalyst. Catalytic converters work in three stages, the reduction catalyst, oxidization catalyst and the control system (see Figs. 1 and 2). Meanwhile, in order to ignite, stabilize and operate under steady-state conditions, the thermal criticality of a burner based on combustion in inert porous media like catalytic converter must be determined [10].

Mathematically speaking, thermal ignition and heat transfer in inert porous media constitutes a nonlinear reaction diffusion problem and the long-time behaviour of the solutions in space will provide us an insight into inherently complex physical process of thermal runaway in the system, [3, 8]. The theory of nonlinear reaction diffusion equations is quite elaborate and their solution in rectangular, cylindrical and spherical coordinate remains an extremely important problem of practical relevance in the engineering sciences, [1, 9]. Several numerical approaches have developed in the last few decades, e.g. finite differences, spectral method, shooting method, etc. to tackle this problem. More recently, the ideas on classical analytical meth-

A Reduction catalyst
B Oxidation catalyst
C Honeycomb

Fig. 1. A picture showing the operation of a catalytic converter in an exhaust pipe

Fig. 2. A ceramic honeycomb catalyst structure of a catalytic converter

ods have experienced a revival, in connection with the proposition of novel hybrid numerical–analytical schemes for nonlinear differential equations. One such trend is related to Hermite–Padé approximation approach, [5, 7, 11]. This approach, over the last few years, proved itself as a powerful benchmarking tool and a potential alternative to traditional numerical techniques in various applications in sciences and engineering. This semi-numerical approach is also extremely useful in the validation of purely numerical scheme.

In this paper, we intend to construct approximate solution for a steady-state reaction diffusion equation that models the thermal operation of a catalytic converter in an exhaust pipe using perturbation technique together with a special type of Hermite–Padé approximants. The mathematical formulation of the problem is established and solved in sections two and three. In section four we introduce and apply some rudiments of Hermite–Padé approximation technique. Both numerical and graphical results are presented and discussed quantitatively with respect to various parameters embedded in the system in section five.

2 Mathematical Model

We modelled the thermal operation in a catalytic converter as a steady-state hydrodynamically and thermally developed unidirectional flow of a viscous combustible reacting fluid in the z-direction inside a cylindrical pipe of uniform cross-section with impermeable isothermal wall at $r = a$, filled with a homogeneous and isotropic porous medium as illustrated in Fig. 1 below.

Neglecting reactant consumption, the governing momentum and energy balance equations are

$$\frac{d^2u}{dr^2} + \frac{1}{r}\frac{du}{dr} - \frac{u}{K} - \frac{1}{\mu}\frac{dP}{dz} = 0 \tag{1}$$

$$\frac{d^2T}{dr^2} + \frac{1}{r}\frac{dT}{dr} + \frac{QC_0A}{k}e^{-\frac{E}{RT}} + \frac{\mu}{k}\left(\frac{du}{dr}\right)^2 + \frac{\mu u^2}{Kk} = 0 \tag{2}$$

Equation (1) is a well-known Brinkman momentum equation (Brinkman [2]) while the additional viscous dissipation term in (2) is due to Al-Hadhrami et al. [1] and is valid in the limit of very small and very large porous medium permeability.

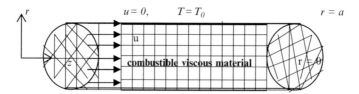

Fig. 3. Geometry of the problem

The appropriate boundary conditions are

$$u = 0, \ T = T_0, \quad \text{on} \quad r = a, \tag{3}$$

$$\frac{du}{dr} = 0, \ \frac{dT}{dr} = 0, \quad \text{on} \quad r = 0, \tag{4}$$

where T is the absolute temperature, P the fluid pressure, T_0 the geometry wall temperature, k the thermal conductivity of the material, K the porous medium permeability parameter, Q the heat of reaction, A the rate constant, E the activation energy, R the universal gas constant, C_0 the initial concentration of the reactant species, a the pipe radius, $(r, \ z)$ the distance measured in the radial and axial directions, respectively, and μ is the combustible material dynamic viscosity coefficient. Let $M = -(a/U\mu)(dP/dz)$ be a constant axial pressure gradient parameter and U the fluid characteristic velocity (Fig. 3). We introduce the following dimensionless variables into (1)–(4);

$$\theta = \frac{E(T - T_0)}{RT_0^2}, \quad \varepsilon = \frac{RT_0}{E}, \quad \bar{r} = \frac{r}{a}, \quad \lambda = \frac{QEAa^2 C_0 e^{-\frac{E}{RT_0}}}{T_0^2 Rk},$$

$$\bar{z} = \frac{z}{a}, \quad W = \frac{u}{UM}, \quad \delta = \frac{\mu M^2 U^2 e^{\frac{E}{RT_0}}}{QAa^2 C_0}, \quad \beta = \sqrt{\frac{1}{Da}}, \quad Da = \frac{K}{a^2}, \tag{5}$$

and obtain the dimensionless governing equation together with the corresponding boundary conditions as (neglecting the bar symbol for clarity);

$$\frac{d^2W}{dr^2} + \frac{1}{r}\frac{dW}{dr} - \beta^2 W + 1 = 0, \tag{6}$$

$$\frac{d^2\theta}{dr^2} + \frac{1}{r}\frac{d\theta}{dr} + \lambda(e^{\left(\frac{\theta}{1+\varepsilon\theta}\right)} + \delta\left(\frac{dW}{dr}\right)^2 + \delta\beta^2 W^2) = 0, \tag{7}$$

$$W(1) = \theta(1) = 0, \tag{8a}$$

$$\frac{dW}{dr}(0) = 0, \quad \frac{d\theta}{dr}(0) = 0, \tag{8b}$$

where λ, ε, δ, β, Da represent the Frank–Kamenetskii parameter, activation energy parameter, the viscous heating parameter, the porous medium shape factor parameter and the Darcy number respectively. In the following sections, (6–8) are solved using both perturbation and multivariate series summation techniques.

3 Perturbation Method

It is very easy to obtain the solution for the fluid velocity profile exactly, however, due to the nonlinear nature of the temperature field (7), it is convenient to form a power series expansion in the Frank–Kamenetskii parameter λ, i.e.,

$$\theta(r) = \sum_{i=0}^{\infty} \theta_i \lambda^i. \tag{9}$$

Substituting the solution series (9) into (7) and collecting the coefficients of like powers of λ, we obtained and solved the equations of the coefficients of solution series iteratively. The solution for the velocity and temperature fields are given as

$$W(r; \beta > 0) = \frac{1}{\beta^2}\left(1 - \frac{I_0(\beta r)}{I_0(\beta)}\right), \tag{10a}$$

$$W(r; \beta \to 0) = -\frac{1}{4}(r^2 - 1) - \frac{\beta^2}{64}(r^2 - 1)(r^2 - 3)$$
$$- \frac{\beta^4}{2304}(r^2 - 1)(r^4 - 8r^2 + 19) + O(\beta^6), \tag{10b}$$

$$\theta(r) = -\frac{1}{7372800}\lambda(r^2 - 1)\Big(188\delta\beta^6 r^8 + 13508\delta\beta^4 r^6 - 2078\delta\beta^6 r^6 + 256008\delta\beta^2 r^4$$
$$+ 8938\delta\beta^6 r^4 - 98508\delta\beta^4 r^4 + 1152008\delta r^2 - 896008\delta\beta^2 r^2 + 225508\delta\beta^4 r^2$$
$$- 180788\delta\beta^6 r^2 + 256008\delta\beta^2 - 206508\delta\beta^4 + 1843200 + 22438\delta\beta^6 + 1152008\delta\Big)$$
$$+ O(\lambda^2) \tag{11}$$

Using a computer symbolic algebra package (MAPLE), we obtained the first 21 terms of the above solution series (11) as well as the series for the fluid maximum temperature $\theta_{max} = \theta(r = 0; \lambda, \varepsilon, \beta, \delta)$. We are aware that the power series solution in (11) is valid for large Darcy number ($\beta \to 0$) and very small Frank–Kamenetskii parameter values ($\lambda \to 0$). However, using Hermite–Padé approximation technique, we have extended the usability of the solution series beyond small parameter values as illustrated in the following section.

4 Thermal Criticality and Bifurcation Study

The concept of thermal criticality or non-existence of steady-state solution to nonlinear reaction diffusion problems for certain parameter values is extremely important from application point of view. This characterizes the thermal stability properties of the materials under consideration and the onset of thermal runaway phenomenon. In order to determine the appearance of thermal runaway in the system together with the evolution of temperature field as the exothermic reaction rate increases (i.e. $\lambda > 0$), we employ a special type of Hermite–Padé approximation technique. Suppose that the partial sum

$$U_{N-1}(\lambda) = \sum_{i=0}^{N-1} a_i \lambda^i = U(\lambda) + O(\lambda^N) \quad \text{as } \lambda \to 0, \tag{12}$$

is given. We are concerned with the bifurcation study by analytic continuation as well as the dominant behaviour of the solution by using partial sum (12). We expect that the accuracy of the critical parameters will ensure the accuracy of the solution. It is well known that the dominant behaviour of a solution of a differential equation can often be written as Guttamann [4],

$$U(\lambda) \approx \begin{cases} H(\lambda_c - \lambda)^\alpha & \text{for} \quad \alpha \neq 0, 1, 2, \ldots \\ H(\lambda_c - \lambda)^\alpha \ln |\lambda_c - \lambda| & \text{for} \quad \alpha = 0, 1, 2, \ldots \end{cases} \quad \text{as } \lambda \to \lambda_c, \tag{13}$$

where H is some constant and λ_c is the critical point with the exponent α. However, we shall make the simplest hypothesis in the contest of nonlinear problems by assuming the $U(\lambda)$ is the local representation of an algebraic function of λ. Therefore, we seek an expression of the form

$$F_d(\lambda, U_{N-1}) = A_{0N}(\lambda) + A_{1N}^d(\lambda)U^{(1)} + A_{2N}^d(\lambda)U^{(2)} + A_{3N}^d(\lambda)U^{(3)}, \tag{14}$$

such that

$$A_{0\mathrm{N}}(\lambda) = 1, \quad A_{iN}(\lambda) = \sum_{j=1}^{d+i} b_{ij}\lambda^{j-1}, \tag{15}$$

and

$$F_d(\lambda, U) = O(\lambda^{N+1}) \quad \text{as } \lambda \to 0, \tag{16}$$

where $d \geq 1$, $i = 1, 2, 3$. The condition (15) normalizes the F_d and ensures that the order of series A_{iN} increases as i and d increase in value. There are thus $3(2 + d)$ undetermined coefficients b_{ij} in the expression (15). The requirement (16) reduces the problem to a system of N linear equations for the unknown coefficients of F_d. The entries of the underlying matrix depend only on the N given coefficients a_i. Henceforth, we shall take

$$N = 3(2 + d), \tag{17}$$

so that the number of equations equals the number of unknowns. (16) is a new special type of Hermite–Padé approximants. Both the algebraic and differential approximants forms of (16) are considered. For instance, we let

$$U^{(1)} = U, \ U^{(2)} = U^2, \ U^{(3)} = U^3,$$

(18)

and obtain a cubic Padé approximant. This enables us to obtain solution branches of the underlying problem in addition to the one represented by the original series. In the same manner, we let

$$U^{(1)} = U, \ U^{(2)} = DU, \ U^{(3)} = D^2U,$$

(19)

in (15), where D is the differential operator given by $D = d/d\lambda$. This leads to a second order differential approximants. It is an extension of the integral approximants idea by Hunter and Baker [5] and enables us to obtain the dominant singularity in the flow field i.e. by equating the coefficient $A_{3N}(\lambda)$ in the (16) to zero. Meanwhile, it is very important to know that the rationale for chosen the degrees of A_{iN} in (15) in this particular application is based on the simple technique of singularity determination in second order linear ordinary differential equation with polynomial coefficients as well as the possibility of multiple solution branches for the nonlinear problem [12]. In practice, one usually finds that the dominant singularities are located at zeroes of the leading polynomial $A_{3N}^{(d)}$ coefficients of the second order linear ordinary differential equation. Hence, some of the zeroes of $A_{3N}^{(d)}$ may provide approximations of the singularities of the series U and we expect that the accuracy of the singularities will ensure the accuracy of the approximants.

The critical exponent α_N can easily be found by using Newton's polygon algorithm. However, it is well known that, in the case of algebraic equations, the only singularities that are structurally stable are simple turning points. Hence, in practice, one almost invariably obtains $\alpha_N = 1/2$. If we assume a singularity of algebraic type as in (13), then the exponent may be approximated by

$$\alpha_N = 1 - \frac{A_{2N}(\lambda_{CN})}{DA_{3N}(\lambda_{CN})}.$$

(20)

5 Results and Discussion

The bifurcation procedure above is applied on the first 21 terms of the solution series and we obtained the results shown in Tables 1 and 2 below:

The result in Table 1 shows the rapid convergence of our procedure for the dominant singularity (i.e. λ_c) together with its corresponding critical exponent α_c with gradual increase in the number of series coefficients utilized in the approximants. In Table 2, we noticed that the magnitude of thermal criticality at very large activation energy ($\varepsilon = 0$) decreases with a decrease in the porous medium permeability ($\beta > 0$). This shows clearly that reducing

Table 1. Computations showing the procedure rapid convergence for $\varepsilon = 0.0$, $\delta = 0.0$

d	N	$\theta_{\max}$	λ_c	α_{cN}
1	9	1.386540593950578	2.0000471922705	0.499999
3	15	1.386294361119890	2.0000000000000	0.500000
5	21	1.386294361119890	2.0000000000000	0.500000

Table 2. Computations showing thermal ignition criticality for different parameter values $(\delta,\ \beta,\ \varepsilon)$

δ	β	ε	$\theta_{\max}$	λ_c	α_{cN}
1.0	0.0	0.1	1.8491492	2.2068382	0.500000
1.0	0.5	1.0	1.8508176	2.2062313	0.500000
0.0	0.0	0.0	1.3865405	2.0000000	0.500000
1.0	0.0	0.0	1.4157385	1.9454358	0.500000
1.0	0.1	0.0	1.4158015	1.9453932	0.500000
1.0	0.3	0.0	1.4162698	1.9450999	0.500000
1.0	0.5	0.0	1.4170165	1.9447578	0.500000

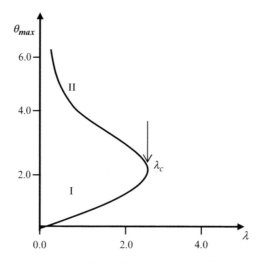

Fig. 4. A slice of approximate bifurcation diagram in the $(\lambda,\ \theta_{\max}(\beta,\ \varepsilon))$ plane

the permeability of a porous medium will enhance the early appearance of ignition in a reactive viscous flow of a combustible fluid. It is noteworthy that a decrease in the combustible fluid activation energy (i.e. $\varepsilon > 0$) will lead to an increase in the magnitude of thermal ignition criticality, hence, delaying the appearance of thermal runaway in the system. A slice of the bifurcation diagram for $0 \le \varepsilon \ll 1$ is shown in Fig. 4. In particular, for every $\beta \ge 0$, there is a critical value λ_c (a turning point) such that, for $0 \le \lambda < \lambda_c$ there are two

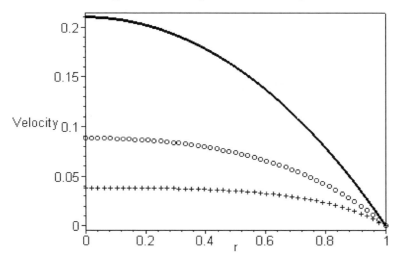

Fig. 5. Fluid velocity profile: $------\beta = 1$; ooooooo $\beta = 3.0$; ++++++$\beta = 5.0$

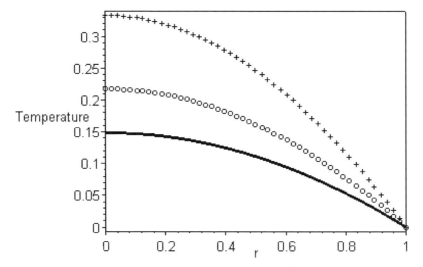

Fig. 6. Fluid temperature profile: $\beta = 1$; $\varepsilon = 0$; $\delta = 1$; $-------\lambda = 0.5$; oooooo $\lambda = 0.7$; ++++++$\lambda = 1$

solutions (labeled I and II) and the solution II diverges to infinity as $\lambda \to 0$. The fully developed dimensionless velocity distribution is shown in Fig. 5. We observed that the magnitude of the fluid velocity increases and tend to that of Poiseuille flow with a gradual increase in the porous medium permeability (i.e. $\beta \to 0$). Similarly, an increase in the fluid temperature is observed with increasing values of λ due to a combined effects of viscous dissipation and exothermic reaction as shown in Fig. 6.

6 Conclusion

The thermal stability of a reactive viscous fluid flowing through a porous-saturated pipe is investigated using perturbation technique together with a special type of Hermite–Padé approximants. We obtained accurately the steady-state thermal ignition criticality conditions as well as the solution branches. It is observed that a reduction in porous medium permeability will enhance complete combustion, hence, improving the effectiveness of engineering equipments like the catalytic converter used in an automobile's exhaust system. Finally, the above analytical and computational procedures are advocated as effective tool for investigating several other parameter dependent nonlinear boundary-value problems.

Nomenclature

a	Pipe radius
A	rate constant
C_0	concentration of the reactant
Da	Darcy number
E	activation energy
k	thermal conductivity
K	permeability
P	fluid pressure
Q	heat of reaction
R	universal gas constant
T_0	wall temperature
T	absolute temperature
W	fluid velocity
z	axial distance
r	radial distance

Greek symbols

μ	fluid dynamics viscosity
λ	Frank–Kamenetskii
ε	activation energy parameter
δ	viscous heating parameter
β	porous medium shape factor
θ	dimensionless temperature

References

1. Al-Hadhrami, A. K., Elliott, L., and Ingham, D. B., 2003, A new model for viscous dissipation in porous media across a range of permeability values, Transport in Porous Media 53, 117–122.
2. Brinkman, H. C., 1947, On the permeability of media consisting of closely packed porous particles, Appl. Sci. Res. A1, 81–86.

3. Frank Kamenetskii, D. A., 1969, Diffusion and Heat Transfer in Chemical kinetics. Plenum Press, New York.
4. Guttamann, A. J., 1989, Asymptotic analysis of power–series expansions, Phase Transitions and Critical Phenomena, C. Domb and J. K. Lebowitz, eds. Academic Press, New York, pp. 1–234.
5. Hunter, D. L. and Baker, G. A., 1979, Methods of series analysis III: Integral approximant methods, Phys. Rev. B 19, 3808–3821.
6. Makinde, O. D., 1999, Extending the utility of perturbation series in problems of laminar flow in a porous pipe and a diverging channel, J. Austral. Math. Soc. Ser. B 41, 118–128.
7. Makinde, O. D., 2004, Exothermic explosions in a slab: a case study of series summation technique, Inter. Comm. Heat & Mass Transfer. 31, 1227–1231.
8. Makinde, O. D., 2005, Strong exothermic explosions in a cylindrical pipe: A case study of series summation technique, Mech. Res. Comm. Vol. 32, 191–195.
9. Makinde, O. D., 2006, Thermal ignition in a reactive viscous flow through a channel filled with a porous medium, ASME - J. Heat Transfer, 128, 601–604.
10. Makinde, O. D., 2007, Solving microwave heating model in a slab using Hermite–Pade approximation technique, App. Therm. Eng., 27, 599–603.
11. Tourigny, Y. and Drazin, P. G., 2000, The asymptotic behaviour of algebraic approximants, Proc. Roy. Soc. London A456, 1117–1137.
12. Vainberg, M. M. and Trenogin V. A., 1974, Theory of branching of solutions of nonlinear equations, Noordoff, Leyden.

Modelling Transmission Dynamics of Childhood Diseases in the Presence of a Preventive Vaccine: Application of Adomian Decomposition Technique

O.D. Makinde

Summary. In recent time, diligent vaccination campaigns have resulted in high levels of permanent immunity against the childhood disease among the population, e.g. measles, mumps, rubella, poliomyelitis, etc. In this paper, a SIR model that monitors the temporal dynamics of a childhood disease in the presence of a preventive vaccine is developed. The qualitative analysis reveals the vaccination reproductive number for disease control and eradication. Adomian decomposition method is also employed to compute an approximation to the solution of the non-linear system of differential equations governing the problem. Graphical results are presented and discussed quantitatively to illustrate the solution.

Keywords: Childhood disease model; Preventive vaccine; Stability analysis; Adomian decomposition

1 Introduction

Diseases such as measles, mumps, chicken pox, poliomyelitis, etc. to which children are born susceptible, and usually contract within the first five years are generally refereed to childhood diseases. Young children are in particularly close contact with their peers, at school and playground; hence such diseases can spread quickly. Meanwhile, the development of vaccines against infectious childhood diseases has been a boon to mankind and protecting children from diseases that can be prevented by vaccination is a primary goal of health administrators. For instance, measles is an acute, highly communicable viral disease with prodromal fever and can be prevented by the *MMR* (Measles| Mumps|Rubella) vaccine. The primary reason for continuing high childhood measles morbidity and mortality in sub-Saharan Africa is the failure to deliver at least one dose of measles vaccine to all infants (see Table 1).

Figure 1 shows the 2002 UNICEF measles campaign in a remote village in central Senegal during which *Serigne Dame Léye – Headman of Ngouye*

Table 1. Estimated measles deaths for 2004 with uncertainty bounds by World Bank geographical region [4]

Region	Estimated measles deaths	Uncertainty bounds
Sub-Saharan Africa	216,000	[216,000 - 279,000]
South Asia	202,000	[145,000 - 264,000]
East Asia & Pacific	32,000	[21,000 - 47,000]
Middle East & North Africa	4,000	[2,000 - 5,000]
Europe & Central Asia	<1,000	[–]
Latin America & Caribbean	<1,000	[–]
High income countries	<1,000	[–]

Fig. 1. Picture showing children after receiving their individual vaccinations against measles – Measles Campaign 2002 by UNICEF

Diaraf remarked "We used to bury two or three children every week because of measles. This does not happen anymore because our children are immunized".

Another illustrative example of childhood disease with preventive vaccine is poliomyelitis [4]. This is a highly contagious, incurable viral infection of the nervous system which can cause crippling paralysis or even death within hours of infection. At its peak, polio paralyzed and killed up to half a million people every year, before Jonas Salk invented a vaccine in 1955, (see Fig. 2).

According to Fig. 3, with only four polio endemic countries left in the world, with constant vaccination strategy-polio transmission could be stopped in every country by the end of 2006. The world could then be certified polio-free by end-2010.

Since vaccination is considered to be the most effective strategy against childhood diseases, the development of a framework that would predict the optimal vaccine coverage level needed to prevent the spread of these diseases is crucial. The SIR model is a standard compartmental model that has been used to describe many epidemiological diseases [8–11]. The way several childhood diseases spread through a population fits into this framework. The model has

Fig. 2. (a) Members of a mobile Polio immunization team in Kano, Nigeria. **(b)** A child receives a dose of oral polio vaccine

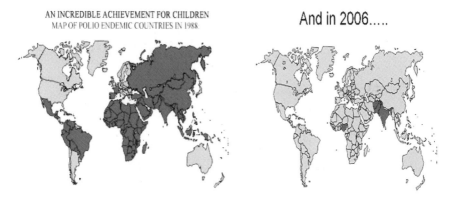

Fig. 3. The map showing the World Polio endemic countries [4]

a susceptible group designated by S, an infected group I, and a removed group R, denoting vaccinated as well as recovered people with permanent immunity. This model assumes that the efficacy of the vaccine is 100% and the natural death rates μ in the classes remain unequal to births, so that the population size N is realistically not constant. Citizens are born into the population at a constant birth rate π with extremely very low childhood disease mortality rate. We denote the fraction of citizens vaccinated at birth each year as P (with $0 < P < 1$) and assume the rest are susceptible. A susceptible individual will move into the infected group through contact with an infected individual, approximated by an average contact rate β. An infected individual recovers at a rate γ, and enters removed group. The removed group also contains people who are vaccinated. The differential equations for the SIR model are

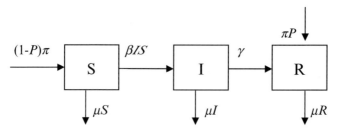

Fig. 4. Flow chart for the SIR model

$$\frac{dS}{dt} = (1 - P)\pi N - \beta\frac{SI}{N} - \mu S, \tag{1}$$

$$\frac{dI}{dt} = \beta\frac{SI}{N} - (\gamma + \mu)I, \tag{2}$$

$$\frac{dR}{dt} = P\pi N + \gamma I - \mu R. \tag{3}$$

We also have the relationship $N = S + I + R$ and assume μ, π, β, γ, μ are all positive constant parameters. Adding (1)–(3), we obtain

$$\frac{dN}{dt} = (\pi - \mu)N, \tag{4}$$

so that we are now dealing with a varying total population, [6].

A summary of the process is drawn in a flow chart in Fig. 1 below: (Fig. 4)

The groups can be scaled by population N using the new variables, $s = S/N$, $i = I/N$, and $r = R/N$. The population is now normalised, meaning $s + i + r = 1$, and we have the new system,

$$\frac{ds}{dt} = (1 - P)\pi - \beta si - \pi s, \tag{5}$$

$$\frac{di}{dt} = \beta si - (\gamma + \pi)i, \tag{6}$$

$$\frac{dr}{dt} = P\pi + \gamma i - \pi r. \tag{7}$$

2 Qualitative Analysis

Since r does not appear in the (5) and (6), we can analyse the system qualitatively by studying the subsystem in the closed set $\Gamma = \{(s, i) \in \Re_+ \,|\, 0 \leq s + i \leq 1\}$. A qualitative investigation of the subsystem described by (5) and (6) reveals that the long-term behaviour, falls into two categories: endemic or die out. When the disease dies out naturally, the solution asymptotically approaches a disease free equilibrium E_0 of the form,

$$E_0 = (1 - P, 0), \tag{8}$$

The threshold that determines the stability of this equilibrium is the vaccination reproduction number,

$$R_v = \frac{\beta(1-P)}{\gamma + \pi},$$ (9)

The disease free equilibrium is locally stable if $R_v < 1$. Global asymptotic stability for disease free equilibrium is also achieved using a Bendixson–Dulac argument for $R_v < 1$ i.e. there are no periodic solutions [5]. Equation (9) also reveals that there is a critical vaccination proportion $P_c = (\beta - \gamma - \pi)/\beta$ above which the disease free equilibrium is stable i.e. $P > P_c$. Thus, in order to successfully prevent disease, the vaccination proportion should be large enough. When the disease free equilibrium is unstable, there exists an endemic equilibrium E_u of the form,

$$E_u = \left(\frac{1-P}{R_v}, \frac{\pi}{\beta}(R_v - 1) \right).$$ (10)

From equation (10) it very obvious that E_u will only exist provided $R_v > 1$. The eigenvalues $(\delta_{1,2})$ of the Jacobian matrix evaluated at the endemic equilibrium E_u is given as

$$\delta_{1,2} = -\frac{\pi}{2}R_v \pm \frac{1}{2}\sqrt{\pi^2 R_v^2 - 4R_v\pi(\gamma + \pi)}.$$ (11)

The endemic equilibrium E_u is locally asymptotically stable provided

$$1 < R_v \le \frac{4(\gamma + \pi)}{\pi},$$ (12)

i.e. the eigenvalues are complex with negative real part and E_u can be classified as a spiral sink. This behavior can be interpreted as follows; for initial low levels of infectives, the numbers of susceptibles build. Then, the number of infectives begins to increase until that process is faster than the number of susceptibles being added to the population. Eventually there are too few people to infect, the outbreak ends, and the number of susceptibles begins to increase again.

3 Adomian Decomposition Technique

In order to explicitly construct approximate non-perturbative solutions of the system described by (5)–(7), Adomian decomposition method well addressed in [1–3, 7] is employed. The advantage of this method is that it provides a direct scheme for solving the problem, i.e., without the need for linearization, perturbation, massive computation and any transformation.

The equivalent canonical form of this system is as follows:

$$s(t) = s(0) + (1 - P)\pi t - \beta \int_0^t sidt - \pi \int_0^t sdt, \tag{13}$$

$$i(t) = i(0) + \beta \int_0^t sidt - (\gamma + \pi) \int_0^t idt, \tag{14}$$

$$r(t) = r(0) + P\pi t + \gamma \int_0^t idt - \pi \int_0^t rdt. \tag{15}$$

As usual in Adomian decomposition method the solutions of (13)–(15) are considered to be as the sum of the following series

$$s = \sum_{n=0}^{\infty} s_n, \quad i = \sum_{n=0}^{\infty} i_n, \quad r = \sum_{n=0}^{\infty} r_n. \tag{16}$$

Then we approximate the nonlinear terms in the system as follows:

$$si = \sum_{n=0}^{\infty} F_n(s_0, \ldots, s_n, i_0, \ldots, i_n), \tag{17}$$

where

$$F_n = \frac{1}{n!} \left[\frac{d^n \left(\sum_{k=0}^{\infty} s_k \lambda^k \right) \left(\sum_{k=0}^{\infty} i_k \lambda^k \right)}{d\lambda^n} \right]_{\lambda=0}. \tag{18}$$

The nonlinear functions F_n are called Adomian's polynomials. Substituting (16)–(18) into (13)–(15), we get:

$$\sum_{n=0}^{\infty} s_n = s(0) + (1 - P)\pi t - \beta \int_0^t \sum_{n=0}^{\infty} F_n dt - \pi \int_0^t \sum_{n=0}^{\infty} s_n dt, \tag{19}$$

$$\sum_{n=0}^{\infty} i_n = i(0) + \beta \int_0^t \sum_{n=0}^{\infty} F_n dt - (\gamma + \pi) \int_0^t \sum_{n=0}^{\infty} i_n dt, \tag{20}$$

$$\sum_{n=0}^{\infty} r_n = r(0) + P\pi t + \gamma \int_0^t \sum_{n=0}^{\infty} i_n dt - \pi \int_0^t \sum_{n=0}^{\infty} r_n dt. \tag{21}$$

From (19)–(21) we define the following scheme:

$$s_0 = s(0) + (1 - P)\pi t, \quad i_0 = i(0), r_0 = r(0) + P\pi t, \tag{22}$$

$$s_{n+1} = -\beta \int_0^t F_n dt - \pi \int_0^t s_n dt, \quad \text{(for } n \geq 0\text{)}, \tag{23}$$

$$i_{n+1} = \beta \int_0^t F_n dt - (\gamma + \pi) \int_0^t i_n dt, \quad \text{(for } n \geq 0\text{)}, \tag{24}$$

$$r_{n+1} = \gamma \int_0^t i_n dt - \pi \int_0^t r_n dt, \quad \text{(for } n \geq 0\text{)}. \tag{25}$$

Using (18), we compute some of the Adomian polynomials as follows:

$$F_0 = s_0 i_0, \quad F_1 = s_0 i_1 + s_1 i_0, \quad F_2 = s_0 i_2 + s_1 i_1 + s_2 i_0,$$
$$F_3 = s_0 i_3 + s_1 i_2 + s_2 i_1 + s_3 i_0, \quad F_4 = s_0 i_4 + s_1 i_3 + s_2 i_2 + s_3 i_1 + s_4 i_0,$$
$$F_5 = s_0 i_5 + s_1 i_4 + s_2 i_3 + s_3 i_2 + s_4 i_1 + s_5 i_0, \ldots \tag{26}$$

Substituting (22)–(26) into (19)–(21), and using MAPLE we obtained a few terms approximation to the solutions as

$$s_N = \sum_{n=0}^{N} s_n, \quad i_N = \sum_{n=0}^{N} i_n, \quad r_N = \sum_{n=0}^{N} r_n, \tag{27}$$

where

$$s(t) = \lim_{N \to \infty}(s_N), \quad i(t) = \lim_{N \to \infty}(i_N), \quad r(t) = \lim_{N \to \infty}(r_N). \tag{28}$$

The decomposition method yields rapidly convergent series solutions by using a few iterations for both linear and non-linear deterministic equations. For the convergence of the Adomian decomposition method the reader is referred to [1].

4 Numerical Results and Discussion

In this section, we monitor the effect of vaccination on the dynamics of a childhood disease described by the SIR model (5)–(7) using Adomian decomposition technique. For illustration purposes the parameter values in Table 2 below are used.

Figure 5 depicts case 1 and shows the impact of high vaccination coverage on the disease free initial population groups. As expected, the population of the susceptible group decreases with time while that of the removed group gradually increases due to inclusion of vaccinated susceptible group. It is very

Table 2. Effect of vaccination coverage at various parameter values

Case	$s(0)$	$i(0)$	$r(0)$	β	γ	π	P	R_v	Comments
1	1	0	0	0.8	0.03	0.4	0.9	0.18604	E_0 stable (disease eradication)
2	0.8	0.2	0	0.8	0.03	0.4	0.9	0.18604	E_0 stable (disease eradication)
3	0.8	0.2	0	0.8	0.03	0.4	0.3	1.30223	E_u stable (no eradication)
4	0.8	0.2	0	0.8	0.03	0.4	0.0	1.86046	E_u stable (no eradication)

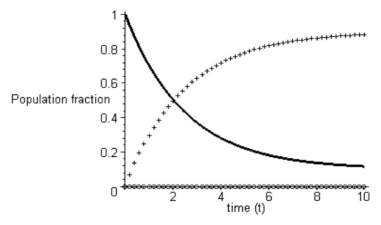

Fig. 5. Population fraction vs. time for case 1: $P = 0.9$; ———, Susceptible fraction; oooooo, infectives fraction; $+++++$, removed fraction

interesting to note that the entire population generally remains disease free with all the time. Figure 6 depicts case 2 and illustrates the impact of high vaccination coverage on the initial population groups with low level of infective group. The populations of the susceptible and infective groups decrease with time while that of the removed group increases due to inclusion of vaccinated and recovered people with permanent immunity and the disease outbreak ends. Case 3 is represented in Fig. 7 and illustrates the effect of low vaccination coverage on the initial population groups with low levels of infective group. The population of the susceptible group decreases with time. A small increase in the population of removed group is also noticed. However, it is noteworthy that the population of infective group will never disappear with time and the endemic situation persists. This confirmed that a disease free equilibrium couldn't be achieved once the vaccination coverage level is lower than a certain threshold and the endemic equilibrium remains stable.

Finally, case 4 is shown in Fig. 8 and illustrates the impact of initial low levels of infective group on the vaccination free population. As expected, the population of susceptible group decreases while that of infective group temporally increases. The disease rapidly spread to the entire population. The only contribution to removed group is the very small proportion of recovered people with permanent immunity.

5 Conclusions

An epidemiological model for the transmission dynamics of a childhood disease in the presence of a preventive vaccine was qualitatively and quantitatively studied. It is observed that the disease free equilibrium is stable provided the vaccination coverage level exceeds a certain threshold (P_c). Adomian

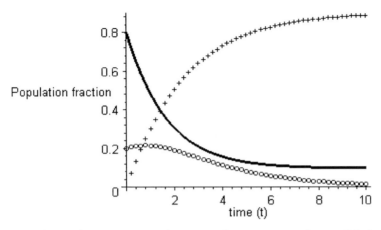

Fig. 6. Population fraction vs. time for case 2: $P = 0.9$; ———, Susceptible fraction; oooooo, infectives fraction; $+ + + + ++$, removed fraction

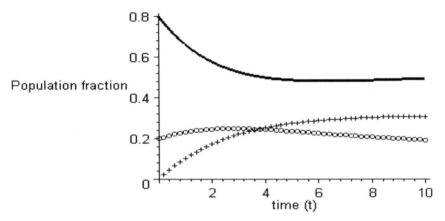

Fig. 7. Population fraction vs. time for case 3: $P = 0.3$; ———, Susceptible fraction; oooooo, infectives fraction; $+ + + + ++$, removed fraction

decomposition method is employed to construct an approximate solution to the problem. The method avoids the difficulties and massive computational work that usually arise from the parallel techniques and finite-difference method.

Appendix

According to the values introduced in the Table 2 the following approximate solutions can are derived;

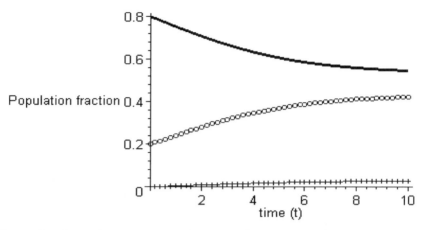

Fig. 8. Population fraction vs. time for case 4: $P = 0$; ———, Susceptible fraction; oooooo, infectives fraction; $+ + + + ++$, removed fraction

Case 1

$$s(t) = 1.0 - 0.36t + 0.72 \times 10^{-1}t^2 - 0.96 \times 10^{-2}t^3 + 0.96 \times 10^{-3}t^4$$
$$- 0.768 \times 10^{-4}t^5 - 0.5688888892 \times 10^{-6}t^6$$
$$r(t) = 0.36t - 0.72 \times 10^{-1}t^2 + 0.96 \times 10^{-2}t^3 - 0.96 \times 10^{-3}t^4$$
$$+ 0.768 \times 10^{-4}t^5 - 0.512 \times 10^{-5}t^6$$

Case 2

$$s(t) = 0.8 - 0.408t + 0.1008^*t^2 - 0.8223999996 \times 10^{-2}t^3$$
$$- 0.1811776 \times 10^{-2}t^4 + 0.2838500158 \times 10^{-3}t^5$$
$$- 0.4866281149 \times 10^{-4}t^6 - 0.1973168518 \times 10^{-5}t^7$$
$$+ 0.1567280763 \times 10^{-7}t^8 + 0.4557699387 \times 10^{-9}t^9$$
$$- 0.1747626667 \times 10^{-11}t^{10}$$
$$i(t) = 0.2 + 0.42 \times 10^{-1}t - 0.2823 \times 10^{-1}t^2 - 0.1169699999 \times 10^{-2}t^3$$
$$+ 0.2759918751 \times 10^{-2}t^4 - 0.3762609484 \times 10^{-3}t^5$$
$$+ 0.4741940899 \times 10^{-4}t^6 + 0.1990139977 \times 10^{-5}t^7$$
$$- 0.1540349563 \times 10^{-7}t^8 - 0.4575903832 \times 10^{-9}t^9$$
$$+ 0.1747626667 \times 10^{-11}t^{10}$$

$$r(t) = 0.366t - 0.7257 \times 10^{-1}t^2 + 0.93937 \times 10^{-2}t^3$$
$$- 0.94814275 \times 10^{-3}t^4 + 0.9241093251 \times 10^{-4}t^5$$
$$- 0.4445486401 \times 10^{-5}t^6 - .1697145904 \times 10^{-7}t^7 - 0.269312 \times 10^{-9}t^8$$
$$+ 0.1820444445 \times 10^{-11}t^9$$

Case 3

$$s(t) = 0.8 - 0.168t + 0.336 \times 10^{-1}t^2 - 0.2464 \times 10^{-2}t^3 - 0.125216 \times 10^{-3}t^4$$
$$+ 0.22308159 \times 10^{-5}t^5 - 0.1932440964 \times 10^{-3}t^6$$
$$- 0.7803698863 \times 10^{-4}t^7 + 0.4251830616 \times 10^{-5}t^8$$
$$+ 0.1094303622 \times 10^{-5}t^9 - 0.293723614 \times 10^{-7}t^{10}$$

$$i(t) = 0.2 + 0.42 \times 10^{-1}t - 0.903 \times 10^{-2}t^2 - 0.721699999 \times 10^{-3}t^3$$
$$+ 0.44919875 \times 10^{-3}t^4 - 0.308446284 \times 10^{-4}t^5$$
$$+ 0.1860204709 \times 10^{-3}t^6 + 0.7869655811 \times 10^{-4}t^7$$
$$- 0.41594566 \times 10^{-5}t^8 - 0.1098674509 \times 10^{-5}t^9$$
$$+ 0.293723614 \times 10^{-7}t^{10}$$

$$r(t) = 0.126t - 0.2457 \times 10^{-1}t^2 + 0.31857 \times 10^{-2}t^3 - 0.32398275 \times 10^3t^4$$
$$+ 0.286138125 \times 10^8t^5 + 0.1534736535 \times 10^{-5}t^6$$
$$- 0.6595694941 \times 10^{-6}t^7 - 0.9237401597 \times 10^{-7}t^8$$
$$+ 0.4370887113 \times 10^{-8}t^9$$

Case 4

$$s(t) = 0.8 - 0.48 \times 10^{-1}t + 0.416 \times 10^{-3}t^3 + 0.26864 \times 10^{-4}t^4$$
$$- 0.7711584 \times 10^{-5}t^5 - 0.2076444349 \times 10^{-3}t^6$$
$$- 0.1402303147 \times 10^{-3}t^7 + 0.1075760762 \times 10^{-4}t^8$$
$$+ 0.4557699387 \times 10^{-5}t^9 - 0.1747626667 \times 10^{-6}t^{10}$$

$$i(t) = 0.2 + 0.42 \times 10^{-1}t + 0.57 \times 10^{-3}t^2 - 0.4977 \times 10^{-3}t^3$$
$$- 0.1496125 \times 10^{-4}t^4 + 0.68491314 \times 10^{-5}t^5 + 0.19838225 \times 10^{-3}t^6$$
$$+ 0.141400832 \times 10^{-3}t^7 - 0.1048829562 \times 10^{-4}t^8$$
$$- 0.4575903832 \times 10^{-5}t^9 + 0.1747626667 \times 10^{-6}t^{10}$$

$$r(t) = 0.6 \times 10^{-2}t - 0.57 \times 10^{-3}t^2 + 0.817 \times 10^{-4}t^3 - 0.1190275 \times 10^{-4}t^4$$
$$+ 0.86245251 \times 10^{-6}t^5 + 0.3573295998 \times 10^{-5}t^6$$
$$- 0.1170517333 \times 10^{-5}t^7 - 0.269312 \times 10^{-6}t^8$$
$$+ 0.1820444445 \times 10^{-7}t^9$$

References

1. K. Abboui, Y. Cherruault, New ideas for proving convergence of decomposition methods, Comput. Appl. Math. 29 (7) (1995) 103–105.
2. G. Adomian, Solving frontier problems of physics: The Decomposition method, Kluwer, Dordecht, (1994).
3. G. Adomian, G. E. Adomian, A global method for solution of complex systems, Math. Model 5 (1984) 521–568.
4. http://www.unicef.org/media/files/globalplan.pdf and http://www.unicef.org/infobycountry
5. F. Brauer, C. Castillo-Chavez, Mathematical models in population biology and epidemiology, Springer-Verlag, (2001).
6. S. Busenberg, P. van den Driessche, Analysis of a disease transmission model in a population with varying size, J. Math. Biol. 28 (1990) 257–270.
7. V. Daftardar-Gejji, H. Jafari: Adomian decomposition-a tool for solving a system of fractional differential equation, J. Math. Anal. Appl. 301 (2) (2005) 508–518.
8. H. W. Hethcote, The mathematics of infectious diseases, Siam Rev. 42 (4) (2000) 599–653.
9. O. D. Makinde, Adomian decomposition approach to a SIR epidemic model with constant vaccination strategy, Appl. Math. Comput. 184 (2007) 842–848.
10. C. Piccolo III, L. Billings, The effect of vaccinations in an immigrant model, Math. Comput. Modell. 42 (2005) 291–299.
11. H. L. Smith, Subharmonic bifurcation in SIR epidemic model, J. Math. Biol. 17 (1983) 163–177.

A New MPFA Formulation for Subsurface Flow Problems on Unstructured Grids: Derivation of the Discrete Problem

A. Njifenjou and I.M. Nguena

Summary. The accurate computation of solutions for multiphase flow problems in a geologically complex subsurface is a great scientific and environmental challenge. Some authors have developed finite volume methods of new generation (the so-called Multi-Point Flux Approximation methods, commonly called MPFA methods) for addressing this kind of problems. This communication presents some finite-volume based flexible MPFA methods which display strong capabilities to handle flow problems in geologically complex reservoirs. It is well known that the flux continuity across grid-block interfaces combined with the local mass conservation (i.e., mass conservation at grid-block scale) make the finite volume methods a powerful computational tool for flow problems. These properties are met by the variants of MPFA methods presented in this paper. In addition, these new variants provide some added values as (1) an approximate pressure which is continuous in the whole domain, (2) sharp results are provided even on relatively coarse grids, (3) large scale flow problems governed by a symmetric and nonsymmetric full effective permeability tensor can be addressed. A Numerical implementation of these variants of MPFA methods has been done for solving a real-life problem, namely the Darcy flow in the Andra Couplex 1 test case.

1 Introduction and the Model Problem

The multiphase flow in geologically complex media is one of the most important issues related to environmental problems. A better understanding of this phenomenon by simulations requires that challenging numerical schemes are designed and implemented. In this connection, several numerical methods are proposed by many authors. Among these methods, let us mention the most important ones: (1) The Galerkin finite element which unfortunately displays limitations about the local conservativity which is an essential condition in fluid flow simulation, (2) The mixed and mixed hybrid finite element methods are two avatars of a same approach which is based upon a pressure-velocity formulation. This formulation meets the local mass conservativity and leads to satisfactory results. However some draw-backs concerning this approach have been pointed out like the violation of the discrete maximum principle near the drillings for flow within aquifers (see [8]), (3) The finite volume, due to its great flexibility toward complex applications and its local mass conservativity, is considered today as the most comfortable numerical way for addressing fluid dynamic problems. Let us mention some recent significant contributions for finite volume solutions to flow problems in anisotropic non-homogeneous media: (1) The Multi-Point Flux Approximation method (see for instance [1, 2, 5, 7, 10, 12, 13]), (2) An extended K-orthogonal grid method has been developed for addressing diffusion problems with full matrix coefficients (see for instance [4, 6]).

The aim of this paper is to present two variants of the MPFA methods involving some novelties as:

(1) The construction of an approximate pressure respecting the continuity condition across the mesh interfaces (recall that, except [12] and [13], the MPFA solution of above mentioned authors does not meet the interelements continuity condition). The prize to pay is that one should solve a larger discrete system since discrete unknowns are located at cell points and corner points of the primary mesh.
(2) The capability of addressing flow problems governed by non symmetric full tensors. Note that when dealing with large scale flow equations, the effective permeabilities governing the flow may be nonsymmetric (see [14] for instance).

For presenting our MPFA finite volume formulation, let us introduce a 2D flow problem which consists in finding a pressure U which satisfies the following steady-state diffusivity equation associated with a Dirichlet boundary condition:

$$- \, div(D \; grad \; U) = f \quad \text{in} \quad \Omega \tag{1}$$

$$U = 0 \quad \text{on} \quad \Gamma, \tag{2}$$

where f is a given function (commonly called source/sink term), Ω is a given open polygonal domain inside which the diffusivity equation is valid and Γ denotes the boundary of Ω. $D = D(x)$, with $x = (x_1, x_2) \in \Omega$, is a full matrix (describing the spatial variation of the permeability tensor) and satisfies the uniform ellipticity, i.e.,

$$\exists \gamma \in \mathbb{R}_+^* \;\; such \;\; that \; \forall \varepsilon \in \mathbb{R}^2 \; \varepsilon^T D(x) \varepsilon \geq \gamma \, |\varepsilon|^2 \quad a.e. \; in \; \Omega, \tag{3}$$

where $| \, . \, |$ denotes the euclidian norm in R^2, $D_{ij}(.) \in L^\infty(\Omega)$ represents the components of the permeability tensor D.

This paper is organized as follows. In the Sect. 2 we deal with the spatial discretization of the model problem. Section 3 presents in details two new variants of MPFA method for solving the system (1)–(2). In the Sect. 4, some numerical simulations are performed for homogeneous and nonhomogeneous anisotropic media. Section 5 is devoted to conclusions and perspectives of the work.

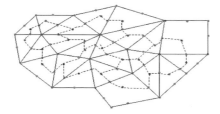

Fig. 1. Detail of primary (*full lines*) and secondary (*dotted lines*) grids for Dirichlet boundary conditions

2 Spatial Discretization

One starts with a primary grid consisting of a collection of polygonal cells made up of full lines. Each polygonal cell is naturally associated with a set of corner points. Inside each cell, a point is arbitrary fixed and defines what we call in the sequel a "cell point." The value of the numerical solution of the system (1)–(2) is computed either at the corner points or cell points (of the primary grid). We denote $x^{(i)}$ either a cell or a corner point, and u_i the corresponding value of the numerical solution.

Let us introduce the notions of cluster and interaction region which play a key role in the discretization process. A collection of mesh cells with one common corner point is called a cluster. The number of cells in a cluster defines the degree of the corresponding corner point. Associated with each cluster is an interaction region which is defined as follows: on each cell edge, choose an arbitrary point $e^{(i)}$ and draw straight dotted lines from this point to the cell points in the two neighboring cells. Inside the cluster, the dotted lines will define a polygon called an interaction region. The set of interaction regions defines the secondary mesh or dual mesh (see Fig. 1).

Each discrete equation is associated with a cell point or a corner point, and is derived from the mass balance equation in a suitable control volume. More precisely, expressing the mass balance in each cell from the primary and secondary mesh, and using convenient quadrature formulas leads to the discrete problem. A detailed description of our technique is presented in the following sections.

3 Finite Volume Formulation

3.1 Description of New Variants of MPFA Methods

Our aim in this section is to present two novel multipoint flux approximation methods designed for flow computations over unstructured irregular grids as

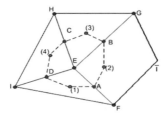

Fig. 2. A cluster (*full lines*) with its interaction region (*dotted lines*) involved in a local computation of the flux

the one shown in Fig. 2. We start with the mass balance equation in every cell C from the primary or secondary mesh

$$ - \int_C div \, [D \, grad \, u] \; dx = \int_C f(x) \, dx. $$

Applying the Ostrogradski theorem to the left hand side of this equality yields what follows:

$$ - \int_{\Gamma_C} [D \, grad \, U] \cdot n \, ds = \int_C f(x) \, dx, $$

where Γ_C is the boundary of C, and where the symbol $\cdot$ denotes the standard scalar product in $\mathbb{R}^2$. The first integral term in this equality may be transformed as follows:

$$ - \int_{\Gamma_C} [D \, grad \, U] \cdot n \, ds = - \sum_{[EA]} \int_{[EA]} [D \, grad \, U] \cdot n_{EA} \, ds $$

leading to the following approximation relation:

$$ - \sum_{[EA]} \int_{[EA]} [D \, grad \, U] \cdot n_{EA} \, ds \approx - \sum_{[EA]} EA \left[(D \, grad \, U \cdot n_{EA})(x^A) \right], \quad (4) $$

where $[EA]$ is a partial edge from the cell C and where n_{EA} is the unit normal vector to $[EA]$ exterior to C, $x^A = (x_1^A, x_2^A)$ are the spatial coordinates of the point A. One can perform the flux approximation in two manners (1) demanding that the flux continuity be satisfied on each partial edge; (2) demanding that the flux continuity be satisfied on each entire edge. In each grid-block one may decide that the degrees of freedom of the approximate pressure U are cell center and corner values of U. This choice will lead to a globally continuous and piecewise linear approximate solution. However another choice is possible as shown in [11] in a square primary grid. In what follows we present in details our strategy.

3.1.1 Derivation of the Flux Approximation Imposing Continuity per Partial Edge

We focus in what follows on the flux computation across the partial edges, under a continuity condition.

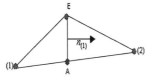

Fig. 3. Flux molecule for formulating the flux continuity over the partial edge $[EA]$

For performing the flux across the partial edge $[EA]$ (see Fig. 3), one should calculate $\frac{\partial U}{\partial x_1}$ and $\frac{\partial U}{\partial x_2}$ at the point A (whose coordinates are $x^A = (x_1^A, x_2^A)$) in view to apply the approximation (4). The following change is made on the space variables:

$$
\begin{cases}
x_1 = x_1^A + (x_1^{(1)} - x_1^A)X_1 + (x_1^E - x_1^A)X_2 \\[2mm]
x_2 = x_2^A + (x_2^{(1)} - x_2^A)X_1 + (x_2^E - x_2^A)X_2
\end{cases}.
$$

Denoting $det_{(1)}$ the determinant of the preceding system, it is easily seen that

$$
det_{(1)} = \left(x_1^{(1)} - x_1^A\right)\left(x_2^E - x_2^A\right) - \left(x_1^E - x_1^A\right)\left(x_2^{(1)} - x_2^A\right)
$$

$$
\frac{\partial X_1}{\partial x_1} = \frac{\left(x_2^E - x_2^A\right)}{det_{(1)}}, \quad \frac{\partial X_1}{\partial x_2} = \frac{\left(x_1^A - x_1^E\right)}{det_{(1)}},
$$

$$
\frac{\partial X_2}{\partial x_1} = \frac{\left(x_2^A - x_2^{(1)}\right)}{det_{(1)}}, \quad \frac{\partial X_2}{\partial x_2} = \frac{\left(x_1^{(1)} - x_1^A\right)}{det_{(1)}}.
$$

Since that A is a point from the triangle $AE(1)$ and accounting with the change of spatial variables, the partial derivatives of U at x^A are given by

$$
\left(\frac{\partial U}{\partial x_1}\right)_{(1)}(x^A) = \frac{\partial U}{\partial X_1}\frac{\partial X_1}{\partial x_1} + \frac{\partial U}{\partial X_2}\frac{\partial X_2}{\partial x_1},
$$

$$
\left(\frac{\partial U}{\partial x_2}\right)_{(1)}(x^A) = \frac{\partial U}{\partial X_1}\frac{\partial X_1}{\partial x_2} + \frac{\partial U}{\partial X_2}\frac{\partial X_2}{\partial x_2}.
$$

Then these partial derivatives are approximated as follows:

$$
\frac{\partial U}{\partial X_1}\frac{\partial X_1}{\partial x_1} + \frac{\partial U}{\partial X_2}\frac{\partial X_2}{\partial x_1} \approx (U_{(1)} - U_A)\frac{(x_2^E - x_2^A)}{det_{(1)}} + (U_E - U_A)\frac{(x_2^A - x_2^{(1)})}{det_{(1)}},
$$

$$
(5)
$$

$$
\frac{\partial U}{\partial X_1}\frac{\partial X_1}{\partial x_2} + \frac{\partial U}{\partial X_2}\frac{\partial X_2}{\partial x_2} \approx (U_{(1)} - U_A)\frac{(x_1^A - x_1^E)}{det_{(1)}} + (U_E - U_A)\frac{(x_1^{(1)} - x_1^A)}{det_{(1)}}.
$$

$$
(6)
$$

On the other hand, the normal vector $n_{(1)}$ is defined as

$$n_{(1)} = \varepsilon \left(\frac{x_2^E - x_2^A}{AE}, \frac{x_1^A - x_1^E}{AE} \right)^t,$$

where $\varepsilon = \pm 1$ is chosen such that $n_{(1)}$ is steered toward the outside of the triangle $AE(1)$ (see Fig. 3).
We deduce that

$$[D \ grad \ U] \cdot n_{(1)} = \varepsilon \left[\frac{x_2^E - x_2^A}{AE} \ \frac{x_1^A - x_1^E}{AE} \right] \left(\begin{bmatrix} D_{11} & D_{12} \\ D_{21} & D_{22} \end{bmatrix} \begin{bmatrix} \frac{\partial U}{\partial x_1} \\ \frac{\partial U}{\partial x_2} \end{bmatrix} \right).$$

Setting

$$\alpha_{(1)}^E = \varepsilon \left(D_{11} \frac{x_2^E - x_2^A}{AE} + D_{21} \frac{x_1^A - x_1^E}{AE} \right), \ \beta_{(1)}^E = \varepsilon \left(D_{12} \frac{x_2^E - x_2^A}{AE} + D_{22} \frac{x_1^A - x_1^E}{AE} \right)$$

it follows that

$$(D \ grad \ U) \cdot n_{(1)} = \alpha_{(1)}^E \frac{\partial U}{\partial x_1} \left(x^A \right) + \beta_{(1)}^E \frac{\partial U}{\partial x_2} \left(x^A \right).$$

Recall that approximations of partial derivatives of U at x^A are given by relations (5)–(6), when considering that A is a point of the triangle $EA(1)$. Similarly, considering that A is also a point from the triangle $EA(2)$, we have

$$\left(\frac{\partial U}{\partial x_1} \right)_{(2)} \left(x^A \right) = \left(U_{(2)} - U_A \right) \frac{(x_2^E - x_2^A)}{det_{(2)}} + (U_E - U_A) \frac{(x_2^A - x_2^{(2)})}{det_{(2)}}$$

$$\left(\frac{\partial U}{\partial x_2} \right)_{(2)} \left(x^A \right) = \left(U_{(2)} - U_A \right) \frac{(x_1^A - x_1^E)}{det_{(2)}} + (U_E - U_A) \frac{(x_1^{(2)} - x_1^A)}{det_{(2)}},$$

where we have set

$$det_{(2)} = (x_1^{(2)} - x_1^A)(x_2^E - x_2^A) - (x_1^E - x_1^A)(x_2^{(2)} - x_2^A).$$

Therefore, the flux continuity equation on $[EA]$ reads

$$\alpha_{(1)}^E \left(\frac{\partial U}{\partial x_1} \right)_{(1)} + \beta_{(1)}^E \left(\frac{\partial U}{\partial x_2} \right)_{(1)} + \alpha_{(2)}^E \left(\frac{\partial U}{\partial x_1} \right)_{(2)} + \beta_{(2)}^E \left(\frac{\partial U}{\partial x_2} \right)_{(2)} = 0 \ . \quad (7)$$

Setting for $i = 1, 2$

$$C_{11}^{(i)E} = \frac{x_2^E - x_2^A}{det_{(i)}}, \quad C_{12}^{(i)E} = \frac{x_2^A - x_2^{(i)}}{det_{(i)}}, \quad C_{21}^{(i)E} = \frac{x_1^A - x_1^E}{det_{(i)}}, \quad C_{22}^{(i)E} = \frac{x_1^{(i)} - x_1^A}{det_{(i)}}$$

and for $i, j = 1, 2$

$$S_{1j}^{(i)E} = \alpha_{(i)}^E C_{1j}^{(i)E} + \beta_{(i)}^E C_{2j}^{(i)E}$$

one deduces from the discrete version of the flux continuity equation (7) that

$$U_A = \frac{S_{11}^{(1)E} U_{(1)} + (S_{12}^{(1)E} + S_{12}^{(2)E}) U_E + S_{11}^{(2)E} U_{(2)}}{S_{11}^{(1)E} + S_{12}^{(1)E} + S_{12}^{(2)E} + S_{11}^{(2)E}} \equiv \gamma_{(1)}^E U_{(1)} + \gamma_E^E U_E + \gamma_{(2)}^E U_{(2)}.$$

Setting also

$$\Phi_{(1)} = U_{(1)}(S_{11}^{(1)E} - \gamma_{(1)}^E(S_{11}^{(1)E} + S_{12}^{(1)E})) + U_E(S_{12}^{(1)E} - \gamma_E^E(S_{11}^{(1)E} + S_{12}^{(1)E}))$$

$$-\gamma_{(2)}^E U_{(2)}(S_{11}^{(1)E} + S_{12}^{(1)E})$$

the flux across $[EA]$ may be written finally as follows:

$$\Phi_{[EA]}^{(1)} = - AE\ \Phi_{(1)} . \tag{8}$$

One should note that the methodology presented above is completely general. This means it applies to all the interior partial edges of cells from the primary and secondary grids. On the other hand, for a boundary partial edge $[EA]$, there is no need to write the continuity equation since U_A is given by the Dirichlet condition. Thus, the global discrete system may be easily derived following what precedes.

Remark 3.1. In the case of a Dirichlet–Neumann boundary problem, one should deal with the case where a partial edge $[EA]$ is included in a Neumann's boundary. The flux across $[EA]$ is exactly given by the Neumann condition.

3.1.2 Derivation of the Flux Approximation Imposing Continuity per Edge

In the previous subsection, the mass balance is performed under the constraint of the flux continuity on partial edges. Unfortunately, it may happen when dealing with the primary mesh, that the expression $S_{11}^{(1)E} + S_{12}^{(1)E} + S_{12}^{(2)E} + S_{11}^{(2)E}$ be null, and therefore it is not possible to compute the flux across $[EA]$, namely $\Phi_{[EA]}^{(1)}$. To overcome this difficulty, one may write the flux continuity equation on the entire edge $[EF]$ (see Fig. 4).

For this purpose, let us set

$$\Phi_{(1)}^{[EF]} = AE \left[\alpha_{(1)}^E \left(\frac{\partial U}{\partial x_1} \right)_{(1)} + \beta_{(1)}^E \left(\frac{\partial U}{\partial x_2} \right)_{(1)} \right]$$

$$+ AF \left[\alpha_{(1)}^F \left(\frac{\partial U}{\partial x_1} \right)_{(1)} + \beta_{(1)}^F \left(\frac{\partial U}{\partial x_2} \right)_{(1)} \right]$$

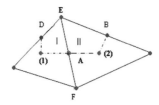

Fig. 4. Flux molecule for formulating the flux continuity over the edge $[EF]$

$$\Phi_{(2)}^{[EF]} = AE \left[\alpha_{(2)}^{E} \left(\frac{\partial U}{\partial x_1} \right)_{(2)} + \beta_{(2)}^{E} \left(\frac{\partial U}{\partial x_2} \right)_{(2)} \right]$$
$$+ AF \left[\alpha_{(2)}^{F} \left(\frac{\partial U}{\partial x_1} \right)_{(2)} + \beta_{(2)}^{F} \left(\frac{\partial U}{\partial x_2} \right)_{(2)} \right].$$

Therefore the flux continuity equation over $[EF]$ reads

$$\Phi_{(1)}^{[EF]} + \Phi_{(2)}^{[EF]} = 0 . \tag{9}$$

The last equation permits to express the numerical potential U_A in terms of $U_{(1)}$, $U_{(2)}$, U_E, and U_F. To solve this equation, one may consider two cases: the case where the elements I and II (see Fig. 4) are from the primary mesh, and the case where these elements are from the secondary mesh.

First case: the elements I and II are from the primary grid. Solving the flux continuity equation (9), where U_A is the unknown, leads to the following expression of the flux $\Phi_{(1)}^{[EF]}$:

$$\Phi_{(1)}^{[EF]} = -EF \left[C_1(U_{(1)} - U_{(2)}) + R\,C_2(U_E - U_F) \right], \tag{10}$$

where we have set

$$C_1 = \frac{S_{11}^{(1)E} S_{11}^{(2)E}}{S_{11}^{(1)E} + S_{11}^{(2)E}}, \quad C_2 = \frac{S_{12}^{(1)E} S_{11}^{(2)E} - S_{11}^{(1)E} S_{12}^{(2)E}}{S_{11}^{(1)E} + S_{11}^{(2)E}}, \quad \text{and} \quad R = \frac{EA}{EF}. \tag{11}$$

Second case: the elements I and II are from the secondary grid. In this case, the flux across $[EF]$ reads as follows:

$$\Phi_{(1)}^{[EF]} = \theta_1 U_{(1)} + \theta_2 U_{(2)} + AE\ S_{12}^{(1)E}\ U_E + AF\ S_{12}^{(1)F}\ U_F , \tag{12}$$

where we have set

$$\theta_1 = -\frac{1}{2} \left(AE\ S_{11}^{(1)E} + AF\ S_{11}^{(1)F} + AE\ S_{12}^{(1)E} + AF\ S_{12}^{(1)F} \right) \tag{13}$$

$$\theta_2 = \frac{1}{2} \left(AE\ S_{11}^{(1)E} + AF\ S_{11}^{(1)F} - AE\ S_{12}^{(1)E} - AF\ S_{12}^{(1)F} \right). \tag{14}$$

Remark 3.2. Due to the discontinuity of the permeability (located on interfaces of the primary grid), the expression $S_{11}^{(1)E} + S_{12}^{(1)E} + S_{12}^{(2)E} + S_{11}^{(2)E}$ may be equal to zero over the discontinuity interface. An example for rectangular grids is provided in [13]. On the other hand, this kind of singularity should not occur for partial or entire edges computation of flux in the dual grid. Indeed, when dealing with the dual grid, the expression $S_{11}^{(1)E} + S_{12}^{(1)E} + S_{12}^{(2)E} + S_{11}^{(2)E}$ is reduced to $S_{11}^{(1)E} + S_{11}^{(2)E}$ which is never null.

3.1.3 Combining the Two Approaches

One can prove that $S_{11}^{(1)E} + S_{12}^{(1)E} + S_{12}^{(2)E} + S_{11}^{(2)E}$ could be null in the scheme (8) only when dealing with cells from the primary grid. It is therefore suitable to use this scheme for cells of the secondary grid. The schemes (10)–(14) apply for cells from the primary grid and those from the secondary grid as well.

3.2 Construction of the Discrete System

3.2.1 Dealing with the Elements of the Primary Mesh

We suppose that the primary mesh consists of polygons. Therefore each mesh element e_i is associated with n_i edges. The data structure used to draw the mesh allow us to obtain from each edge all the corresponding mesh elements.

For a given mesh element e_i from the primary mesh, one carries out the computation of the total flux across its boundary in the following way. One considers all the facets of this element and for each facet, one calculates the flow exchanged by using the relations (8) or (10) depending on the chosen variant of the proposed method. The element is cut out in triangles whose vertices are the cell point and the two extremities of a facet. The integral in the right hand side of the balance equation over the element is obtained by summing the integrals calculated over these triangles.

3.2.2 Dealing with the Elements of the Dual Mesh

First of all, let us recall that the cell point associated with every element of the dual mesh is a vertex of the primary mesh. In order to do the flux balance in an element of the dual mesh (denoted ed_i), one has to consider the set of facets from the primary mesh elements admitting ed_i as one of the vertices. Let us call this set bd_i. To illustrate, let us consider the dual mesh element centered on the point C (see Fig. 5). Let us call it ed_1. Therefore bd_1 is the following set:

$$bd_1 = \{[CA], [CB], [CD], [CE], [CF], [CG]\}.$$

Each facet a_i from this set admits C as an end. The other end of a_i is the center of another element of the dual mesh, having a common edge with ed_1.

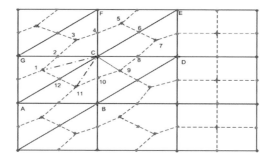

Fig. 5. A local numbering of nodes for a local flux balance

The vertices of the two half-edges of the common frontier are the center of the edge a_i and the center of the two elements of the primary mesh sharing the edge a_i. For instance, let us consider the facet [CD] of the set bd_1. The vertex D of the primary mesh is the center of another element of the dual mesh. The frontier between this element and ed_1 is made by the two half-edges (9,8) and (8,7). Using the relation (8) or (10) together with triangles (C,7,8) and (D,7,8), one can compute the flux through the first half-edge. Using the same equation and the triangles (C,8,9) and (D,8,9), it is also possible to compute the flux across the second half-edge.

In order to compute the right hand side of the balance equation, each element is divided into triangles defined by its center, the middle of each edge of bd_i and the centers of the primary mesh elements sharing this edge. As an example, one can consider the triangles $(12, C, 1)$ and $(11, C, 12)$ in Fig. 5. The integral in the right hand side of the balance equation is obtained by summing integrals over these triangles.

3.3 Definition of an Approximate Solution in Terms of Function

Recall that the proposed variants of MPFA method lead to solving a square system of equations involving all the discrete unknowns U_m representing the approximate pressure at cell centers and cell corners (with respect to the primary grid). Once these unknowns are computed, one utilizes them as interpolation data in the construction of an approximate solution denoted U_h defined as a piecewise linear function (more details are given below).

The computed quantities U_m actually correspond to the values of U_h at cell centers and cell corners. Thus these quantities satisfy the relation

$$U_m = U_h(x^{(m)}),$$

where $x^{(m)}$ is a node, i.e., a cell center or a corner point.

We should do some comments about the definition of U_h. Without any loose of generality, let us consider a *triangular* grid cell of the primary grid. To define U_h at every point of that grid cell, we divide it into three triangular

Fig. 6. A grid block divided into three triangles for a piecewise linear approximation of the solution

elements constructed by joining the cell center to the three cell corners (see Fig. 6).

Definition 3.3. *Let $x^{(i)}$, $x^{(j)}$ and $x^{(k)}$ denote the vertices of a triangular element T obtained from the division of a primary grid cell. The approximate solution U_h of the diffusion problem (1)–(2) is defined in T as follows:*

$$U_h(x) = \alpha \cdot (x - x^{(i)}) + U_i \,, \tag{15}$$

where $x = (x_1, x_2)^t$, $\alpha = (\alpha_1, \alpha_2)^t$, $x^{(i)} = (x_1^{(i)}, x_2^{(i)})^t$, and $U_i = U_h(x^{(i)})$, with $(.,.)^t$ denoting the transposition operator. The components of the vector α are easily calculated thanks to the fact that $U_j = U_h(x^{(j)})$ and $U_k = U_h(x^{(k)})$ are given (from the solution of the finite volume discrete system).

Proposition 3.4. *The approximate solution U_h is a continuous function in $\overline{\Omega}$ (closure of Ω). Moreover U_h is in the space $H_0^1(\Omega)$.*

Before carrying out the proof of this Proposition, let us recall that for an open subset D of $\mathbb{R}^2$, the spaces $H^1(D)$ and $H_0^1(D)$ are defined as follows:

$$H^1(D) = \left\{ v \in L^2(D) ; \ \frac{\partial v}{\partial x_i} \in L^2(D) \quad for \ \ i = 1, 2 \right\} \tag{16}$$

$$H_0^1(D) = \{ v \in H^1(D); \quad v = 0 \quad on \ \Gamma \} \,, \tag{17}$$

where $L^2(D)$ is the space (of classes) of functions v such that $\int_D v^2 dx$ converges, and where $\frac{\partial v}{\partial x_i}$ denotes partial derivatives in a distributional sense (see for instance [3] for more details). The mapping

$$v \longmapsto \left[\int_\Omega |grad \, v|^2 \, dx \right]^{\frac{1}{2}} \tag{18}$$

defines the well-known $H_0^1(\Omega) - norm$.

Proof. One easily checks that U_h is continuous on grid blocks boundaries. This follows from the fact that U_h is linear in each triangular element T (inside grid blocks) and is continuous at the corner points (of the primary grid). So U_h is continuous over $\overline{\Omega}$. Since the restriction of U_h in each triangular element T is in $H^1(T)$ and U_h takes zero value on the boundary Γ of Ω, then U_h is in $H_0^1(\Omega)$. □

4 Numerical Experiments

In this section, we present some numerical results following two purposes. The first one is to compare our MPFA solution with the one found in the literature concerning a real-life problem, namely the Andra Couplex 1. The second objective is to compare our MPFA solution with the one given by the so-called MPFA O-method (see [1]).

4.1 The Andra Couplex 1 Test Case [9]

This Andra Couplex test case consists to compute a simplified 2D far field model used in nuclear waste management simulation. From the mathematical point of view, one should solve a system of an elliptic equation (hydrodynamic problem) and a hyperbolic equation (transport problem, with a very concentrated nature of the source). We should focus in what follows on the hydrodynamic problem for testing our MPFA formulation.

4.1.1 Preliminaries

The repository lies at a depth of 450 m (meters) inside a clay layer which has above it a layer of limestone and a layer of marl and below it is a layer of dogger limestone. The water flows slowly (creeping flow) through these porous media and convects the radioactive materials once the containers leak. There is also a dilution effect which in mathematical terms is similar to diffusion. The problem involves three main technical difficulties:

(1) The diffusion constants are very different from one layer to another (maximal ratio equal to 8×10^6).
(2) There is a large contrast between the width (25,000 m) and the height (695 m).
(3) The imposed boundary conditions create a flow which is neither parallel nor orthogonal to layers interfaces.

4.1.2 The Geometry

In this First test case, the computation is restricted to a 2D section of the disposal site. The geometry is summarized in Fig. 7.

4.1.3 The Flow Computation

The permeability tensor D, assumed constant in each layer is given in Table 1 below.

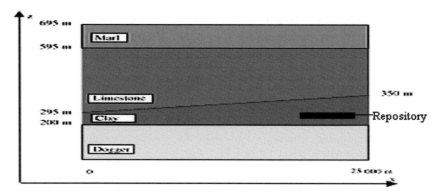

Fig. 7. Geometry of Andra Couplex 1 computational domain

	Marl	Limestone	Clay	Dogger
K (m/year)	3.1536e-5	6.3072	3.1536e-6	25.2288

Table 1. Permeability tensor in the four rock layers

On the boundary, conditions are:

$H = 289$	on $\{25000\} \times (0, 200)$,
$H = 310$	on $\{25000\} \times (350, 595)$,
$H = 180 + 160x/25000$	on $(0, 25000) \times \{695\}$,
$H = 200$	on $\{0\} \times (295, 595)$,
$H = 286$	on $\{0\} \times (0, 200)$,
$\frac{\partial H}{\partial n} = 0$	elsewhere.

Fig. 8. Pressure computation. *Left*: Mixed hybrid finite element solution from [9] on a relatively fine grid (54,432 elements involving 130,000 unknowns). *Right*: MPFA solution on a relatively coarse grid (23,280 elements corresponding to 34,500 unknowns)

Let us mention that the quadruple precision were utilized when implementing the mixed hybrid finite element (MHFE) as this method did not converge for usual linear solver performing with the double precision (Fig. 8). The proposed MPFA combined with linear solver has displayed a capability to yield satisfactory results based upon double precision (Fig. 9). The MPFA methods remain challenging even when they are implemented on coarse grids for problems with strong discontinuities (Fig. 10).

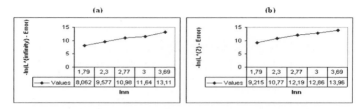

Fig. 9. (a) Log of L^∞-error for the proposed method, with the convergence rate which is equal to 2.69. (b) Log of L^2 -error for the proposed method, with the convergence rate which is equal to 2.55

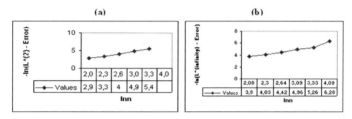

Fig. 10. (a) Log of L^∞-error for the O-method, with the convergence rate which is equal to 1.22. (b) Log of L^2-error for the O-method, with the convergence rate which is equal to 2

4.2 Second Test Problem

We consider the following elliptic problem governed by a full matrix of diffusion:

$$-div(D\ grad\ u) = f \quad \text{in} \quad]0,1[^2 \quad \text{with} \quad D = \begin{bmatrix} 1 & 10 \\ 10 & 1000 \end{bmatrix}$$

which possesses the exact solution $\varphi\,(x_1,x_2) = \sin(\pi x_1)\sin(\pi x_2)$.

5 Conclusions and Perspectives

We have presented in this work a new MPFA method for addressing flow problems in anisotropic nonhomogeneous media. This method has displayed a large capability for computing, with a satisfactory accuracy, the flows within nonhomogeneous media involving strong anisotropies (see numerical experiments of 5). A numerical experiment have been performed with an example extracted from the literature, namely the Andra Couplex 1 test case.

Theoretical investigations concerning the stability and the convergence of this method have been done for rectangular grids (see [13] and [11]). We have been investigating the case of unstructured irregular grids.

References

1. Aavatsmark I., Barkve T., Boe O., Mannseth T. (1996), A class of discretization methods for structured and unstructured grids in anisotropic, inhomogeneous media, 5th European Conference on the Mathematics of Oil Recovery, September 1996.
2. Aavatsmark I., Barkve T., Boe O., Mannseth T. (1998), Discretization on unstructured grids for inhomogeneous, anisotropic media. Part I: derivation of the methods, SIAM J. Sci. Comp., vol. 19, no. 5, pp. 1700–1716.
3. Brézis H. (1983), Analyse fonctionnelle, théorie et applications, Paris, Masson.
4. Du Q., Gunzburger M.D., Ju L. (2003), Voronoï-based finite volume methods, optimal voronoï meshes, and PDE on the sphere, Comput. Methods Appl. Mech. Engrg., vol. 192, pp. 3933–3957.
5. Eigestad G.T., Aadland T., Aavastmark I., Klausen R.A., Nordbotten J.M. (2004), Recent advances for MPFA methods, Ecmor IX, 2004.
6. Eymard R., Gallouët T., Herbin R. (2002), A finite volume scheme for anisotropic diffusion problems, C. R. Acad. Sci., Paris (Note presented by P.G. Ciarlet).
7. Eigestad G.T., Klausen R.A. (2005), On the convergence of the multi-point flux approximation O-method; Numerical experiments for discontinuous permeability, Numer. Methods Partial Diff. Eqns., vol. 21, pp. 1079–1098.
8. Hoteit H. (2002), Simulation d'écoulements et de transports de polluants en milieu poreux: Application à la modé lisation de la sûreté des dépôts de déchets radioactifs, PhD thesis, Université de Rennes 1 (France).
9. Erhel J., Hoteit H., Simulation dun stockage profond de dchets radioactifs. Etude des cas Couplex. Chaire UNESCO - Calcul numrique intensif, Tunis (Tunisie), March 2004.
10. Mishev I.D. (2003), Analysis of a new mixed finite volume method, Computational Methods in Applied Mathematics, vol. 3, no. 1, pp. 189–201.
11. Njifenjou A., Kinfack A.J., Convergence analysis of a MPFA method for flow problems in anisotropic heterogeneous porous media. Accepted for publication in International Journal on Finite Volumes.
12. Nguena I.M., Njifenjou A. (2005), A new finite volume formulation for diffusion problems in anisotropic non-homogeneous media, 4th International Symposium on Finite Volume for Complex Applications, Marraketch (Morocco), July 4–8, 2005, pp. 435–446.
13. Njifenjou A., Nguena I.M. (2006), A finite volume approximation for second order elliptic problems with a full matrix on quadrilateral grids: derivation of the scheme and a theoretical analysis, Int. J. Finite Volumes, vol. 3, no. 2 (electronic publication).
14. Renard P. (2005), Averaging methods for permeability fields, development, protection, management and sequestration of subsurface. Fluid Flow and Transport in Porous and Fractured Media, Summer School, Cargèse (France), July 27, 2005.

Analysis of a New MPFA Formulation for Flow Problems in Geologically Complex Media

A. Njifenjou and M. Mbehou

Summary. This work analyzes some mathematical aspects of a new Multi-Point Flux Approximation (MPFA) formulation for flow problems. This MPFA formulation has been developed in [12, 13] for quadrilateral grids and [10] for unstructured grids. Our MPFA formulation displays capabilities for handling flow problems in geologically complex media modelled by spatially varying full permeability tensor. However in this work, we focus our attention on the case of anisotropic homogeneous porous media. In this framework, the proposed MPFA formulation leads to a well-posed discrete problem which is a linear system whose associated matrix is symmetric and positive definite, even if the permeability tensor governing the flow is only positive definite. Following the spirit of the finite element theory, we have introduced the concept of globally continuous and piecewise linear approximate solution. The convergence analysis of this solution is strongly based upon another concept: the weak approximate solution. Stability and convergence results for the weak approximate solution are proven for L^2- and L^∞-norm, and for a discrete energy norm as well. These results permit to prove some error estimates related to the globally continuous and piecewise linear approximate solution.

1 Introduction and the Model Problem

Due to their local mass conservation properties, their flexibility for complex geometries and the flux continuity, the mixed finite element (MFE) methods have been widely used for modelling subsurface flow problems. Note that the local mass conservation and the flux continuity ensure the global mass conservation. Moreover, the MFE methods allow an accurate computation of the flow velocity and also handle well discontinuous coefficients. A computational drawback of these methods is the need to solve an algebraic system of saddle point type.

The multipoint flux approximation (MPFA) methods (see for instance [1, 2, 6, 13]) have been developed these last years, as finite volume methods of new generation for addressing flow problems governed by full permeability tensors, over distorted meshes. The MPFA methods combine the advantages of the MFE methods, i.e., local and global mass conservation principles, accuracy for rough grids and discontinuous full permeability tensors. In addition, the gridding flexibility makes the MPFA methods more competitive than the classical finite volume methods. Several MPFA methods lead to cell-centered stencil, for the pressures, which is relatively easy to solve. However a recent variant of MPFA (see [12, 13]) giving higher order approximate solutions, leads to cell-centered and vertex-centered stencils for the pressures. Since a variational formulation of MPFA methods is an ongoing research issue (at our knowledge), there exist only limited theoretical results in the literature concerning the well posedness and convergence of these methods (see [4, 7, 8, 13]).

This work is a contribution to the theoretical analysis of the MPFA formulation. It is also the continuation of the work done in [13] and is enriched with the concept of globally continuous and locally linear approximate solution following the spirit of the finite element theory [15].

For presenting our MPFA finite volume formulation, let us consider the 2D flow problem consisting in finding a function U (i.e. the pressure) which satisfies the following partial differential equation associated with a Dirichlet boundary condition:

$$-div(D \; grad \; U) = f \quad in \quad \Omega \tag{1}$$

$$U = 0 \quad on \quad \Gamma \tag{2}$$

where f is a given function (commonly called source/sink term), Ω is a given open square domain and Γ denotes its boundary. $D = D(x)$, with $x = (x_1, x_2) \in \Omega$, is a full symmetric matrix describing the spatial variation of the permeability tensor which satisfies the uniform ellipticity i.e.

$$\exists \gamma_{\min}, \gamma_{\max} \in \mathbb{R}_+^* \quad such \; that \quad \forall \xi \in \mathbb{R}^2, \; \xi \neq 0$$

$$\gamma_{\min} \; |\xi|^2 \leq \xi^T D(x) \xi \leq \gamma_{\max} \; |\xi|^2 \quad a.e. \; in \; \Omega \tag{3}$$

where $| \, . \, |$ denotes the euclidian norm in $\mathbb{R}^2$, $D_{ij}(.)$ are the components of D and are $L^\infty (\Omega)$ functions.

This paper is organized as follows. The second section deals with an MPFA finite volume formulation of the model problem. Within this section we bring an affirmative answer to the well posedness issue concerning the discrete problem. In the third section we introduce a notion of approximate solution in terms of globally continuous and piecewise linear functions. In the fourth section, we investigate the theoretical properties (stability and error estimates in convenient discrete norms) for the solution of the discrete problem. Based upon the discrete solution properties, convergence results are given in terms of error estimates for the approximate solution in the fifth section. The sixth section is devoted to conclusions and perspectives of this work.

2 An MPFA Formulation

In what follows we are dealing with the case of flow phenomena governed by piecewise constant full permeability tensors. This assumption is not restrictive. Indeed when the permeability tensor is spatially varying, Petroleum Engineers looks for its mean value over each grid-block from the primary grid. Therefore from the practical point of view this assumption is very realistic (see [5, 14] for instance).

2.1 The Discrete Problem

Our purpose in this section is to describe the matrix form of our MPFA finite volume formulation for (1)–(2). Let us suppose that the spatial domain Ω is $]0, 1[\times]0, 1[$. However the method applies to bounded polygonal domains as shown in [10]. We assume that Ω is covered with a square primary grid denoted $\mathcal{P}$ whose size is $h = \frac{1}{N}$, where N is a given positive integer. On the other hand, we denote K_{ij} the grid block defined by: $K_{ij} = \left[x_1^{i-\frac{1}{2}}, x_1^{i+\frac{1}{2}}\right] \times \left[x_2^{j-\frac{1}{2}}, x_2^{j+\frac{1}{2}}\right]$ where $x_1^{i+\frac{1}{2}} = x_1^{i-\frac{1}{2}} + h$, $x_2^{j+\frac{1}{2}} = x_2^{j-\frac{1}{2}} + h$, for $i, j = 1, \ldots, N$ with $x_1^{\frac{1}{2}} = x_2^{\frac{1}{2}} = 0$.

Recall that $L^2(\Omega)$ is the space (of classes) of functions v such that $\int_\Omega v^2 dx$ is a finite quantity, and for any positive integer m the so-called Sobolev space $H^m(\Omega)$ is defined by:

$$H^m(\Omega) = \left\{ v \in L^2(\Omega) \,;\; \frac{\partial^{|\alpha|} v}{\partial x_1^{\alpha_1} \partial x_2^{\alpha_2}} \in L^2(\Omega)\,, \text{ with } \forall 0 \leq |\alpha| = \alpha_1 + \alpha_2 \leq m \right\}$$

where the partial derivatives are taken in the distributional sense. We denote $\|\cdot\|_{m,\Omega}$ the standard norm of $H^m(\Omega)$ and we adopt the convention that $H^0(\Omega) = L^2(\Omega)$, which implies that $\|\cdot\|_{0,\Omega} = \|\cdot\|_{L^2(\Omega)}$.

From the boundary-value problem theory (see for instance [3]), the system (1)–(2) possesses a unique solution in $H^2(\Omega)$ under the assumption (3) and the condition $f \in L^2(\Omega)$.

In what follows we assume that the solution of (1)–(2) is sufficiently regular for our purpose (more precisions will be given later about the solution regularity). We should look for a finite volume formulation of the problem (1)–(2) in terms of a linear system which is derived from the elimination of auxiliary unknowns, namely interface pressures, from flux balance equations over grid-blocks. This linear system involves $\{u_{i,j}\}_{1\leq i,j\leq N}$ and $\left\{u_{i+\frac{1}{2},j+\frac{1}{2}}\right\}_{1\leq i,j\leq N-1}$ as discrete unknowns which are expected to be reasonable approximations of $\{\varphi_{i,j}\}_{1\leq i,j\leq N}$ (cell center pressures) and $\left\{\varphi_{i+\frac{1}{2},j+\frac{1}{2}}\right\}_{1\leq i,j\leq N-1}$ (cell corner pressures) respectively, where $\varphi_{i,j} = \varphi\left(x_1^i, x_2^j\right)$ and $\varphi_{i+\frac{1}{2},j+\frac{1}{2}} = \varphi\left(x_1^{i+\frac{1}{2}}, x_2^{j+\frac{1}{2}}\right)$, with:

$$x_1^i = \frac{x_1^{i-\frac{1}{2}} + x_1^{i+\frac{1}{2}}}{2}, \qquad x_2^j = \frac{x_2^{j-\frac{1}{2}} + x_2^{j+\frac{1}{2}}}{2} \qquad 1 \le i,j \le N \qquad (4)$$

We also adopt the following conventions:

$$x_1^0 = x_1^{\frac{1}{2}}, \quad x_1^{N+1} = x_1^{N+\frac{1}{2}}, \quad x_2^0 = x_2^{\frac{1}{2}}, \quad x_2^{N+1} = x_2^{N+\frac{1}{2}} \qquad (5)$$

It is easy to check that the exact solution φ satisfy the following systems (see [9]):

$$\frac{2D_{22}^{ij}D_{22}^{ij+1}}{D_{22}^{ij}+D_{22}^{ij+1}}\left[\varphi_{i,j} - \varphi_{i,j+1}\right] + \left(\frac{D_{22}^{ij}D_{21}^{ij+1}+D_{22}^{ij+1}D_{21}^{ij}}{D_{22}^{ij}+D_{22}^{ij+1}}\right)\left[\varphi_{i-\frac{1}{2},j+\frac{1}{2}} - \varphi_{i+\frac{1}{2},j+\frac{1}{2}}\right]$$

$$+\frac{2D_{22}^{ij}D_{22}^{ij-1}}{D_{22}^{ij}+D_{22}^{ij-1}}\left[\varphi_{i,j} - \varphi_{i,j-1}\right] + \left(\frac{D_{22}^{ij}D_{12}^{ij-1}+D_{22}^{ij-1}D_{12}^{ij}}{D_{22}^{ij}+D_{22}^{ij-1}}\right)\left[\varphi_{i+\frac{1}{2},j-\frac{1}{2}} - \varphi_{i-\frac{1}{2},j-\frac{1}{2}}\right]$$

$$+\frac{2D_{11}^{ij}D_{11}^{i+1j}}{D_{11}^{ij}+D_{11}^{i+1j}}\left[\varphi_{i,j} - \varphi_{i+1,j}\right] + \left(\frac{D_{11}^{ij}D_{21}^{i+1j}+D_{11}^{i+1j}D_{21}^{ij}}{D_{11}^{ij}+D_{11}^{i+1j}}\right)\left[\varphi_{i+\frac{1}{2},j-\frac{1}{2}} - \varphi_{i+\frac{1}{2},j+\frac{1}{2}}\right] \quad (6)$$

$$+\frac{2D_{11}^{ij}D_{11}^{i-1j}}{D_{11}^{ij}+D_{11}^{i-1j}}\left[\varphi_{i,j} - \varphi_{i-1,j}\right] + \left(\frac{D_{11}^{ij}D_{12}^{i-1j}+D_{11}^{i-1j}D_{12}^{ij}}{D_{11}^{ij}+D_{11}^{i-1j}}\right)\left[\varphi_{i-\frac{1}{2},j+\frac{1}{2}} - \varphi_{i+\frac{1}{2},j+\frac{1}{2}}\right]$$

$$\approx \int_{K_{ij}} f(x)dx \qquad\qquad \forall 1 \le i,\ j \le N$$

$$\left(\frac{D_{11}^{ij+1}D_{21}^{i+1j+1}+D_{11}^{i+1j+1}D_{21}^{ij+1}}{D_{11}^{ij+1}+D_{11}^{i+1j+1}}\right)\left[\varphi_{i,j+1} - \varphi_{i+1,j+1}\right]$$

$$+\left(\frac{\left(D_{12}^{i+1j+1}-D_{12}^{ij+1}\right)\left(D_{21}^{ij+1}-D_{21}^{i+1j+1}\right)}{2\left(D_{11}^{ij+1}+D_{11}^{i+1j+1}\right)} + \frac{D_{22}^{ij+1}+D_{22}^{i+1j+1}}{2}\right)\left[\varphi_{i+\frac{1}{2},j+\frac{1}{2}} - \varphi_{i+\frac{1}{2},j+\frac{3}{2}}\right]$$

$$+\left(\frac{D_{11}^{i+1j}D_{21}^{ij}+D_{11}^{ij}D_{21}^{i+1j}}{D_{11}^{ij}+D_{11}^{i+1j}}\right)\left[\varphi_{i+1,j} - \varphi_{i,j}\right]$$

$$+\left(\frac{\left(D_{12}^{i+1j}-D_{12}^{ij}\right)\left(D_{21}^{ij}-D_{21}^{i+1j}\right)}{2\left(D_{11}^{ij}+D_{11}^{i+1j}\right)} + \frac{D_{22}^{ij}+D_{22}^{i+1j}}{2}\right)\left[\varphi_{i+\frac{1}{2},j+\frac{1}{2}} - \varphi_{i+\frac{1}{2},j-\frac{1}{2}}\right]$$

$$+\left(\frac{D_{22}^{i+1j}D_{12}^{i+1j+1}+D_{22}^{i+1j+1}D_{12}^{i+1j}}{D_{22}^{i+1j}+D_{22}^{i+1j+1}}\right)\left[\varphi_{i+1,j} - \varphi_{i+1,j+1}\right] \qquad (7)$$

$$+\left(\frac{\left(D_{21}^{i+1j+1}-D_{21}^{i+1j}\right)\left(D_{12}^{i+1j}-D_{12}^{i+1j+1}\right)}{2\left(D_{22}^{i+1j}+D_{22}^{i+1j+1}\right)} + \frac{D_{11}^{i+1j}+D_{11}^{i+1j+1}}{2}\right)\left[\varphi_{i+\frac{1}{2},j+\frac{1}{2}} - \varphi_{i+\frac{3}{2},j+\frac{1}{2}}\right]$$

$$+\left(\frac{D_{22}^{ij+1}D_{12}^{ij}+D_{22}^{ij}D_{12}^{ij+1}}{D_{22}^{i+1j}+D_{22}^{i+1j+1}}\right)\left[\varphi_{i,j+1} - \varphi_{i,j}\right]$$

$$+\left(\frac{\left(D_{21}^{ij+1}-D_{21}^{ij}\right)\left(D_{12}^{ij}-D_{12}^{ij+1}\right)}{2\left(D_{22}^{ij}+D_{22}^{ij+1}\right)} + \frac{D_{11}^{ij}+D_{11}^{ij+1}}{2}\right)\left[\varphi_{i+\frac{1}{2},j+\frac{1}{2}} - \varphi_{i-\frac{1}{2},j+\frac{1}{2}}\right]$$

$$\approx \int_{K_{i+\frac{1}{2},j+\frac{1}{2}}} f(x)dx \qquad \forall\ 1 \le i,\ j \le N-1$$

where we have set

$$\varphi_{i+\frac{1}{2},\frac{1}{2}} = \varphi_{i+\frac{1}{2},N+\frac{1}{2}} = \varphi_{\frac{1}{2},j+\frac{1}{2}} = \varphi_{N+\frac{1}{2},j+\frac{1}{2}} = 0 \qquad \forall 0 \le i, j \le N \quad (8)$$

and

$$\varphi_{i,0} = \varphi_{0,j} = \varphi_{i,N+1} = \varphi_{N+1,j} = 0 \qquad \forall 1 \le i, j \le N \qquad (9)$$

The discrete problem consists in finding $\{u_{i,j}\}_{1 \le i,j \le N}$ and $\left\{u_{i+\frac{1}{2},j+\frac{1}{2}}\right\}_{1 \le i,j \le N-1}$, real quantities such that:

$$\frac{2D_{22}^{ij}D_{22}^{ij+1}}{D_{22}^{ij}+D_{22}^{ij+1}}\left[u_{i,j}-u_{i,j+1}\right] + \left(\frac{D_{22}^{ij}D_{21}^{ij+1}+D_{22}^{ij+1}D_{21}^{ij}}{D_{22}^{ij}+D_{22}^{ij+1}}\right)\left[u_{i-\frac{1}{2},j+\frac{1}{2}}-u_{i+\frac{1}{2},j+\frac{1}{2}}\right]$$

$$+\frac{2D_{22}^{ij}D_{22}^{ij-1}}{D_{22}^{ij}+D_{22}^{ij-1}}\left[u_{i,j}-u_{i,j-1}\right] + \left(\frac{D_{22}^{ij}D_{12}^{ij-1}+D_{22}^{ij-1}D_{12}^{ij}}{D_{22}^{ij}+D_{22}^{ij-1}}\right)\left[u_{i+\frac{1}{2},j-\frac{1}{2}}-u_{i-\frac{1}{2},j-\frac{1}{2}}\right]$$

$$+\frac{2D_{11}^{ij}D_{11}^{i+1j}}{D_{11}^{ij}+D_{11}^{i+1j}}\left[u_{i,j}-u_{i+1,j}\right] + \left(\frac{D_{11}^{ij}D_{21}^{i+1j}+D_{11}^{i+1j}D_{21}^{ij}}{D_{11}^{ij}+D_{11}^{i+1j}}\right)\left[u_{i+\frac{1}{2},j-\frac{1}{2}}-u_{i+\frac{1}{2},j+\frac{1}{2}}\right] \quad (10)$$

$$+\frac{2D_{11}^{ij}D_{11}^{i-1j}}{D_{11}^{ij}+D_{11}^{i-1j}}\left[u_{i,j}-u_{i-1,j}\right] + \left(\frac{D_{11}^{ij}D_{12}^{i-1j}+D_{11}^{i-1j}D_{12}^{ij}}{D_{11}^{ij}+D_{11}^{i-1j}}\right)\left[u_{i-\frac{1}{2},j+\frac{1}{2}}-u_{i+\frac{1}{2},j+\frac{1}{2}}\right]$$

$$= \int_{K_{ij}} f(x)dx \qquad \forall 1 \le i, j \le N$$

$$\left(\frac{D_{11}^{ij+1}D_{21}^{i+1j+1}+D_{11}^{i+1j+1}D_{21}^{ij+1}}{D_{11}^{ij+1}+D_{11}^{i+1j+1}}\right)\left[u_{i,j+1}-u_{i+1,j+1}\right]$$

$$+\left(\frac{(D_{12}^{i+1j+1}-D_{12}^{ij+1})(D_{21}^{ij+1}-D_{21}^{i+1j+1})}{2(D_{11}^{ij+1}+D_{11}^{i+1j+1})}+\frac{D_{22}^{ij+1}+D_{22}^{i+1j+1}}{2}\right)\left[u_{i+\frac{1}{2},j+\frac{1}{2}}-u_{i+\frac{1}{2},j+\frac{3}{2}}\right]$$

$$+\left(\frac{D_{11}^{i+1j}D_{21}^{ij}+D_{11}^{ij}D_{21}^{i+1j}}{D_{11}^{ij}+D_{11}^{i+1j}}\right)\left[u_{i+1,j}-u_{i,j}\right]$$

$$+\left(\frac{(D_{12}^{i+1j}-D_{12}^{ij})(D_{21}^{ij}-D_{21}^{i+1j})}{2(D_{11}^{ij}+D_{11}^{i+1j})}+\frac{D_{22}^{ij}+D_{22}^{i+1j}}{2}\right)\left[u_{i+\frac{1}{2},j+\frac{1}{2}}-u_{i+\frac{1}{2},j-\frac{1}{2}}\right]$$

$$+\left(\frac{D_{22}^{i+1j}D_{12}^{i+1j+1}+D_{22}^{i+1j+1}D_{12}^{i+1j}}{D_{22}^{i+1j}+D_{22}^{i+1j+1}}\right)\left[u_{i+1,j}-u_{i+1,j+1}\right]$$

$$+\left(\frac{(D_{21}^{i+1j+1}-D_{21}^{i+1j})(D_{12}^{i+1j}-D_{12}^{i+1j+1})}{2(D_{22}^{i+1j}+D_{22}^{i+1j+1})}+\frac{D_{11}^{i+1j}+D_{11}^{i+1j+1}}{2}\right)\left[u_{i+\frac{1}{2},j+\frac{1}{2}}-u_{i+\frac{3}{2},j+\frac{1}{2}}\right]$$

$$+ \left(\frac{D_{22}^{ij+1} D_{12}^{ij} + D_{22}^{ij} D_{12}^{ij+1}}{D_{22}^{i+1j} + D_{22}^{i+1j+1}} \right) [u_{i,j+1} - u_{i,j}]$$

$$+ \left(\frac{(D_{21}^{ij+1} - D_{21}^{ij})(D_{12}^{ij} - D_{12}^{ij+1})}{2(D_{22}^{ij} + D_{22}^{ij+1})} + \frac{D_{11}^{ij} + D_{11}^{ij+1}}{2} \right) \left[u_{i+\frac{1}{2},j+\frac{1}{2}} - u_{i-\frac{1}{2},j+\frac{1}{2}} \right] \quad (11)$$

$$= \int_{K_{i+\frac{1}{2},j+\frac{1}{2}}} f(x) dx \quad \forall 1 \leq i, j \leq N - 1$$

where we have set

$$u_{i+\frac{1}{2},\frac{1}{2}} = u_{i+\frac{1}{2},N+\frac{1}{2}} = u_{\frac{1}{2},j+\frac{1}{2}} = u_{N+\frac{1}{2},j+\frac{1}{2}} = 0 \quad \forall \ 0 \leq i, j \leq N \quad (12)$$

and

$$u_{i,0} = u_{0,j} = u_{i,N+1} = u_{N+1,j} = 0 \quad \forall \ 1 \leq i, j \leq N \quad (13)$$

If the preceding discrete problem gets a unique solution, one can deduce the approximate values of the potential at the midpoints of interfaces using the following relations which expresses the flux continuity at grid-block interfaces. For $0 \leq i \leq N$ and $1 \leq j \leq N$:

$$u_{i+\frac{1}{2},j} = \frac{1}{2(D_{11}^{i,j} + D_{11}^{i+1,j})} \left\{ 2D_{11}^{i,j} u_{ij} + 2D_{11}^{i+1,j} u_{i+1,j} \right.$$

$$\left. + \left[D_{12}^{i+1,j} - D_{12}^{i,j} \right] \left[u_{i+\frac{1}{2},j+\frac{1}{2}} - u_{i+\frac{1}{2},j-\frac{1}{2}} \right] \right\} \quad (14)$$

For $1 \leq i \leq N$ and $0 \leq j \leq N$:

$$u_{i,j+\frac{1}{2}} = \frac{1}{2(D_{22}^{i,j} + D_{22}^{i,j+1})} \left\{ 2D_{22}^{i,j} u_{ij} + 2D_{22}^{i,j+1} u_{i,j+1} \right.$$

$$\left. + \left[D_{21}^{i,j+1} - D_{21}^{i,j} \right] \left[u_{i+\frac{1}{2},j+\frac{1}{2}} - u_{i-\frac{1}{2},j+\frac{1}{2}} \right] \right\} \quad (15)$$

Therefore, one can deduce the fluxes over the grid-block interfaces from the following relations

$$q_{i+\frac{1}{2},j} = D_{12}^{ij} \left[u_{i+\frac{1}{2},j-\frac{1}{2}} - u_{i+\frac{1}{2},j} \right] + 2D_{11}^{ij} \left[u_{ij} - u_{i+\frac{1}{2},j} \right]$$

$$+ D_{12}^{ij} \left[u_{i+\frac{1}{2},j} - u_{i+\frac{1}{2},j+\frac{1}{2}} \right] \quad (16)$$

$$q_{i,j+\frac{1}{2}} = -D_{21}^{ij} \left[u_{i+\frac{1}{2},j+\frac{1}{2}} - u_{i,j+\frac{1}{2}} \right] + 2D_{22}^{ij} \left[u_{ij} - u_{i,j+\frac{1}{2}} \right]$$

$$- D_{21}^{ij} \left[u_{i,j+\frac{1}{2}} - u_{i-\frac{1}{2},j+\frac{1}{2}} \right] \quad (17)$$

2.2 Existence and Uniqueness for a Solution of the Discrete Problem

We are going to deal now with the existence and uniqueness of a solution for the discrete problem (10)–(11). Before giving the two main results of this subsection, let us shortly comment about this discrete problem. Its matrix form may be expressed as follows:

$$
\begin{pmatrix} A & B \\ B^T & C \end{pmatrix} \begin{pmatrix} U_{cc} \\ U_{vc} \end{pmatrix} = \begin{pmatrix} F_{cc} \\ F_{vc} \end{pmatrix}
\tag{18}
$$

where we have set:

$$
U_{cc} = \{u_{i,j}\}_{1 \leq i,j \leq N} \qquad \text{and} \qquad U_{vc} = \left\{ u_{i+\frac{1}{2}, j+\frac{1}{2}} \right\}_{1 \leq i,j \leq N-1}
\tag{19}
$$

and where:

F_{cc} is a sub-vector with N^2 components defined by the right hand side of (10) only as we account with (12) and (13).

F_{vc} is a sub-vector with $(N-1)^2$ components defined by the right hand side of (11) only as we account with (12) and (13).

A is a $N \times N$ symmetric positive definite matrix, associated to the classical grid-centered finite volume when D is diagonal i.e. $D_{12} = D_{21} = 0$.

C is a $(N-1) \times (N-1)$ symmetric positive definite matrix, associated with the classical vertex-centered finite volume when D is diagonal matrix.

B is a $N \times (N-1)$ matrix and B^T is its transpose.

Let us give now the two main results of this subsection.

Proposition 2.1. *The discrete problem consisting to find* $\{u_{i,j}\}_{1 \leq i,j \leq N}$ *and* $\left\{ u_{i+\frac{1}{2}, j+\frac{1}{2}} \right\}_{1 \leq i,j \leq N-1}$ *such that the equations (10)–(11) are satisfied under the conditions (12) and (13), possesses a unique solution.*

Proposition 2.2. *The matrix* $\begin{pmatrix} A & B \\ B^T & C \end{pmatrix}$ *associated with the discrete problem (10)–(11) is symmetric and positive definite.*

The proof of Proposition 2.1 and 2.2, can be found in [9].

3 The Approximate Solution in Terms of Piecewise Linear Function

Solving the discrete problem (10)–(11) leads to determining all the discrete unknowns at grid-block centers and grid-block corners (with respect to the primary grid). In what follows, we make use of the simplified notation u_m representing either u_{ij}, $u_{i+\frac{1}{2}j+\frac{1}{2}}$, $u_{i+\frac{1}{2}j}$ or $u_{ij+\frac{1}{2}}$.

Fig. 1. A (primary) grid block divided into four triangular elements T for a piecewise linear approximation of the solution. The symbol • represents a degree of freedom (which is a nodal value) of the approximate solution over triangular elements

We start by dividing each grid-block (of the primary grid $\mathcal{P}$) into four triangular elements, whose generic name is T, constructed by joining each grid-block center to the corresponding grid-block corners (see Fig. 1). By doing so, one generates over Ω a new grid denoted $\mathcal{T}$. Let $U_{\mathcal{T}}^h$ be the piecewise linear approximate solution associated with the new grid $\mathcal{T}$. The quantities u_m actually correspond here to the values of $U_{\mathcal{T}}^h$ at grid-block centers and grid-block corners. Thus these quantities satisfy the following relation

$$u_m = U_{\mathcal{T}}^h(x^{(m)})$$

where $x^{(m)}$ is a node i.e. a grid-block center or a grid-block corner.

Definition 3.1. *Let $x^{(i)}$, $x^{(j)}$ and $x^{(k)}$ denote the vertices of a triangular element $T \in \mathcal{T}$. The approximate solution $U_{\mathcal{T}}^h$ of the flow problem (1)–(2) is defined in T as follows:*

$$U_{\mathcal{T}}^h(x) = \alpha \cdot (x - x^{(i)}) + u_i$$

where $x = (x_1, x_2)^t$, $\alpha = (\alpha_1, \alpha_2)^t$, $x^{(i)} = (x_1^{(i)}, x_2^{(i)})^t$ and $u_i = U_h(x^{(i)})$, with $(.,.)^t$ denoting the transposition operator. The components of the vector α are easily calculated due to the fact that $u_j = U_{\mathcal{T}}^h(x^{(j)})$ and $u_k = U_{\mathcal{T}}^h(x^{(k)})$ are given (from the solution of the discrete problem).

Proposition 3.2. *The approximate solution $U_{\mathcal{T}}^h$ is a continuous function in $\overline{\Omega}$ (closure of Ω). Moreover $U_{\mathcal{T}}^h$ belongs to the space $H_0^1(\Omega)$.*

Before carrying out the proof of this proposition, let us recall that $H_0^1(\Omega)$ is defined as follows:

$$H_0^1(\Omega) = \{v \in H^1(\Omega); \quad v = 0 \quad \text{on} \quad \Gamma \} \tag{20}$$

The mapping

$$v \longmapsto \left[\int_\Omega |grad\ v|^2\ dx \right]^{\frac{1}{2}} \tag{21}$$

defines the well-known $H_0^1(\Omega) - norm$.

Proof. One easily checks that U_T^h is continuous on grid-block boundaries. This follows from the fact that U_T^h is linear in each triangular element $T \in \mathcal{T}$ and is continuous at the corner points (of the primary grid). So U_T^h is continuous over $\overline{\Omega}$. Since the restriction of U_T^h in each triangular element T is in $H^1(T)$ and U_T^h takes zero value on the boundary Γ of Ω, it is clear that U_T^h is in $H_0^1(\Omega)$. $\square$

4 Stability and Error Estimates for the Solution of the Discrete Problem

4.1 Preliminaries: Notion of "Weak Approximate Solution"

We start by considering an additive grid $\mathcal{L}$ associated with the primary grid (see Fig. 2). The elements of $\mathcal{L}$ are made of rhombi L completely imbedded in $\overline{\Omega}$. We denote Γ_L the boundary of $L \in \mathcal{L}$ and $\mathbf{E}(\mathcal{L})$ the space of functions v defined almost everywhere in Ω such that v is constant in every $L \in \mathcal{L}$ and zero elsewhere. This space is obviously non-empty since it contains the null function.

Let us endow $\mathbf{E}(\mathcal{L})$ with the following discrete energy norm. For all $v \in \mathbf{E}(\mathcal{L})$ we set:

$$\|v\|_{1,h} = \left[\sum_{s \in S} (\varDelta_s v) \right]^{\frac{1}{2}} \tag{22}$$

where

$$(\varDelta_s v) = \sum_{\substack{L,\, K \in \mathcal{L} \text{ such that} \\ \Gamma_K \cap \Gamma_L = \{s\}}} |v_L - v_K|^2 \tag{23}$$

and where S is the set of vertices.

Note that a vertex $s \in S$ could belong to the boundary Γ of the domain Ω. In this case, there exists a unique element L of $\mathcal{L}$ such that s belongs to the boundary Γ_L of L. It is therefore natural to define $(\varDelta_s v)$ in this case by

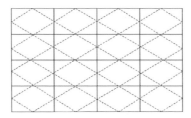

Fig. 2. An example of grid $\mathcal{L}$ made of lozenges associated with a primary rectangular grid

$(\Delta_s v) = |v_L|^2$. The norm defined by (22) could be viewed as a discrete version of the classical $H_0^1(\Omega)$ norm.

Let us introduce the space

$$C_0\left(\overline{\Omega}\right) = \{v : \overline{\Omega} \longrightarrow \mathbb{R} \text{ is continuous, and } v = 0 \text{ on } \Gamma\}$$

and the following operator:

$$\Pi : C_0\left(\overline{\Omega}\right) \longrightarrow E\left(\mathcal{L}\right)$$

$$v \longmapsto \Pi v$$

with:

$$[\Pi v]\left(x\right) = \begin{cases} v(x_L), & \text{if } x \in Int(L), \text{with} L \in \mathcal{L} \\ 0 \text{ elsewhere} \end{cases}$$

where $L \in \mathcal{L}$ and where $x_L = (x_1^L, x_2^L)$ are the coordinates of the center of L.

Since the approximate solution U_h of the diffusion problem (1)–(2) is in $C^0\left(\overline{\Omega}\right)$ (see Subsection 3.2 for the definition of U_h), we have:

$$[\Pi U_h]\left(x\right) = \begin{cases} U_L, & \text{if } x \in Int(L), \text{ with } L \in \mathcal{L} \\ 0 & \text{elsewhere} \end{cases} \tag{24}$$

Definition 4.1. *Let v be a function of $E(\mathcal{L})$. $v_{|\Omega}$ is a weak approximate solution for the diffusion problem (1)–(2) if there exists an approximate solution V of (1)–(2) in the sense of Definition 3.1 such that $v = \Pi V$.*

Remark 4.2. According to this definition, ΠU_T^h, is a weak approximate solutions of (1)–(2). Moreover it is denoted U_h in the sequel for the sake of simplicity of notations.

Note that U_L represents $u_{i,j}$ or $u_{i+\frac{1}{2},j+\frac{1}{2}}$ depending on whether x_L is a grid-block center point or a grid-block corner from the primary grid.

4.2 Stability of the Weak Approximate Solution

We are going to prove here the stability of the weak approximate solution in the sense of the discrete energy norm (22). The main ingredient for the proof of this result is a discrete version of the Poincaré inequality which reads as follows.

Lemma 4.3. *(discrete version of Poincaré inequality)*
   There exists a strictly positive number P such that

$$\|v\|_{L^2(\Omega)} \leq P \|v\|_{1,h} \qquad \forall\, v \in \mathbf{E}(\mathcal{L})$$

where we have set

$$\|v\|_{L^2(\Omega)} = \left(\int_\Omega v^2 dx\right)^{\frac{1}{2}}$$

Proof. See [9].   □

Proposition 4.4. *(Stability result)*
The weak approximate solution u_h of the diffusion problem (1)–(2) satisfies the following inequality:

$$\|u_h\|_{1,h} \le C \, \|f\|_{L^2(\Omega)}$$

where C is a strictly positive real number not depending on the spatial discretization.

For the proof of this proposition one may see [9].

Remark 4.5. This stability result implies the L^2–stability of the weak approximate solution. This follows from Lemma 4.3.

4.3 Error Estimates for the Weak Approximate Solution

Let us define a function ε_h almost everywhere in $\mathbb{R}^2$ in the following way:

$$\varepsilon_h\left(x\right) = \begin{cases} \varepsilon_L & \text{if } x \in Int(L) \\ 0 & \text{elsewhere} \end{cases} \qquad \text{with} \quad L \in \mathcal{L} \qquad (25)$$

where we have set $\varepsilon_L = \varphi_L - u_L$ for all $L \in \mathcal{L}$. Note that the element L of the additive mesh $\mathcal{L}$ is necessary centered on a point whose cartesian coordinates are of the form $\left(x_1^i, x_2^j\right)$ or $\left(x_1^{i+\frac{1}{2}}, x_2^{j+\frac{1}{2}}\right)$. ε_L is the generic name for $\varepsilon_{i,j}$ or $\varepsilon_{i+\frac{1}{2},j+\frac{1}{2}}$.

One can check easily that the following quantities $\{\varepsilon_{i,\,j}\}_{1 \le i,\,j \le N}$ and $\left\{\varepsilon_{i+\frac{1}{2},\,j+\frac{1}{2}}\right\}_{1 \le i,\,j \le N-1}$ are a solution of a discrete problem of the form (10)–(13). Therefore,

$$\frac{2D_{22}^{ij}D_{22}^{ij+1}}{D_{22}^{ij}+D_{22}^{ij+1}} \left[\varepsilon_{i,j} - \varepsilon_{i,j+1}\right] + \frac{D_{22}^{ij}D_{21}^{ij+1}+D_{22}^{ij+1}D_{21}^{ij}}{D_{22}^{ij}+D_{22}^{ij+1}} \left[\varepsilon_{i-\frac{1}{2},j+\frac{1}{2}} - \varepsilon_{i+\frac{1}{2},j+\frac{1}{2}}\right]$$

$$+ \frac{2D_{22}^{ij}D_{22}^{ij-1}}{D_{22}^{ij}+D_{22}^{ij-1}} \left[\varepsilon_{i,j} - \varepsilon_{i,j-1}\right] + \frac{D_{22}^{ij}D_{12}^{ij-1}+D_{22}^{ij-1}D_{12}^{ij}}{D_{22}^{ij}+D_{22}^{ij-1}} \left[\varepsilon_{i+\frac{1}{2},j-\frac{1}{2}} - \varepsilon_{i-\frac{1}{2},j-\frac{1}{2}}\right]$$

$$+ \frac{2D_{11}^{ij}D_{11}^{i+1j}}{D_{11}^{ij}+D_{11}^{i+1j}} \left[\varepsilon_{i,j} - \varepsilon_{i+1,j}\right] + \frac{D_{11}^{ij}D_{21}^{i+1j}+D_{11}^{i+1j}D_{21}^{ij}}{D_{11}^{ij}+D_{11}^{i+1j}} \left[\varepsilon_{i+\frac{1}{2},j-\frac{1}{2}} - \varepsilon_{i+\frac{1}{2},j+\frac{1}{2}}\right] \quad (26)$$

$$+ \frac{2D_{11}^{ij}D_{11}^{i-1j}}{D_{11}^{ij}+D_{11}^{i-1j}} \left[\varepsilon_{i,j} - \varepsilon_{i-1,j}\right] + \frac{D_{11}^{ij}D_{12}^{i-1j}+D_{11}^{i-1j}D_{12}^{ij}}{D_{11}^{ij}+D_{11}^{i-1j}} \left[\varepsilon_{i-\frac{1}{2},j+\frac{1}{2}} - \varepsilon_{i+\frac{1}{2},j+\frac{1}{2}}\right]$$

$$= \sum_{e \in E_{i,j}} hR_{i,j}^e \qquad \forall \ 1 \le i,\, j \le N$$

$$\left(\frac{D_{11}^{ij+1}D_{21}^{i+1j+1}+D_{11}^{i+1j+1}D_{21}^{ij+1}}{D_{11}^{ij+1}+D_{11}^{i+1j+1}}\right)[\varepsilon_{i,j+1}-\varepsilon_{i+1,j+1}]$$

$$+\left(\frac{\left(D_{12}^{i+1j+1}-D_{12}^{ij+1}\right)\left(D_{21}^{ij+1}-D_{21}^{i+1j+1}\right)}{2\left(D_{11}^{ij+1}+D_{11}^{i+1j+1}\right)}+\frac{D_{22}^{ij+1}+D_{22}^{i+1j+1}}{2}\right)\left[\varepsilon_{i+\frac{1}{2},j+\frac{1}{2}}-\varepsilon_{i+\frac{1}{2},j+\frac{3}{2}}\right]$$

$$+\left(\frac{D_{11}^{i+1j}D_{21}^{ij}+D_{11}^{ij}D_{21}^{i+1j}}{D_{11}^{ij}+D_{11}^{i+1j}}\right)[\varepsilon_{i+1,j}-\varepsilon_{i,j}]$$

$$+\left(\frac{\left(D_{12}^{i+1j}-D_{12}^{ij}\right)\left(D_{21}^{ij}-D_{21}^{i+1j}\right)}{2\left(D_{11}^{ij}+D_{11}^{i+1j}\right)}+\frac{D_{22}^{ij}+D_{22}^{i+1j}}{2}\right)\left[\varepsilon_{i+\frac{1}{2},j+\frac{1}{2}}-\varepsilon_{i+\frac{1}{2},j-\frac{1}{2}}\right]$$

$$+\left(\frac{D_{22}^{i+1j}D_{12}^{i+1j+1}+D_{22}^{i+1j+1}D_{12}^{i+1j}}{D_{22}^{i+1j}+D_{22}^{i+1j+1}}\right)[\varepsilon_{i+1,j}-\varepsilon_{i+1,j+1}]$$

$$+\left(\frac{\left(D_{21}^{i+1j+1}-D_{21}^{i+1j}\right)\left(D_{12}^{i+1j}-D_{12}^{i+1j+1}\right)}{2\left(D_{22}^{i+1j}+D_{22}^{i+1j+1}\right)}+\frac{D_{11}^{i+1j}+D_{11}^{i+1j+1}}{2}\right)\left[\varepsilon_{i+\frac{1}{2},j+\frac{1}{2}}-\varepsilon_{i+\frac{3}{2},j+\frac{1}{2}}\right]$$

$$(27)$$

$$+\left(\frac{D_{22}^{ij+1}D_{12}^{ij}+D_{22}^{ij}D_{12}^{ij+1}}{D_{22}^{i+1j}+D_{22}^{i+1j+1}}\right)[\varepsilon_{i,j+1}-\varepsilon_{i,j}]$$

$$+\left(\frac{\left(D_{21}^{ij+1}-D_{21}^{ij}\right)\left(D_{12}^{ij}-D_{12}^{ij+1}\right)}{2\left(D_{22}^{ij}+D_{22}^{ij+1}\right)}+\frac{D_{11}^{ij}+D_{11}^{ij+1}}{2}\right)\left[\varepsilon_{i+\frac{1}{2},j+\frac{1}{2}}-\varepsilon_{i-\frac{1}{2},j+\frac{1}{2}}\right]$$

$$=\sum_{e\in E_{i+\frac{1}{2},j+\frac{1}{2}}}hR^e_{i+\frac{1}{2},j+\frac{1}{2}}\quad\forall\ 1\le i,j\le N-1$$

with, due to (8), (9), (12) and (13)

$$\varepsilon_{i+\frac{1}{2},\frac{1}{2}}=\varepsilon_{i+\frac{1}{2},N+\frac{1}{2}}=\varepsilon_{\frac{1}{2},j+\frac{1}{2}}=\varepsilon_{N+\frac{1}{2},j+\frac{1}{2}}=0\quad\forall\ 0\le i,j\le N\quad(28)$$

$$\varepsilon_{i,0}=\varepsilon_{0,j}=\varepsilon_{i,N+1}=\varepsilon_{N+1,j}=0\quad\forall\ 1\le i,j\le N\quad(29)$$

We have the following result.

Theorem 4.6. (*Error estimates in following norms:* $\mathbf{L}^\infty(\Omega)$ *and* $\|.\|_{1,h}$)
Assume that the permeability tensor D in the flow problem (1)–(2) is a symmetric positive definite full matrix with piecewise constant coefficients. Assume also that the unique variational solution φ of (1)–(2) satisfies $\varphi\in C^2\left(\overline\Omega\right)$ and consider the space $\mathbf{E}(\mathcal{L})$ made up of functions v defined almost everywhere in Ω such that v is constant in each element of the mesh $\mathcal{L}$ (see Fig. 2 for the definition of $\mathcal{L}$) and zero elsewhere. Recall that $\Pi\varphi$ is a function of $\mathbf{E}(\mathcal{L})$ defined as follows:

$$\Pi\varphi_{|L}(x)=\varphi_L\equiv value\ of\ \varphi\ at\ the\ center\ of\ L,\ for\ all\ L\ \in\mathcal{L}$$

and (of course) zero elsewhere.

Then, the function $\varepsilon_h = \varphi_h - u_h$, where ε_h is defined by (25) and $u_h = \Pi U_h$, satisfies the following inequalities:

$$(i) \qquad\qquad \|\varepsilon_h\|_{1,h} \leq Ch$$

$$(ii) \qquad\qquad \|\varepsilon_h\|_{L^\infty(\Omega)} \leq Ch^{\frac{1}{2}}$$

Moreover, if D is uniformly constant and the unique variational solution φ of (1)–(2) lies in $C^3\left(\overline{\Omega}\right)$ then:

$$(i) \qquad\qquad \|\varepsilon_h\|_{1,h} \leq Ch^2$$

$$(ii) \qquad\qquad \|\varepsilon_h\|_{L^\infty(\Omega)} \leq Ch^{\frac{3}{2}}$$

where C represents miscellaneous strictly positive constants without dependence on h.

The proof of this theorem can found in [9, 13]. From Lemma 4.3 (discrete version of Poincaré inequality), one gets:

Corollary 4.7. *(Error estimate in $L^2(\Omega) - norm$)*
ε_h satisfies the following inequality
First case: D is piecewise constant and the unique variational solution φ of (1)–(2) lies in $C^2\left(\overline{\Omega}\right)$.

$$\|\varepsilon_h\|_{L^2(\Omega)} \leq Ch.$$

Second case: D is uniformly constant and the unique variational solution φ of (1)–(2) belongs to $C^3\left(\overline{\Omega}\right)$.

$$\|\varepsilon_h\|_{L^2(\Omega)} \leq Ch^2.$$

where C represents miscellaneous strictly positive constants without dependence on h.

5 Convergence Results

In what follows, C denotes miscellaneous constants without dependence on h and Λ is the classical Lagrange interpolation operator associated with the nodes of the grid $\mathcal{T}$.

Let us carry out the error estimates for the approximate solution introduced in the section 3. The notations introduced in the previous sections are conserved here. In this connection, we recall that φ denotes the exact solution of the boundary-value problem (1)–(2).

We give in this section the error estimates for the piecewise linear approximate solution introduced in section 3.

Proposition 5.1. *First case: D is piecewise constant and the unique variational solution φ of (1)–(2) lies in $C^2\left(\overline{\Omega}\right)$. Then the linear approximate solution U_T^h satisfy the following inequality:*

$$\left\|\varphi - U_T^h\right\|_{0,\Omega} \leq C\,h$$

Second case: D is uniformly constant and the unique variational solution φ of (1)–(2) is in $C^3\left(\overline{\Omega}\right)$. Then the linear approximate solution U_T^h satisfy the following inequalities:

$$\left\|\varphi - U_T^h\right\|_{0,\Omega} \leq C\,h^2.$$

The proof of this proposition can be found in [9].

6 Conclusions and Perspectives

We have presented in this work the formulation of a MPFA finite volume scheme for flow problems in anisotropic heterogeneous media. The well posedness of the discrete problem was not an obvious issue and was solved affirmatively. The stability and the convergence of the weak approximate solution have been shown for L^2 and L^∞−norm in the one hand and in a discrete energy norm in the other hand. It is also proven that the error estimates of the weak approximate solution play a key role for the derivation of the error estimates of the piecewise linear approximate solution introduced in the third section.

The analysis of the presented MPFA finite volume method on unstructured irregular meshes is our objective today as we have successfully implemented it for such meshes.

References

1. I. Aavatsmark. (2002), *An introduction to multipoint flux approximations for quadrilateral grids*, Comput. Geosci. 6, pp. 405–432.
2. I. Aavatsmark, T. Barkve, O. Boe, T. Mannseth. (1998), *Discretization on unstructured grids for inhomogeneous, anisotropic media. part I: derivation of the methods*, Siam Vol. 19, No. 5, pp. 1700–1716.
3. H. Brézis. (1983), *Analyse fonctionnelle, théorie et applications*, Ed. Masson, Paris.
4. S. -H. Chou, D. Y. Kwak, K. Y. Kim. (2001), *A general framework for constructing and analyzing mixed finite volume methods on quadrilateral grids: the overlapping covolume case*, SIAM J. Numer. Anal. Vol. 39, pp. 1170–1196. (electronic).
5. L. J. Durlofsky. (2005), *Upscaling and Gridding of Geologically Complex Systems*, Department of Petroleum Engineering, Stanford University Chevron Texaco E & P Technology Company.

6. M. G. Edwards. (2002), *Unstructured, control-volume distributed, full-tensor finite-volume schemes with flow based grids*, Comput. Geosci. 6, pp. 433–452.
7. R. A. Klausen, T. F. Russell. *Relationships among some locally conservative discretization methods which handle discontinuous coefficients.* To appear in Computational Geoscience.
8. I. D. Mishev. (2003), *Analysis of a new mixed finite volume method*, Comput. Methods Appl. Math. Vol. 3, No. 1, pp. 189–201.
9. A. Njifenjou, A. J. Kinfack. *Convergence analysis of a MPFA method for flow problems in anisotropic heterogeneous porous media.* Accepted for publication in International Journal on Finite Volume (in press).
10. A. Njifenjou, I. Moukouop Nguena. (2006) *A new MPFA formulation for subsurface flow problems on unstructured grids: derivation of the discrete problem*, International Workshop on Mathematical Modelling, Simulation, Visualization and E-learning, Bellagio (Italy), November 20–26.
11. A. Njifenjou, B. Mampassi, S. Njipouakouyou. (2006), *A finite volume formulation and its connection with mixed hybrid finite element method*, Far East J. Appl. Math. Vol. 24, No. 3, pp. 267–280.
12. I. Nguena Moukouop, A. Njifenjou. (2005), *A new finite volume formulation for diffusion problems in anisotropic non-homogeneous media*, Proceedings of the 4th International Symposium on Finite Volume for Complex Applications, Marraketch (Morocco), July 4–8, pp. 435–446, (Hermes edition).
13. A. Njifenjou, I. M. Nguena. (2006), *A finite volume approximation for second order elliptic problems with a full matrix on quadrilateral grids: derivation of the scheme and a theoretical analysis*, Int. J. Finite Vol., Vol.3, No. 2, (electronic publication).
14. P. Renard. (2005), *Averaging methods for permeability fields, Development, Protection, Management and Sequestration of Subsurface. Fluid Flow and Transport in Porous and Fractured Media*, Summer School, Cargèse (France), July 27.
15. P. A. Raviart, J. M. Thomas. (1983), *Introduction à l'analyse numérique des équations aux dérivées partielles*, Collection Mathématiques Appliquées pour la Maîtrise, under the direction of P.G. Ciarlet et J.L. Lions. Editor: Masson.

A Small Eddy Correction Algorithm for the Primitive Equations of the Ocean

T. Tachim Medjo and R. Temam

Summary. Considering the interaction between the baroclinic and barotropic flows and using the idea of the Newton iteration, a small eddy correction method is proposed for approximating and numerically solving the primitive equations of the ocean. We assume that the barotropic approximation to the solution is known. Formally applying the Newton iterative procedure to the baroclinic flow equation, we then generate approximate systems. It is shown that the first step leads to the well known quasi-geostrophic equations. The convergence analysis is presented and the results show that the small eddy correction method can greatly improve the accuracy of the quasi-geostrophic approximate solution. More precisely, we prove that the approximate system derived from the procedure converges to the primitive equations of the ocean and we estimate the rate of convergence as a function of the aspect ratio of the ocean. Some numerical simulations of a wind-driven circulation problem are presented to illustrate the method.

1 Introduction

In dynamical systems theory the objective is to study the long-term behavior of solutions of an evolution equation. When the equation is dissipative all solutions converge as $t \longmapsto \infty$ to a complicated set $\mathcal{A}$, the global attractor, which may be a fractal set. This set embodies the large-time dynamics of the equations, corresponding to all sorts of regimes, including turbulent ones. Although this set may be fairly complicated, in general it has finite dimension, see e.g., [7, 23, 26]. Despite the considerable increase in the available computing power during the past few years, the numerical approximation of the global attractor remains a difficult task specially for important systems such as the Navier–Stokes (NS) equations or the primitive equations (PEs) of the ocean. For the NS flows, there are some approaches to deriving simplified behavioral laws for the smallest structure set in motion with the aim of reducing the computational cost, see e.g., [4, 19]. In the nonlinear Galerkin (NLG) method introduced in [19], the small scales are given as a function of the large scales and the nonlinear interaction between the large and the small scales is only approximately modelled. In [11], the authors presented a small eddy correction method for the 2D NS equations. It is shown that the first step of this iterative method leads to the standard Galerkin method and the second step yields the nonlinear Galerkin method.

Although the source of the extensive scale variability differs for the NS and the PEs models (the scale variability in the NS system is mainly the result of the nonlinear term, while the sources are more varied for the PEs model), there exists an energy cascade that is similar for the two models, and for which one can apply the main principle of description given by Charney [3].

Every mode undergoes constraints due to wind. Even for a fairly constant wind, there is still an infinite number of modes stimulated by the boundary conditions. These modes will exhibit different behaviors with respect to the stimulations based on their position in the spectrum. They can be grouped into three categories:

- At the largest scales, geophysical flows such as the ocean and the atmosphere are essentially two-dimensional (barotropic component). These barotropic modes transmit their energy in the two following ways:
  (1) At modes of greater dimension, through an inverse kinetic energy cascade. The surplus energy is then dissipated by the boundary conditions.
  (2) At modes of smaller dimension, through an enstrophy barotropic cascade.
- At the medium scales, we have the baroclinic modes. These modes will redistribute their energy as a baroclinic energy cascade, thus transporting the energy to the viscous dispersal area. This cascade is similar to the energy cascade predicted by Kolmogorov for the Navier–Stokes system, [4, 21].
- At the very small scales, the energy provided by the surface forces is insufficient to oppose the viscous dispersion constraint.

Given the similarities with the NS system and inspired by the results obtained for the 2D NS equations with the NLG method, we present in this article a small eddy correction method for the PEs of the ocean. Considering the interaction between the baroclinic and barotropic flows and using the idea of the Newton iteration, a small eddy correction method is proposed for approximating and numerically solving the PEs of the ocean. We assume that the barotropic approximation to the solution is known. Formally applying the Newton iterative procedure to the baroclinic flow equation, we then generate approximate systems. It is shown that the first step leads to the well known quasi-geostrophic equations. The convergence analysis is presented and the results show that the small eddy correction method can greatly improve the accuracy of the quasi-geostrophic approximate solution. More precisely, we prove that the approximate system derived from the procedure converges to the original PEs with a rate of convergence $O(\delta^{(3/2)^l})$, where δ is the shape ratio of the ocean and l is the number of small eddy iterations.

The article is organized as follows. In Sect. 2, we recall the PEs of the ocean and their mathematical setting. Section 3 is devoted to the existence and uniqueness of strong solutions when the aspect ratio is small enough and the initial condition and body forces satisfy some restrictive conditions. For the latest results concerning the existence and uniqueness of solutions of the primitive equations in space dimension 2 and 3, see [2, 14, 15] and the review articles [22, 28]. For flows in shallow domains see [17] which however does not apply here as the boundary conditions are different; also because our approach here differs from that of [17] because the vertical height has been normalized to 1, thus introducing the shape factor δ in the equations. Note that the issues of existence and uniqueness of solutions are not our main objective here and they are presented as a step toward the introduction of the *small-eddy algorithm*. Section 4 presents the small eddy correction method and studies its boundedness. Section 5 is devoted to the convergence of the small eddy correction models to the PEs of the ocean as the aspect ratio goes to zero. We derive an estimate on the rate of convergence as a power of the aspect ratio of the ocean. Although the approach used here bear some similarities with [11], there are several differences between the work of [11] and the one presented here. First, our model is more complicated. In fact, the PEs of the ocean possess some specific difficulties to circumvent, for instance the nonlocal constraint (incompressible condition) and the integral expression of the vertical velocity lead to a strong nonlinear term

$$\left(\int_z^0 \mathrm{div} v ds \right) \frac{\partial v}{\partial z}. \tag{1}$$

More importantly, in [11] the authors used the eigenvalues of the Stokes operator to split the solution between the large and small scales, while in this article the large scale is the depth average (barotropic mode) of the solution and the small scale is the deviation (baroclinic mode), a decomposition commonly used in ocean modeling. To illustrate the method, some numerical simulations are presented in the last section of this article. See also related but different ideas on multilevel methods in [5, 17, 18].

2 A Navier–Stokes Type Equation and its Mathematical Setting

2.1 Governing Equations

Throughout this article, we use Δ, ∇, div to denote the two-dimensional gradient, Laplacian and divergence operators on the horizontal plane. The nondimensional domain $\mathcal{M}$ occupied by the fluid is given by

$$\mathcal{M} = \mathcal{O} \times (-h, 0), \tag{2}$$

where $\mathcal{O} \subset R^2$ is a smooth convex, bounded open set of R^2 with boundary $\partial \mathcal{O}$, and $h > 0$ is a constant.

The boundary of $\mathcal{M}$ consists of

$$\partial \mathcal{M} = \Gamma_i \cup \Gamma_l \cup \Gamma_b, \tag{3}$$

where

$$
\begin{aligned}
\Gamma_i(z = 0) &= \text{ upper boundary of the ocean (interface with air)}, \\
\Gamma_l &= \text{ lateral boundary, i.e., } \Gamma_l = \partial \mathcal{O} \times (-h, 0), \\
\Gamma_b(z = -h) &= \text{ bottom of the ocean.}
\end{aligned} \tag{4}
$$

We first recall the set of equations which describe the motion and state of an idealized ocean. In the nondimensional form, the equations read

$$
\left\{
\begin{aligned}
&\frac{\partial u}{\partial t} - \frac{1}{R_{e_1}} \Delta u - \frac{1}{\delta^2 R_{e_2}} \frac{\partial^2 u}{\partial z^2} + \nabla \mathcal{L}(u) + f k_0 \times u + (u \cdot \nabla)u \\
&\quad + W(u)\frac{\partial u}{\partial z} + \text{grad } p + \text{grad} \int_z^0 \rho ds = F_1, \\
&\text{div} \int_{-h}^0 u \, dz = 0, \\
&\frac{\partial \rho}{\partial t} - \frac{1}{R_{t_1}} \Delta \rho - \frac{1}{\delta^2 R_{t_2}} \frac{\partial^2 \rho}{\partial z^2} + (u \cdot \nabla)\rho + W(u)\frac{\partial \rho}{\partial z} = F_2.
\end{aligned}
\right. \tag{5}
$$

The boundary conditions are

$$\frac{\partial u}{\partial z} = 0, \quad \frac{\partial \rho}{\partial z} = 0 \text{ on } \Gamma_i \cup \Gamma_b, \quad u = 0, \quad \rho = 0 \text{ on } \Gamma_l, \tag{6}$$

and the initial conditions are

$$(u, \rho) = \mathbf{a} = (a^1, a^2) \text{ at } t = 0. \tag{7}$$

In (5), the unknown functions are the horizontal velocity $u = (u_1, u_2)$ and the density ρ of the fluid. The constants $R_{e_1} = \frac{\mu}{L_1 U_1} > 0$, $R_{e_2} = \frac{\nu}{L_1 U_1} > 0$, $R_{t_1} = \frac{\mu_T}{L_1 U_1} > 0$ and $R_{t_2} = \frac{\nu_T}{L_1 U_1} > 0$ are nondimensional Reynolds numbers, $\delta = \frac{H_1}{L_1}$ is the aspect ratio, $p = p(x, y)$ is the surface pressure of the fluid, F_1, F_2 are the volume forces and k_0 is the unit vector in the vertical direction. Here U_1 is the reference value for the horizontal velocity, L_1 is the reference value for the horizontal length scale, H_1 is the reference value for the vertical length scale, μ and ν are the effective molecular dissipations in the horizontal and vertical directions, μ_T and ν_T reflect the heat diffusion [16–18].

The Coriolis parameter f is defined by $f = f_0 + \beta y$, where $\beta > 0$, $f_0 > 0$ are positive constants. The operator W is defined by

$$W(u) = \int_z^0 \text{div} u \, ds, \tag{8}$$

and it represents the vertical velocity of the fluid [16].

The operator $\mathcal{L}$ is defined by

$$\mathcal{L}(u) = \int_z^0 \mathcal{L}_1(u)ds, \ \mathcal{L}_1(u) = -\frac{1}{R_{e_1}}\Delta W(u) - \frac{1}{\delta^2 R_{e_2}}\frac{\partial^2 W(u)}{\partial z^2}. \tag{9}$$

Remark 2.1. Hereafter, for simplicity we will consider only homogeneous boundary conditions and we will assume that the function τ_0 in (5) is identically zero, that is $\tau_0 = 0$. Let us mention that with a boundary data of the form $\tau_0 = c\sin \pi y$ that is commonly used in oceanography to simulate the double-gyre phenomena [6, 8, 29], the boundary conditions in (5) present a discontinuity on $\partial\mathcal{O} \times \{z = 0\}$. This discontinuity may affect the accuracy/stability of the numerical schemes if special care is not taken. To overcome this problem, it is common in oceanography to take $\tau_0 = 0$ and to compensate with a body force $\mathcal{F} = g(z)\tau_0$ in the momentum equations. The function $g(z)$ is chosen such that the forcing F is nonzero only on a thin layer from the surface of the ocean, the goal being to reproduce the effect of the wind-stress on the ocean circulation.

2.2 Mathematical Setting

In this section we first define the function spaces suitable for the mathematical setting of (5), see [1] and [12, 16] for details. Let $\vec{n}$ be the unit outward normal vector to the boundary $\partial\mathcal{O}$ of the domain $\mathcal{O}$. We denote by $H^s(\mathcal{O})$ (resp. $H^s(\mathcal{M})$), for $s \in R$, the Sobolev spaces constructed on $L^2(\mathcal{O})$ (resp. $L^2(\mathcal{M})$), and by $H_0^s(\mathcal{O})$ (resp. $H_0^s(\mathcal{M})$,) for $s > 1/2$, the closure of $C_c^\infty(\mathcal{O})$ (resp. $C_c^\infty(\mathcal{M})$) in $H^s(\mathcal{O})$ (resp. $H^s(\mathcal{M})$), the space of infinitely differentiable functions with compact support in $\mathcal{O}$ (resp. $\mathcal{M}$).

The Velocity Function Spaces

Motivated by the boundary conditions for the velocity field, we define

$$\mathcal{V}_1 = \{u \in (C^\infty(\mathcal{M}))^2; \ u = 0 \text{ in a neighborhood of } \Gamma_l, \ W(u) \in H_0^1(\mathcal{M}),$$

$$\int_{-h}^0 \text{div } udz = 0\},$$

and denote by H_1 (resp. V_1) the closure of $\mathcal{V}_1$ in $(L^2(\mathcal{M}))^2$ (resp. $(H^1(\mathcal{M}))^2$).

The scalar product in H_1 is denoted by $\langle u, v\rangle = \int_\mathcal{M} u \cdot v d\mathcal{M}$, that on V_1 is denoted by

$$((u, v)) = \int_\mathcal{M} \left(\nabla u \cdot \nabla v + \frac{\partial u}{\partial z} \cdot \frac{\partial v}{\partial z}\right) d\mathcal{M}$$

and the associated norms are denoted by $|\cdot|_{L^2}$ and $\|\cdot\|$ respectively.

We also equip V_1 with the norm

$$\|u\|_w^2 = \|u\|^2 + \|W(u)\|^2.$$

We will denote by $\mathcal{P}_1$ the orthogonal projector of the space $(L^2(\mathcal{M}))^2$ onto the space H_1.

Throughout this article, we will use the notation

$$B_1(u,v) = \mathcal{P}_1\left((u \cdot \nabla)v + W(u)\frac{\partial v}{\partial z}\right), \quad \forall u, v \in V_1. \tag{10}$$

For $u, v \in V_1$, we set

$$a_{11}(u,v) = \frac{1}{R_{e_1}}\langle \nabla u, \nabla v\rangle + \frac{1}{R_{e_1}}\langle \nabla W(u), \nabla W(v)\rangle \equiv \langle A_{11}u, v\rangle,$$

$$a_{12}(u,v) = \frac{1}{R_{e_2}}\left\langle \frac{\partial u}{\partial z}, \frac{\partial v}{\partial z}\right\rangle + \frac{1}{R_{e_2}}\langle \mathrm{div}u, \mathrm{div}v\rangle \equiv \langle A_{21}u, v\rangle,$$

$$e_1(u,v) = \langle fk_0 \times u, v\rangle \equiv \langle E_1 u, v\rangle, \tag{11}$$

$$a_1 = a_{11} + a_{12}, \quad A_1 = A_{11} + A_{12}.$$

The operator A_1 is the isomorphism from V_1 onto the dual V_1' of V_1 defined by:

$$\langle A_1 u, v\rangle_{V_1', V_1} = a_1(u,v), \quad \forall u, v \in V_1,$$

where $\langle \cdot, \cdot \rangle_{V', V}$ is the duality bracket between V_1' and V_1. The operator A_1 is also a linear unbounded operator in H_1 with domain $D(A_1) = (H^2(\mathcal{M}))^2 \cap V_1$. Let us recall (see [28]) that there exists a positive constant c such that

$$\|v\|_{H^2(\mathcal{M})} \le c|A_1 v|_{L^2}, \quad \forall v \in D(A_1). \tag{12}$$

Lemma 2.2. *For all $u, v \in D(A_1)$ such that $\frac{\partial u}{\partial z} = \frac{\partial v}{\partial z} = 0$ on $\Gamma_i \cup \Gamma_b$, we have*

$$\left\langle -\nabla \int_z^0 \Delta W(u)ds, v\right\rangle = \langle \nabla W(u), \nabla W(v)\rangle,$$

$$\left\langle -\nabla \int_z^0 \frac{\partial^2 W(u)}{\partial z^2}ds, v\right\rangle = \left\langle \frac{\partial W(u)}{\partial z}, \frac{\partial W(v)}{\partial z}\right\rangle,$$

$$\langle -\Delta u, A_{12}v\rangle = \frac{1}{R_{e_2}}\left\langle \frac{\partial \nabla u}{\partial z}, \frac{\partial \nabla v}{\partial z}\right\rangle + \frac{1}{R_{e_2}}\langle \nabla\, \mathrm{div}u, \nabla\, \mathrm{div}v\rangle,$$

$$\langle \nabla \mathcal{L}(u), v\rangle = \frac{1}{R_{e_1}}\langle \nabla W(u), \nabla W(v)\rangle + \frac{1}{\delta^2 R_{e_2}}\langle \mathrm{div}u, \mathrm{div}v\rangle \tag{13}$$

$$= \frac{1}{R_{e_1}}\langle \nabla W(u), \nabla W(v)\rangle$$

$$+ \frac{1}{\delta^2 R_{e_2}}\left\langle \frac{\partial W(u)}{\partial z}, \frac{\partial W(v)}{\partial z}\right\rangle.$$

Proof. See [28]. □

Proposition 2.3. *The following properties hold true*

$a_1(u, u) \geq \alpha_1 \|u\|_w^2, \ \forall u \in V_1,$

$\langle B_1(u, v), v \rangle = 0, \quad e_1(u, u) = 0, \ \forall u, v \in V_1,$

$|B_1(u, v)|_{L^2} \leq c\|u\|_w \|v\|^{\frac{1}{2}} |A_1 v|_{L^2}^{\frac{1}{2}}, \ \forall u \in V_1, v \in D(A_1),$

$|B_1(\bar{u}, v)|_{L^2} \leq c\|\bar{u}\| \|v\|^{\frac{1}{2}} |A_1 v|_{L^2}^{\frac{1}{2}}, \ \forall \bar{u} \in MV_1, v \in D(A_1),$

$|\langle B_1(u, v), w \rangle| \leq c\|u\|_w \|v\| \|w\|_{L^2}^{\frac{1}{4}} \|w\|^{\frac{3}{4}}, \ \forall u, v, w \in V_1,$

$|\langle B_1(u^\flat, \bar{v}), w^\flat \rangle| \leq c|u^\flat|_{L^2}^{\frac{1}{2}} \|u^\flat\|^{\frac{1}{2}} \|\bar{v}\| |w^\flat|_{L^2}^{\frac{1}{2}} \|w^\flat\|^{\frac{1}{2}}, \ \forall u^\flat, w^\flat \in NV_1, \bar{v} \in MV_1,$

$$(14)$$

where $c > 0$ and $\alpha_1 > 0$ are constants depending only on $\mathcal{M}, R_{e_1}, R_{e_2}$ and the operators M and N are defined in (24)–(25) below.

Proof. The proof of $(14)_1 - (14)_5$ is given in [16], see also [28]. For $(14)_6$, we have

$$\int_{\mathcal{M}} (u^\flat \nabla \bar{v}) \cdot w^\flat d\mathcal{M} \leq \int_{-1}^0 |u^\flat|_{L^4(\mathcal{O})} |\nabla \bar{v}|_{L^2(\mathcal{O})} |w^\flat|_{L^4(\mathcal{O})} dz$$

$$\leq c|\nabla \bar{v}|_{L^2(\mathcal{O})} \int_{-1}^0 |u^\flat|_{L^2(\mathcal{O})}^{\frac{1}{2}} |\nabla u^\flat|_{L^2(\mathcal{O})}^{\frac{1}{2}} |w^\flat|_{L^2(\mathcal{O})}^{\frac{1}{2}} |\nabla w^\flat|_{L^2(\mathcal{O})}^{\frac{1}{2}} dz$$

$$\leq c|u^\flat|_{L^2}^{\frac{1}{2}} \|u^\flat\|^{\frac{1}{2}} \|\bar{v}\| |w^\flat|_{L^2}^{\frac{1}{2}} \|w^\flat\|^{\frac{1}{2}}$$

$$(15)$$

and $(14)_6$ follows. $\square$

We also define the following function spaces

$$\tilde{\mathcal{V}}_1 = \{u \in (C_0^\infty(\mathcal{O}))^2; \ \text{div } u = 0\},$$

and denote by $\tilde{H}_1$ (resp. $\tilde{V}_1$) the closure of $\tilde{\mathcal{V}}_1$ in $(L^2(\mathcal{O}))^2$ (resp. $(H^1(\mathcal{O}))^2$); we have

$$\tilde{H}_1 = \{u \in (L^2(\mathcal{O}))^2; \ \text{div } u = 0 \text{ in } \mathcal{O}, \ u \cdot \vec{n} = 0 \text{ on } \partial\mathcal{O}\}$$

and

$$\tilde{V}_1 = \{u \in (H_0^1(\mathcal{O}))^2; \ \text{div } u = 0 \text{ in } \mathcal{O}\}.$$

The scalar product in $\tilde{H}_1$ is denoted by $\langle u, v \rangle = \int_{\mathcal{O}} u \cdot v dx dy$, that on $\tilde{V}_1$ is denoted by

$$((u, v)) = \int_{\mathcal{O}} \nabla u \cdot \nabla v dx dy,$$

and the associated norms are also denoted by $|\cdot|_{L^2}$ and $\|\cdot\|$, respectively.

We denote by $\tilde{A}_1$ the Stokes operator, defined as an isomorphism from $\tilde{V}_1$ onto the dual $\tilde{V}_1'$ of $\tilde{V}_1$ such that, for $u \in \tilde{V}_1, \tilde{A}_1 u$ is defined by

$$\langle \tilde{A}_1 u, v \rangle_{\tilde{V}_1', \tilde{V}_1} = ((u,v)), \ \forall u, v \in \tilde{V}_1,$$

where $\langle \cdot, \cdot \rangle_{\tilde{V}_1', \tilde{V}_1}$ is the duality bracket between $\tilde{V}_1'$ and $\tilde{V}_1$. The operator $\tilde{A}_1$ can also be seen as a linear unbounded operator in H_1 with domain $D(\tilde{A}_1) = (H^2(\mathcal{O}))^2 \cap \tilde{V}_1$ when $\partial \mathcal{O}$ is of class C^2. We also denote by $\mathcal{P}_2$ the Leray–Hopf projector, which is the orthogonal projector of the space $(L^2(\mathcal{O}))^2$ onto the space $\tilde{H}_1$. The Stokes operator is related to $\mathcal{P}_2$ by

$$\tilde{A}_1 u = -\mathcal{P}_2(\Delta u), \ \forall u \in D(\tilde{A}_1).$$

We also define the bilinear mapping $\tilde{B}_1$ by

$$\tilde{B}_1(u,v) = \mathcal{P}_2\left((u \cdot \nabla)v\right), \ \forall u, v \in \tilde{V}_1,$$

which maps $\tilde{V}_1$ into $\tilde{V}_1'$.

The following properties hold true, [28].

$$\langle \tilde{B}_1(u,v), v \rangle = 0, \ \forall u, v \in \tilde{V}_1,$$
$$|\tilde{B}_1(u,v)|_{L^2} \le c|u|_{L^2}^{\frac{1}{2}}|\tilde{A}_1 u|_{L^2}^{\frac{1}{2}}\|v\|, \ \forall u \in D(\tilde{A}_1), v \in \tilde{V}_1, \qquad (16)$$

where c is a constant depending only on $\mathcal{O}$.

We define the function spaces X_1 and $\tilde{X}_1$ by

$$\tilde{X}_1 = \left\{ \bar{u} \in L^2(0,T; D(\tilde{A}_1)), \ \frac{d\bar{u}}{dt} \in L^2(0,T; \tilde{H}_1) \right\},$$

$$X_1 = \left\{ u^\flat \in L^2(0,T; D(A_1)), \ \frac{du^\flat}{dt} \in L^2(0,T; H_1) \right\}.$$

The spaces $\tilde{X}_1$ and X_1 are endowed with the norms

$$\|\bar{u}\|_{\tilde{X}_1} = \left(\|\bar{u}\|_{L^2(0,T;D(\tilde{A}_1))}^2 + \left\| \frac{d\bar{u}}{dt} \right\|_{L^2(0,T;\tilde{H}_1)}^2 \right)^{\frac{1}{2}},$$

$$\|u^\flat\|_{X_1} = \left(\|u^\flat\|_{L^2(0,T;D(A_1))}^2 + \left\| \frac{du^\flat}{dt} \right\|_{L^2(0,T;H_1)}^2 \right)^{\frac{1}{2}}.$$

Let us recall that

$$\tilde{X}_1 \subset C(0,T; \tilde{V}_1), \ X_1 \subset C(0,T; V_1),$$

with continuous injections.

The Density Function Spaces

We also define the following function spaces

$$\mathcal{V}_2 = \{\rho \in C^\infty(\mathcal{M}); \ \rho = 0 \text{ in a neighborhood of } \Gamma_l, \ \},$$

and denote by H_2 (resp. V_2) the closure of $\mathcal{V}_2$ in $L^2(\mathcal{M})$ (resp. $H^1(\mathcal{M})$).

The scalar product in H_2 is denoted by $\langle \rho, \phi \rangle = \int_{\mathcal{M}} \rho \cdot \phi \, d\mathcal{M}$, that on V_2 is denoted by

$$((\rho, \phi)) = \int_{\mathcal{M}} \left(\nabla \rho \cdot \nabla \phi + \frac{\partial \rho}{\partial z} \frac{\partial \phi}{\partial z} \right) d\mathcal{M}$$

and the associated norms are denoted by $| \cdot |_{L^2}$ and $\| \cdot \|$, respectively.

Throughout this article, we will use the notation

$$B_2(u, \rho) = (u \cdot \nabla)\rho + W(u)\frac{\partial \rho}{\partial z}, \ \forall u \in V_1, \rho \in V_2. \tag{17}$$

For $\rho, \phi \in V_2$, we set

$$a_{21}(\rho, \phi) = \frac{1}{R_{t_1}} \langle \nabla \rho, \nabla \phi \rangle \equiv \langle A_{21}\rho, \phi \rangle,$$

$$a_{22}(\rho, \phi) = \frac{1}{R_{t_2}} \left\langle \frac{\partial \rho}{\partial z}, \frac{\partial \phi}{\partial z} \right\rangle \equiv \langle A_{22}\rho, \phi \rangle, \tag{18}$$

$$\Lambda_2 \rho = \int_z^0 \text{grad } \rho ds,$$

$$a_2 = a_{21} + a_{22}, \ A_2 = A_{21} + A_{22}.$$

The operator A_2 is the isomorphism from V_2 onto the dual V_2' of V_2 defined by:

$$\langle A_2 \rho, \phi \rangle_{V_2', V_2} = a_2(\rho, \phi), \ \forall \rho, \phi \in V_2,$$

where $\langle \cdot, \cdot \rangle_{V', V}$ is the duality bracket between V_2' and V_2. The operator A_2 is extended to H_2 as a linear unbounded operator with domain $D(A_2) = H^2(\mathcal{M}) \cap V_2$. Let us recall that (see [28]) there exists a positive constant c such that

$$\|\phi\|_{H^2(\mathcal{M})} \leq c|A_2\phi|_{L^2}, \ \forall \phi \in D(A_2). \tag{19}$$

Proposition 2.4. *We have the following inequalities:*

$$a_2(\rho, \rho) \geq \alpha_2 \|\rho\|^2, \ \forall \rho \in V_2,$$

$$\langle B_2(u, \rho), \rho \rangle = 0, \ \forall u \in V_1, \rho \in V_2,$$

$$|B_2(u, \rho)|_{L^2} \leq c\|u\|_w \|\rho\|^{\frac{1}{2}} |A_2\rho|_{L^2}^{\frac{1}{2}}, \ \forall u, \in V_1, \rho \in D(A_2),$$

$$|B_2(\bar{u}, \rho)|_{L^2} \leq c\|\bar{u}\| \|\rho\|^{\frac{1}{2}} |A_2\rho|_{L^2}^{\frac{1}{2}}, \ \forall \bar{u} \in MV_1, \rho \in D(A_2),$$

$$|\langle B_2(u, \rho), \phi \rangle| \le c\|u\|_w \|\rho\| |\phi|_{L^2}^{\frac{1}{4}} \|\phi\|^{\frac{3}{4}}, \ \forall u \in V_1, \rho, \phi \in V_2,$$

$$|\langle B_2(u^b, \bar{\rho}), \phi^b \rangle| \le c|u^b|_{L^2}^{\frac{1}{2}} \|u^b\|^{\frac{1}{2}} \|\bar{\rho}\| |\phi^b|_{L^2}^{\frac{1}{2}} \|\phi^b\|^{\frac{1}{2}},$$

$$\forall u^b \in NV_1, \phi^b \in NV_2, \bar{\rho} \in MV_2, \tag{20}$$

where $c > 0$ and $\alpha_2 > 0$ are constants depending only on $\mathcal{M}, R_{t_1}, R_{t_2}$.

Proof. The proof is similar to that of Proposition 2.2.   □

We introduce the function space $\tilde{\mathcal{V}}_2 = C_0^\infty(\mathcal{O})$, and denote by $\tilde{H}_2$ (resp. $\tilde{V}_2$) the closure of $\tilde{\mathcal{V}}_2$ in $L^2(\mathcal{O})$ (resp. $H^1(\mathcal{O})$).

The scalar product in $\tilde{H}_2$ is denoted by $\langle \rho, \phi \rangle = \int_{\mathcal{O}} \rho \cdot \phi \, dx dy$, that on $\tilde{V}_2$ is denoted by

$$((\rho, \phi)) = \int_{\mathcal{O}} \nabla \rho \cdot \nabla \phi \, dx dy,$$

and the associated norms are also denoted by $|\cdot|_{L^2}$ and $\|\cdot\|$, respectively.

We denote by $\tilde{A}_2$ the operator defined as an isomorphism from $\tilde{V}_2$ onto the dual $\tilde{V}_2'$ of $\tilde{V}_2$ by: $\tilde{A}_2 \rho$ is defined by

$$\langle \tilde{A}_1 \rho, \phi \rangle_{\tilde{V}_2', \tilde{V}_2} = ((\rho, \phi)), \ \forall \rho, \phi \in \tilde{V}_2,$$

where $\langle \cdot, \cdot \rangle_{\tilde{V}_2', \tilde{V}_2}$ is the duality bracket between $\tilde{V}_2'$ and $\tilde{V}_2$. The operator $\tilde{A}_2$ is extended to $\tilde{H}_2$ as a linear unbounded operator with domain $D(\tilde{A}_2) = H^2(\mathcal{O}) \cap \tilde{V}_2$ when $\partial \mathcal{O}$ is of class C^2.

We also define the bilinear mapping $\tilde{B}_2$ by

$$\tilde{B}_2(u, \rho) = (u \cdot \nabla)\rho, \ \forall u \in \tilde{V}_1, \forall \rho \in \tilde{V}_2,$$

which maps $\tilde{V}_1 \times \tilde{V}_2$ into $\tilde{V}_1' \times \tilde{V}_2'$.

The following properties hold true, [28].

$$\langle (u \cdot \nabla)v, v \rangle = 0, \ \forall u, v \in \tilde{V}_1,$$

$$|(u \cdot \nabla)v|_{L^2} \le c|u|_{L^2}^{\frac{1}{2}} |\tilde{A}_1 u|_{L^2}^{\frac{1}{2}} \|v\|, \ \forall u \in D(\tilde{A}_1), \forall v \in \tilde{V}_1,$$

$$\langle (u \cdot \nabla)\rho, \rho \rangle = 0, \ \forall u \in \tilde{V}_1, \forall \rho \in \tilde{V}_2, \tag{21}$$

$$|(u \cdot \nabla)\rho|_{L^2} \le c|u|_{L^2}^{\frac{1}{2}} |\tilde{A}_1 u|_{L^2}^{\frac{1}{2}} \|\rho\|, \ \forall u \in D(\tilde{A}_2), \forall \rho \in \tilde{V}_2,$$

We also define the function spaces Y_1 and Y_2 by

$$\tilde{Y}_1 = \left\{ \bar{\rho} \in L^2(0, T; D(\tilde{A}_2)), \ \frac{d\bar{\rho}}{dt} \in L^2(0, T; \tilde{H}_2) \right\},$$

$$Y_1 = \left\{ \rho^b \in L^2(0, T; D(A_2)), \ \frac{d\rho^b}{dt} \in L^2(0, T; H_2) \right\}.$$

The spaces $\tilde{Y}_1$ and Y_1 are endowed with the norms

$$\|\bar{\rho}\|_{\tilde{Y}_1} = \left(\|\bar{\rho}\|^2_{L^2(0,T;D(\tilde{A}_2))} + \left\| \frac{d\bar{\rho}}{dt} \right\|^2_{L^2(0,T;\tilde{H}_2)} \right)^{\frac{1}{2}},$$

$$\|\rho^b\|_{Y_1} = \left(\|\rho^b\|^2_{L^2(0,T;D(A_2))} + \left\| \frac{d\rho^b}{dt} \right\|^2_{L^2(0,T;H_2)} \right)^{\frac{1}{2}}.$$

Let us recall that

$$\tilde{Y}_1 \subset C(0,T;\tilde{V}_2), \ Y_1 \subset C(0,T;V_2),$$

with continuous and dense injections.

Now we set $V = V_1 \times V_2$, $H = H_1 \times H_2$, $X = X_1 \times Y_1$, $\tilde{V} = \tilde{V}_1 \times \tilde{V}_2$, $\tilde{H} = \tilde{H}_1 \times \tilde{H}_2$ and $\tilde{X} = \tilde{X}_1 \times \tilde{Y}_1$.

For $\mathbf{u} = (u,\rho)$, $\mathbf{v} = (v,\phi) \in V$, we set

$$B(\mathbf{u},\mathbf{v}) = (B_1(u,v), B_2(\rho,\phi)), \ \mathcal{A}_1\mathbf{u} = (A_{11}u, A_{21}\rho),$$

$$\mathcal{A}_2\mathbf{u} = (A_{12}u, A_{22}\rho), \ \mathcal{A} = \mathcal{A}_1 + \mathcal{A}_2, \tag{22}$$

$$E\mathbf{u} = (E_1u, 0), \ \Lambda\mathbf{u} = (\Lambda_2\rho, 0).$$

For $\bar{\mathbf{u}} = (\bar{u},\bar{\rho})$, $\bar{\mathbf{v}} = (\bar{v},\bar{\phi}) \in \tilde{V}$, we set

$$\tilde{B}(\bar{\mathbf{u}},\bar{\mathbf{v}}) = \left(\tilde{B}_1(\bar{u},\bar{v}), \tilde{B}_2(\bar{\rho},\bar{\phi}) \right), \ \tilde{A}\bar{\mathbf{u}} = (\tilde{A}_1u, \tilde{A}_2\rho). \tag{23}$$

We also equip V with the norm

$$\|\mathbf{u}\|^2_w = \|u\|^2_w + \|\rho\|^2,$$

for $\mathbf{u} = (u,\rho) \in V$.

Hereafter, if Z is any other Hilbert space, we will denote by $\langle \cdot, \cdot \rangle_Z$ the scalar product in Z and by $\| \cdot \|_Z$ the associated norm. We will also denote by c a numerical constant that depends only on the data.

We will use the following notations. We define the operators M and N by

$$Mu = \frac{1}{h} \int_{-h}^0 udz, \ Nu = u - Mu. \tag{24}$$

For $u \in L^2(\mathcal{M})$, we set

$$\bar{u} = Mu, \ u^b = Nu = u - Mu. \tag{25}$$

In oceanography, the vertical average $\bar{u}$ is referred to as the barotropic flow and u^b is called the baroclinic flow [8, 29]. The following properties hold.

Proposition 2.5. *The following properties hold true*

$$\langle \bar{u}, v^b \rangle = 0, \ \forall u, v \in L^2(\mathcal{M}),$$

$$|u^b|_{L^2} \le 2h \left| \frac{\partial u}{\partial z} \right|_{L^2}, \ \forall u \in H^1(\mathcal{M}),$$

$$\langle \Delta \bar{u}, v^b \rangle = 0, \ \forall u \in H^2(\mathcal{M}), \forall v \in L^2(\mathcal{M}),$$

$$\left\langle \bar{u}, \Delta v^b + \frac{\partial^2 v^b}{\partial z^2} \right\rangle = 0, \ \forall u \in L^2(\mathcal{M}), \forall v \in H^2(\mathcal{M}) \tag{26}$$
$$\text{satisfying } \frac{\partial v}{\partial z} = 0 \ \text{ on } \Gamma_i \cup \Gamma_b,$$

$$\left\langle \Delta \bar{u}, \Delta v^b + \frac{\partial^2 v^b}{\partial z^2} \right\rangle = 0, \ \forall u \in H^2(\mathcal{M}), \forall v \in H^2(\mathcal{M})$$
$$\text{satisfying } \frac{\partial v}{\partial z} = 0 \ \text{ on } \Gamma_i \cup \Gamma_b.$$

Proof. See [25]. □

Hereafter, we assume that the initial data $\mathbf{a} = \bar{\mathbf{a}} + \mathbf{a}^b$ and the forcing $F = \bar{F} + F^b$ satisfy the following smallness assumptions. There exist positive constants c_1, c_2 and k, with $0 < k < 1$, such that

$$|\nabla \bar{a}|_{L^2}^2 + \int_0^T |\bar{F}|_{L^2}^2 dt \le c_1 \delta^k, \ \left| \frac{\partial \mathbf{a}^b}{\partial z} \right|_{L^2}^2 + \left| \text{div } \mathbf{a}^b \right|_{L^2}^2$$

$$= \left| \frac{\partial \mathbf{a}^b}{\partial z} \right|_{L^2}^2 + \left| \frac{\partial W(\mathbf{a}^b)}{\partial z} \right|_{L^2}^2 \le c_2 \delta^2. \tag{27}$$

Remark 2.6. Although the conditions (27) impose some restrictions on the data, they are still physically relevant as (27) is satisfied, for δ small enough, by all data $\bar{\mathbf{a}}, \mathbf{a}^b, \bar{F}$ that are uniformly bounded (in L^∞) with a bound of order 1; more generally, if we denote by μ the uniform bound of all these quantities, then (27) is satisfied for δ small, $0 < \delta < \delta_*$, with δ_* depending on μ. Moreover, there is no restriction on the size of $F^b = NF$. Furthermore, since $k < 1$, $(27)_1$ is less restrictive than $(27)_2$.

Using the previous notations and setting $\mathbf{u} = (u, \rho)$, it is easy to check that a weak formulation of (5) reads:
Find $\mathbf{u} \in L^2(0, T; V)$ satisfying

$$\frac{d\mathbf{u}}{dt} + \mathcal{A}_1 \mathbf{u} + \delta^{-2} \mathcal{A}_2 \mathbf{u} + E\mathbf{u} + B(\mathbf{u}, \mathbf{u}) + \Lambda \mathbf{u} = F \text{ in } V', \ \mathbf{u}(0) = \mathbf{a}. \tag{28}$$

Taking the vertical average of (28), we derive that the barotropic and baroclinic flows $\bar{\mathbf{u}}$ and $\mathbf{u}^b$ satisfy the following functional equations:

$$\frac{d\bar{\mathbf{u}}}{dt} + \tilde{A}_1\bar{\mathbf{u}} + E\bar{\mathbf{u}} + MB(\bar{\mathbf{u}} + \mathbf{u}^b, \bar{\mathbf{u}} + \mathbf{u}^b) = \bar{F} \text{ in } (MV)', \ \bar{\mathbf{u}}(0) = \bar{\mathbf{a}}, \quad (29)$$

$$\frac{d\mathbf{u}^b}{dt} + A_1\mathbf{u}^b + \delta^{-2}A_2\mathbf{u}^b + E\mathbf{u}^b + NB_2(\bar{\mathbf{u}} + \mathbf{u}^b, \bar{\mathbf{u}} + \mathbf{u}^b) + \Lambda(\bar{\mathbf{u}} + \mathbf{u}^b) = F^b \text{ in } (NV)',$$

$$\mathbf{u}^b(0) = \mathbf{a}^b. \quad (30)$$

Let us note that $M\Lambda(\bar{\mathbf{u}} + \mathbf{u}^b) = 0$ in $(MV)'$ and $N\Lambda(\bar{\mathbf{u}} + \mathbf{u}^b) = \Lambda(\bar{\mathbf{u}} + \mathbf{u}^b)$ in $(NV)'$.

2.2.1 The 2D Navier–Stokes

We first recall the following 2D Navier–Stokes equations (with a Coriolis force) with an associated transport equation:

$$\begin{cases} \dfrac{\partial \bar{v}}{\partial t} - \dfrac{1}{R_{e_1}}\Delta \bar{v} + fk_0 \times \bar{v} + (\bar{U} \cdot \nabla)\bar{v} + (\bar{v} \cdot \nabla)\bar{U} + (\bar{v} \cdot \nabla)\bar{v} \\ \qquad + \text{grad } p = \bar{F}_1, \\ \text{div } \bar{v} = 0, \ \bar{v} = 0, \ \text{ on } \partial\mathcal{O}, \ \bar{v} = \bar{\mathbf{a}}^1 \text{ at } t = 0, \end{cases} \quad (31)$$

$$\begin{cases} \dfrac{\partial \bar{q}}{\partial t} - \dfrac{1}{R_{t_1}}\Delta \bar{q} + (\bar{U} \cdot \nabla)\bar{q} + (\bar{v} \cdot \nabla)\bar{\psi} + (\bar{v} \cdot \nabla)\bar{q} = \bar{F}_2, \\ \bar{q} = 0, \ \text{ on } \partial\mathcal{O}, \ \bar{q} = \bar{\mathbf{a}}^2 \text{ at } t = 0, \end{cases} \quad (32)$$

or equivalently

$$\frac{d\bar{\mathbf{v}}}{dt} + \tilde{A}_1\bar{\mathbf{v}} + E\bar{\mathbf{v}} + \tilde{B}(\bar{\mathbf{U}}, \bar{\mathbf{v}}) + \tilde{B}(\bar{\mathbf{v}}, \bar{\mathbf{U}}) + \tilde{B}(\bar{\mathbf{v}}, \bar{\mathbf{v}}) = \bar{F} \text{ in } (MV)', \ \bar{\mathbf{v}}(0) = \bar{\mathbf{a}},$$
$$(33)$$

where $\bar{\mathbf{v}} = (\bar{v}, \bar{q})$, $\bar{\mathbf{U}} = (\bar{U}, \bar{\psi})$.

In (31), the unknown functions are the velocity $\bar{v}$ and the pressure p. The volume force $\bar{F}_1$ and the initial condition $\bar{\mathbf{a}}^1 = M\mathbf{a}^1$ are given. The Coriolis parameter f is given by $f = f_0 + \beta y$, where $f_0 > 0$, $\beta > 0$ are given constants.

In (32), the unknown function is the density $\bar{q}$. The volume force $\bar{F}_2$ and the initial condition $\bar{\mathbf{a}}^2 = M\mathbf{a}^2$ are given. The scalar function $\bar{\psi}$ is given and $\bar{v} = (\bar{v}_1, \bar{v}_2)$ is the solution to (31) (note that the systems (31) and (32) are decoupled).

We assume the following regularity conditions:

$$\bar{F}_1, \in L^2(0, T; \tilde{H}_1), \ \forall \ T > 0, \ \bar{\mathbf{a}}^1 \in \tilde{V}_1, \ \bar{F}_2, \in L^2(0, T; \tilde{H}_2), \ \bar{\mathbf{a}}^2 \in \tilde{V}_2,$$
$$\bar{U} \in L^\infty(0, T; \tilde{V}_1) \cap L^2(0, T; D(\tilde{A}_1)), \ \bar{\psi} \in L^\infty(0, T; \tilde{V}_2) \cap L^2(0, T; D(\tilde{A}_2)).$$
$$(34)$$

The domain $\mathcal{O}$ occupied by the fluid is a smooth, convex and bounded open set of R^2 with boundary $\partial\mathcal{O}$. Finally the constants $R_{e_1} > 0$, $R_{t_1} > 0$ are the nondimensional Reynolds numbers.

Proposition 2.7. *The Navier–Stokes system (31) has a unique solution $\bar{v} \in L^2(0,T;D(\tilde{A}_1)) \cap L^\infty(0,T;\tilde{V}_1)$. The system (32) has a unique solution $\bar{q} \in L^2(0,T;D(\tilde{A}_2)) \cap L^\infty(0,T;\tilde{V}_2)$. Moreover, we have the estimates:*

$$|\bar{v}(t)|_{L^2}^2 + |\bar{q}(t)|_{L^2}^2 \le e^{M_0(t)} \left(|\bar{\mathbf{a}}|_{L^2}^2 + \int_0^T |\bar{F}|_{L^2}^2 ds \right),$$

$$\int_0^T \left(\|\bar{v}\|^2 + \|\bar{q}\|^2 \right) dt \le e^{M_0(T)} \left(|\bar{\mathbf{a}}|_{L^2}^2 + \int_0^T |\bar{F}|_{L^2}^2 ds \right),$$

$$\|\bar{v}(t)\|^2 + \|\bar{q}(t)\|^2 \le e^{M_1(t)} \left(\|\bar{\mathbf{a}}\|^2 + \int_0^T |\bar{F}|_{L^2}^2 ds \right), \tag{35}$$

$$\int_0^T \left(|\tilde{A}_1 \bar{v}|_{L^2}^2 + |\tilde{A}_2 \bar{q}|_{L^2}^2 \right) dt \le e^{M_1(T)} (\|\bar{\mathbf{a}}\|^2 + \int_0^T |\bar{F}|_{L^2}^2 ds),$$

where

$$M_0(t) = c \int_0^t \left(\|\bar{U}\|^2 + \|\bar{\psi}\|^2 \right) ds,$$

$$M_1(t) = c \int_0^t \left(|\bar{U}|_{L^2} |\tilde{A}_1 \bar{U}|_{L^2} + |\bar{\psi}|_{L^2} |\tilde{A}_2 \bar{\psi}|_{L^2} \right) ds \tag{36}$$

$$+ e^{2M_0(t)} \left(|\bar{\mathbf{a}}|_{L^2}^2 + \int_0^T |\bar{F}|_{L^2}^2 ds \right)^2 + \int_0^t \|\bar{\psi}\|^4 ds,$$

and

$$\bar{\mathbf{a}} = (\bar{\mathbf{a}}^1, \bar{\mathbf{a}}^2) \in \tilde{V}_1 \times \tilde{V}_2, \ \bar{F} = (\bar{F}_1, \bar{F}_2) \in L^2(0,T;\tilde{H}_1) \times L^2(0,T;\tilde{H}_2). \tag{37}$$

Proof. The existence and uniqueness of solutions to (31), (32), are well-known, see e.g., [26]. The estimates (35) are also standard, but for the sake of completeness, we give a sketch of the proof.

For $(35)_1$, multiplying (31) by $\bar{v}$, (32) by $\bar{q}$ and summing yield

$$\frac{1}{2}\frac{d}{dt} \left(|\bar{v}|_{L^2}^2 + |\bar{q}|_{L^2}^2 \right) + \alpha_3 \left(\|\bar{v}\|^2 + \|\bar{q}\|^2 \right)$$

$$\le c|\bar{v}|_{L^2} \|\bar{v}\| \|\bar{U}\| + c|\bar{v}|_{L^2}^{\frac{1}{2}} \|\bar{v}\|^{\frac{1}{2}} \|\bar{\psi}\| |\bar{q}|_{L^2}^{\frac{1}{2}} \|\bar{q}\|^{\frac{1}{2}} + c|\bar{F}_1|_{L^2} |\bar{v}|_{L^2} + c|\bar{F}_2|_{L^2} |\bar{q}|_{L^2}$$

$$\le \frac{\alpha_3}{2} \left(\|\bar{v}\|^2 + \|\bar{q}\|^2 \right) + c|\bar{v}|_{L^2}^2 \|\bar{U}\|^2 + c\|\bar{\psi}\|^2 \left(|\bar{v}|_{L^2}^2 + |\bar{q}|_{L^2}^2 \right) + c|\bar{F}_1|_{L^2}^2 + c|\bar{F}_2|_{L^2}^2, \tag{38}$$

which gives

$$\frac{d}{dt} \left(|\bar{v}|_{L^2}^2 + |\bar{q}|_{L^2}^2 \right) + \alpha_3 \left(\|\bar{v}\|^2 + \|\bar{q}\|^2 \right) \le c \left(\|\bar{U}\|^2 + \|\bar{\psi}\|^2 \right) \left(|\bar{v}|_{L^2}^2 + |\bar{q}|_{L^2}^2 \right) + c|\bar{F}|_{L^2}^2, \tag{39}$$

where $\alpha_3 = Min\left(\frac{1}{R_{e_1}}, \frac{1}{R_{t_1}}\right)$ and $(35)_1$ follows from the Gronwall's Lemma, [26].

For $(35)_2$, multiplying (31) by $\tilde{A}_1\bar{v}$, (32) by $\tilde{A}_2\bar{q}$, summing and using (21) yield

$$\frac{1}{2}\frac{d}{dt}\left(\|\bar{v}\|^2 + \|\bar{q}\|^2\right) + \alpha_3\left(|\tilde{A}_1\bar{v}|_{L^2}^2 + |\tilde{A}_2q|_{L^2}^2\right)$$

$$\leq c|\bar{v}|_{L^2}|\tilde{A}_1\bar{v}|_{L^2} + c|\bar{U}|_{L^2}^{\frac{1}{2}}|\tilde{A}_1\bar{U}|_{L^2}^{\frac{1}{2}}\|\bar{v}\||\tilde{A}_1\bar{v}|_{L^2} + c|\bar{v}|_{L^2}^{\frac{1}{2}}|\tilde{A}_1\bar{v}|_{L^2}^{\frac{3}{2}}\|\bar{U}\|$$

$$+c|\bar{v}|_{L^2}^{\frac{1}{2}}\|\bar{v}\||\tilde{A}_1\bar{v}|_{L^2}^{\frac{3}{2}} + c|\bar{U}|_{L^2}^{\frac{1}{2}}|\tilde{A}_1\bar{U}|_{L^2}^{\frac{1}{2}}\|\bar{q}\||\tilde{A}_2\bar{q}|_{L^2} + c|\bar{v}|_{L^2}^{\frac{1}{2}}|\tilde{A}_1\bar{v}|_{L^2}^{\frac{1}{2}}\|\bar{\psi}\||\tilde{A}_2\bar{q}|_{L^2}$$

$$+c|\bar{v}|_{L^2}^{\frac{1}{2}}|\tilde{A}_1\bar{v}|_{L^2}^{\frac{1}{2}}\|\bar{q}\||\tilde{A}_2\bar{q}|_{L^2} + c|\bar{F}|_{L^2}\left(|\tilde{A}_1\bar{v}|_{L^2} + |\tilde{A}_2\bar{q}|_{L^2}\right)$$

$$\leq \frac{\alpha_3}{2}\left(|\tilde{A}_1\bar{v}|_{L^2}^2 + |\tilde{A}_2\bar{q}|_{L^2}^2\right) + c|\bar{v}|_{L^2}^2 + c|\bar{F}|_{L^2}^2$$

$$+c|\bar{U}|_{L^2}|\tilde{A}_1\bar{U}|_{L^2}\left(\|\bar{v}\|^2 + \|\bar{q}\|^2\right) + c\left(|\bar{v}|_{L^2}^2 + |\bar{q}|_{L^2}^2\right)\left(\|\bar{v}\|^2 + \|\bar{q}\|^2\right)^2 + c\|\bar{\psi}\|^4\|\bar{v}\|^2.$$

$$(40)$$

Let

$$h(t) = c|\bar{U}|_{L^2}|\tilde{A}_1\bar{U}|_{L^2} + c(|\bar{v}|_{L^2}^2 + |\bar{q}|_{L^2}^2)(\|\bar{v}\|^2 + \|\bar{q}\|^2) + c\|\bar{\psi}\|^4.$$

Then

$$\int_0^t h(s)ds \leq c\int_0^t |\bar{U}|_{L^2}|\tilde{A}_1\bar{U}|_{L^2}ds + \sup_s\left(|\bar{v}(s)|_{L^2}^2 + |\bar{q}(s)|_{L^2}^2\right)$$

$$\times \int_0^t \left(\|\bar{v}\|^2 + \|\bar{q}\|^2\right)ds + c\|\bar{\psi}\|^4$$

$$\leq c\int_0^t(|\bar{U}|_{L^2}^2 + |\tilde{A}_1\bar{U}|_{L^2}^2)ds + e^{2M_0(t)}\left(|\bar{\mathbf{a}}|_{L^2}^2 + \int_0^t |\bar{F}|_{L^2}^2 ds\right)^2$$

$$+ \int_0^t \|\bar{\psi}\|^4 ds \equiv M_1(t),$$

$$(41)$$

where $M_0(t)$ is given by (36).

Therefore

$$\frac{1}{2}\frac{d}{dt}\left(\|\bar{v}\|^2 + \|\bar{q}\|^2\right) + \alpha_3(|\tilde{A}_1\bar{v}|_{L^2}^2 + |\tilde{A}_2\bar{q}|_{L^2}^2)$$

$$\leq c|\bar{F}|_{L^2}^2 + c|\bar{v}|_{L^2}^2 + h(t)\left(\|\bar{v}\|^2 + \|\bar{q}\|^2\right),$$

$$(42)$$

and the estimates (35) follow from the standard Gronwall Lemma. $\square$

2.2.2 The 3D Linear System

We also consider the following 3D heat type equations

$$\frac{d\mathbf{v}^\flat}{dt} + \mathcal{A}_1\mathbf{v}^\flat + \delta^{-2}\mathcal{A}_2\mathbf{v}^\flat + E\mathbf{v}^\flat + \Lambda\mathbf{v}^\flat = F^\flat \quad \text{in } (NV)', \ \mathbf{v}^\flat(0) = \mathbf{a}^\flat, \quad (43)$$

which is equivalent to

$$\frac{dv^\flat}{dt} + A_{11}v^\flat + \delta^{-2}A_{12}v^\flat + E_1v^\flat + \Lambda_2\phi^\flat = F_1^\flat \quad \text{in } (NV_1)', \ v^\flat(0) = \mathbf{a}_1^\flat, \quad (44)$$

$$\frac{d\phi^\flat}{dt} + A_{21}\phi^\flat + \delta^{-2}A_{22}\phi^\flat = F_2^\flat \quad \text{in } (NV_2)', \ \phi^\flat(0) = \mathbf{a}_2^\flat, \quad (45)$$

with $\mathbf{v}^\flat = (v^\flat, \phi^\flat)$.

In (43), the unknown function is $\mathbf{v}^\flat = (v^\flat, \phi^\flat)$; the volume force $F^\flat$ and the initial condition $\mathbf{a}^\flat$ are given. The Coriolis parameter f and the aspect ratio $\delta << 1$ are given.

We assume the following regularity condition

$$\mathbf{a}^\flat \in V, \ F^\flat \in L^2(0, T; H). \quad (46)$$

We also assume the additional condition

$$\int_{-h}^{0} F^\flat dz = 0, \ \int_{-h}^{0} \mathbf{a}^\flat dz = 0. \quad (47)$$

Remark 2.8. Condition (47) appears in the equations for the baroclinic flow $\mathbf{u}^\flat$ of the PEs (5). From (47), the flow $\mathbf{v}^\flat$ of (43) must satisfy $\int_{-h}^{0} \mathbf{v}^\flat dz = 0$.

Proposition 2.9. *The heat type equation (43) has a unique solution* $\mathbf{v}^\flat = (v^\flat, \phi^\flat) \in L^2(0, T; D(\mathcal{A})) \cap L^\infty(0, T; V)$. *Moreover, we have the estimates:*

$$\|\mathbf{v}^\flat(t)\|_w^2 = \|v^\flat(t)\|_w^2 + \|\phi^\flat(t)\|^2 \leq c(|\nabla\mathbf{a}^\flat|_{L^2}^2 + c_2 + \int_0^T |F^\flat|_{L^2}^2 ds) \equiv c_3,$$

$$\int_0^T |\mathcal{A}\mathbf{v}^\flat|_{L^2}^2 dt = \int_0^T (|A_1v^\flat|_{L^2}^2 + |A_2\phi^\flat|_{L^2}^2) dt \leq c_3,$$

$$\int_0^T \|\mathbf{v}^\flat\|_w^2 dt = \int_0^T (\|v^\flat\|_w^2 + \|\phi^\flat\|^2) dt \leq \delta^2 c_3,$$

$$|\mathbf{v}^\flat(t)|_{L^2}^2 = |v^\flat(t)|_{L^2}^2 + |\phi^\flat(t)|_{L^2}^2 \leq \delta^2 c_3.$$

$$(48)$$

Proof. Multiplying (45) by $\phi^\flat$ gives

$$|\phi^\flat(t)|^2_{L^2} + c \int_0^t \left(|\nabla\phi^\flat|^2_{L^2} + \delta^{-2} \left| \frac{\partial\phi^\flat}{\partial z} \right|^2_{L^2} \right) dt \le c(|\mathbf{a}^\flat_2|^2_{L^2} + \int_0^t |F^\flat_2|^2_{L^2} ds). \quad (49)$$

We also note that

$$|\nabla\phi^\flat|_{L^2} \le c \left| \frac{\partial}{\partial z} \nabla\phi^\flat \right|_{L^2}, \quad (50)$$

since $\int_{-h}^0 \nabla\phi^\flat dz = 0$.

Now, multiplying (45) by $A_{21}\phi^\flat + \delta^{-2} A_{22}\phi^\flat$ yields (see (13))

$$|\nabla\phi^\flat(t)|^2_{L^2} + \delta^{-2} \left| \frac{\partial\phi^\flat}{\partial z}(t) \right|^2_{L^2}$$

$$+ c\int_0^T \left(|A_{21}\phi^\flat|^2_{L^2} + \delta^{-2} \left| \frac{\partial}{\partial z} \nabla\phi^\flat \right|^2_{L^2} + \delta^{-4} |A_{22}\phi^\flat|^2_{L^2} \right) dt$$

$$\le c|\nabla N\mathbf{a}^2|^2_{L^2} + \frac{c}{\delta^2} \left| \frac{\partial N\mathbf{a}^2}{\partial z} \right|^2_{L^2} + c\int_0^t |F^\flat_2|^2_{L^2} ds$$

$$\le c(|\nabla N\mathbf{a}^2|^2_{L^2} + c_2 + \int_0^t |F^\flat_2|^2_{L^2} ds) \equiv c_3, \quad (51)$$

and (48) follows from (51) (note that $\delta << 1$) for the second component $\phi^\flat$ of $\mathbf{v}^\flat$.

Multiplying (43) by $v^\flat$ gives

$$|v^\flat(t)|^2_{L^2} + c\int_0^T \left(|\nabla v^\flat|^2_{L^2} + |\nabla W(v^\flat)|^2_{L^2} + \delta^{-2} \left| \frac{\partial v^\flat}{\partial z} \right|^2_{L^2} + \delta^{-2} \left| \operatorname{div} v^\flat \right|^2_{L^2} \right) dt$$

$$\le c(|N\mathbf{a}^1|^2_{L^2} + \int_0^T |F^\flat_1|^2_{L^2} ds). \quad (52)$$

We also note that

$$|\nabla v^\flat|_{L^2} \le c \left| \frac{\partial}{\partial z} \nabla v^\flat \right|_{L^2}, \quad (53)$$

since $\int_{-h}^0 \nabla v^\flat dz = 0$.

Now, multiplying (43) by $A_{11}v^\flat + \delta^{-2}A_{12}v^\flat$ yields (see (13))

$$|\nabla v^\flat(t)|^2_{L^2} + |\nabla W(v^\flat)(t)|^2_{L^2} + \delta^{-2}\left|\frac{\partial v^\flat}{\partial z}(t)\right|^2_{L^2} + \delta^{-2}\left|\text{div }v^\flat\right|^2_{L^2}$$

$$+c\int_0^T\left(|A_{11}v^\flat|^2_{L^2} + \delta^{-2}\left|\frac{\partial}{\partial z}\nabla v^\flat\right|^2_{L^2} + \delta^{-2}\left|\nabla\text{div }v^\flat\right|^2_{L^2} + \delta^{-4}|A_{12}v^\flat|^2_{L^2}\right)dt$$

$$+c\int_0^T|A_1 v^\flat|^2_{L^2}dt$$

$$\leq c|\nabla N\mathbf{a}^1|^2_{L^2} + c|\nabla W(N\mathbf{a}^1)|^2_{L^2} + c\delta^{-2}\left|\frac{\partial N\mathbf{a}^1}{\partial z}\right|^2_{L^2} + c\delta^{-2}\left|\text{div }N\mathbf{a}^1\right|^2_{L^2}$$

$$+c\int_0^T|F_1^\flat|^2_{L^2}ds$$

$$\leq c(|\nabla N\mathbf{a}^1|^2_{L^2} + c_2 + \int_0^T|F_1^\flat|^2_{L^2}ds) \leq c_3, \tag{54}$$

and (48) follows from (53) and the fact that $|\Lambda_2\phi^\flat|_{L^2} \leq c|\nabla\phi^\flat|_{L^2}$.   □

3 Existence of Strong Solution to (5)

To (28), we associate the following system:

$$\frac{d\mathbf{u}_0}{dt} + A_1\mathbf{u}_0 + \delta^{-2}A_2\mathbf{u}_0 + E\mathbf{u}_0 + \Lambda\mathbf{u}_0 = F \text{ in } V', \quad \mathbf{u}_0(0) = \mathbf{a}. \tag{55}$$

Taking the vertical average of (55), we derive the following equations for the barotropic component $\bar{\mathbf{u}}_0$ and baroclinic flow $\mathbf{u}_0^\flat$:

$$\frac{d\bar{\mathbf{u}}_0}{dt} + \tilde{A}_1\bar{\mathbf{u}}_0 + E\bar{\mathbf{u}}_0 = \bar{F} \text{ in } (MV)', \quad \bar{\mathbf{u}}_0(0) = \bar{\mathbf{a}}, \tag{56}$$

$$\frac{d\mathbf{u}_0^\flat}{dt} + A_1\mathbf{u}_0^\flat + \delta^{-2}A_2\mathbf{u}_0^\flat + E\mathbf{u}_0^\flat + \Lambda(\bar{\mathbf{u}}_0 + \mathbf{u}_0^\flat) = F^\flat \text{ in } (NV)', \quad \mathbf{u}_0^\flat(0) = \mathbf{a}^\flat. \tag{57}$$

Remark 3.1. Note that (56) (resp. (57)) has a unique solution $\bar{\mathbf{u}}_0$ (resp. $\mathbf{u}_0^\flat$) and estimates similar to (35) (resp. (48)) hold. In particular, we have

$$\|\bar{\mathbf{u}}_0\|^2_{\tilde{X}} \leq \delta^k\alpha_0^2,$$

$$\|\mathbf{u}_0\|^2_X \equiv \|\bar{\mathbf{u}}_0\|^2_{\tilde{X}} + \|\mathbf{u}_0^\flat\|^2_X \leq \alpha_0^2,$$

$$\int_0^T\|\mathbf{u}_0^\flat\|^2_w dt \leq \delta^2\alpha_0^2, \quad |\mathbf{u}_0^\flat(t)|^2_{L^2} \leq \delta^2\alpha_0^2, \tag{58}$$

where

$$\alpha_0^2 \equiv c(c_1 + c_2 + |\nabla \mathbf{a}^\flat|_{L^2}^2 + \int_0^T |F^\flat|_{L^2}^2 dt), \text{ and } k \text{ is as in } (27), 0 < k < 1.$$

Now let us set $\mathbf{v} = \mathbf{u} - \mathbf{u}_0$. Then $\mathbf{v}$ satisfies

$$\frac{d\mathbf{v}}{dt} + \mathcal{A}_1 \mathbf{v} + \delta^{-2} \mathcal{A}_2 \mathbf{v} + E\mathbf{v} + B(\mathbf{v} + \mathbf{u}_0, \mathbf{v} + \mathbf{u}_0) + \Lambda(\mathbf{v} + \mathbf{u}_0) = 0 \text{ in } V', \ \mathbf{v}(0) = 0. \tag{59}$$

Taking the vertical average of (59), we derive the following equations for the barotropic flow $\bar{\mathbf{v}}$ and the baroclinic flow $\mathbf{v}^\flat$.

$$\frac{d\bar{\mathbf{v}}}{dt} + \tilde{\mathcal{A}}_1 \bar{\mathbf{v}} + E\bar{\mathbf{v}} + MB(\mathbf{v} + \mathbf{u}_0, \mathbf{v} + \mathbf{u}_0) = 0 \text{ in } (MV)', \ \bar{\mathbf{v}}(0) = 0, \tag{60}$$

$$\frac{d\mathbf{v}^\flat}{dt} + \mathcal{A}_1 + \delta^{-2} \mathcal{A}_2 \mathbf{v}^\flat + E\mathbf{v}^\flat + NB(\mathbf{v} + \mathbf{u}_0, \mathbf{v} + \mathbf{u}_0) + \Lambda(\mathbf{v} + \mathbf{u}_0) = 0 \text{ in } (NV)',$$

$$\mathbf{v}^\flat(0) = 0. \tag{61}$$

Let us note that

$$MB(\mathbf{v} + \mathbf{u}_0, \mathbf{v} + \mathbf{u}_0) = \tilde{B}(\bar{\mathbf{v}} + \bar{\mathbf{u}}_0, \bar{\mathbf{v}} + \bar{\mathbf{u}}_0) + MB(\mathbf{v}^\flat + \mathbf{u}_0^\flat, \mathbf{v}^\flat + \mathbf{u}_0^\flat),$$

$$NB(\mathbf{v} + \mathbf{u}_0, \mathbf{v} + \mathbf{u}_0) = B(\bar{\mathbf{v}} + \bar{\mathbf{u}}_0, \mathbf{v}^\flat + \mathbf{u}_0^\flat) + B(\mathbf{v}^\flat + \mathbf{u}_0^\flat, \bar{\mathbf{v}} + \bar{\mathbf{u}}_0)$$

$$+ NB(\mathbf{v}^\flat + \mathbf{u}_0^\flat, \mathbf{v}^\flat + \mathbf{u}_0^\flat). \tag{62}$$

We also have

$$MB(\bar{\mathbf{v}}, \mathbf{w}^\flat) = MB(\mathbf{v}^\flat, \bar{\mathbf{w}}) = 0, \ \forall \mathbf{v}^\flat, \mathbf{w}^\flat \in NV, \ \forall \bar{\mathbf{v}}, \bar{\mathbf{w}} \in MV. \tag{63}$$

To solve (60)–(3), we consider the sequences $(\bar{\mathbf{v}}_n)$, $(\mathbf{v}_n^\flat)$ given by the systems:

$$\frac{d\bar{\mathbf{v}}_{n+1}}{dt} + \tilde{\mathcal{A}}_1 \bar{\mathbf{v}}_{n+1} + E\bar{\mathbf{v}}_{n+1} + \tilde{B}(\bar{\mathbf{v}}_{n+1} + \bar{\mathbf{u}}_0, \bar{\mathbf{v}}_{n+1} + \bar{\mathbf{u}}_0)$$
$$+ MB(\mathbf{v}_n^\flat + \mathbf{u}_0^\flat, \mathbf{v}_n^\flat + \mathbf{u}_0^\flat) = 0 \text{ in } (MV)', \tag{64}$$
$$\bar{\mathbf{v}}_{n+1}(0) = 0,$$

$$\frac{d\mathbf{v}_{n+1}^\flat}{dt} + \mathcal{A}_1 \mathbf{v}_{n+1}^\flat + \delta^{-2} \mathcal{A}_2 \mathbf{v}_{n+1}^\flat + E\mathbf{v}_{n+1}^\flat + NB(\mathbf{v}_n + \mathbf{u}_0, \mathbf{v}_n + \mathbf{u}_0)$$
$$+ \Lambda(\mathbf{v}_{n+1} + \mathbf{u}_0) = 0 \text{ in } (NV)', \tag{65}$$
$$\mathbf{v}_{n+1}^\flat(0) = 0,$$

for $(\bar{\mathbf{v}}_0, \mathbf{v}_0^\flat)$ given such that

$$\|\bar{\mathbf{v}}_0\|_{\tilde{X}}^2 \le \delta^k R^2,$$

$$\|\mathbf{v}_0\|_X^2 \equiv \|\bar{\mathbf{v}}_0\|_{\tilde{X}}^2 + \|\mathbf{v}_0^\flat\|_X^2 \le R^2, \tag{66}$$

$$\int_0^T \|\mathbf{v}_0^\flat\|_w^2 dt \le \delta^2 R^2, \ |\mathbf{v}_0^\flat(t)|_{L^2}^2 \le \delta^2 R^2,$$

*where $R = R(\mathbf{a}, g, T, R_{e_1}, R_{e_2}) > 0$ will be defined later (see (87), (83)), and
k is as in (27), $0 < k < 1$.*

Note that for $\mathbf{v}_n^\flat$, $\bar{\mathbf{v}}_n$ given, $\bar{\mathbf{v}}_{n+1}$ and $\mathbf{v}_{n+1}^\flat$ are solutions of a 2D Navier–Stokes equations and a 3D heat equation, respectively. Therefore, the existence and uniqueness of $\bar{\mathbf{v}}_{n+1}$ are $v_{n+1}^\flat$ is well known provided that $v_n^\flat$ are $\bar{\mathbf{v}}_n$ are regular enough. Moreover, estimates similar to (35) and (48) hold for $\bar{\mathbf{v}}_{n+1}$ and $v_{n+1}^\flat$, respectively.

Let us set

$$r_n = MB(\mathbf{v}_n^\flat + \mathbf{u}_0^\flat, \mathbf{v}_n^\flat + \mathbf{u}_0^\flat) + B_1(\bar{\mathbf{u}}_0, \bar{\mathbf{u}}_0),$$
$$s_n = NB(\mathbf{v}_n + \mathbf{u}_0, \mathbf{v}_n + \mathbf{u}_0) + \Lambda(\bar{\mathbf{v}}_{n+1} + \bar{\mathbf{u}}_0 + \mathbf{u}_0^\flat) = B(\bar{\mathbf{v}}_n + \bar{\mathbf{u}}_0, \mathbf{v}_n^\flat + \mathbf{u}_0^\flat)$$

$$+ B(\mathbf{v}_n^\flat + \mathbf{u}_0^\flat, \bar{\mathbf{v}}_n + \bar{\mathbf{u}}_0) + NB(\mathbf{v}_n^\flat + \mathbf{u}_0^\flat, \mathbf{v}_n^\flat + \mathbf{u}_0^\flat) + \Lambda(\bar{\mathbf{v}}_{n+1} + \bar{\mathbf{u}}_0 + \mathbf{u}_0^\flat).$$
$$(67)$$

Then (64) and (65) can be rewritten into the form

$$\frac{d\bar{\mathbf{v}}_{n+1}}{dt} + \tilde{A}_1 \bar{\mathbf{v}}_{n+1} + E\bar{\mathbf{v}}_{n+1} + \tilde{B}(\bar{\mathbf{v}}_{n+1}, \bar{\mathbf{u}}_0) + \tilde{B}(\bar{\mathbf{u}}_0, \bar{\mathbf{v}}_{n+1})$$
$$+ \tilde{B}(\bar{\mathbf{v}}_{n+1}, \bar{\mathbf{v}}_{n+1}) + r_n = 0 \text{ in } (MV)', \qquad (68)$$
$$\bar{\mathbf{v}}_{n+1}(0) = 0,$$

$$\frac{d\mathbf{v}_{n+1}^\flat}{dt} + \mathcal{A}_1 \mathbf{v}_{n+1}^\flat + \delta^{-2} \mathcal{A}_2 \mathbf{v}_{n+1}^\flat + E\mathbf{v}_{n+1}^\flat + \Lambda \mathbf{v}_{n+1}^\flat + s_n = 0 \text{ in } (NV)',$$
$$\mathbf{v}_{n+1}^\flat(0) = 0. \qquad (69)$$

In the next step, we derive some estimates on the barotropic and baroclinic components $\bar{\mathbf{v}}_{n+1}$ and $\mathbf{v}_{n+1}^\flat$, respectively, assuming some regularity conditions on $\bar{\mathbf{v}}_n$, $\mathbf{v}_n^\flat$, $\bar{\mathbf{u}}_0$ and $\mathbf{u}_0^\flat$. The goal is to prove (using a fixed-point argument) that for a suitable choice of R, the sequence $(\mathbf{v}_n)$ is convergent for δ small enough.

Proposition 3.2. *We assume that*

$$\|\bar{\mathbf{v}}_n\|_{\tilde{X}}^2 \le \delta^k R^2,$$

$$\|\mathbf{v}_n\|_X^2 \equiv \|\mathbf{v}_n^\flat\|_X^2 + \|\bar{\mathbf{v}}_n\|_{\tilde{X}}^2 \le R^2, \qquad (70)$$

$$\int_0^T \|\mathbf{v}_n^\flat\|_w^2 dt \le \delta^2 R^2,$$

where k is as in (27), $0 < k < 1$.

Then the following estimates hold true for r_n and s_n :

$$\int_0^T |r_n|_{L^2}^2 dt \le c(\delta R^2 \|\mathbf{v}_n\|_X^2 + \delta R^4 + \delta \alpha_0^4 + \delta^k \alpha_0^4),$$

$$\int_0^T |s_n|_{L^2}^2 dt \le c(\delta R^2 \|\mathbf{v}_n\|_X^2 + \delta R^4 + \delta \alpha_0^4 + \delta^k \alpha_0^2) + c\int_0^T \|\bar{\mathbf{v}}_{n+1}\|^2 dt. \quad (71)$$

Proof. Using (14), (21), (58) and (70) we have

$$|r_n|_{L^2}^2 \leq c\|\mathbf{v}_n^\flat + \mathbf{u}_0^\flat\|_w^2 |\mathcal{A}(\mathbf{v}_n^\flat + \mathbf{u}_0^\flat)|_{L^2} \|\mathbf{v}_n^\flat + \mathbf{u}_0^\flat\| + c|\bar{\mathbf{u}}_0|_{L^2} |\tilde{\mathcal{A}}_1 \bar{\mathbf{u}}_0|_{L^2} \|\bar{\mathbf{u}}_0\|^2, \tag{72}$$

and

$$\int_0^T |r_n|_{L^2}^2 dt \leq c \sup \|\mathbf{v}_n^\flat + \mathbf{u}_0^\flat\|_w^2 \int_0^T |\mathcal{A}(v_n^\flat + \mathbf{u}_0^\flat)|_{L^2} \|\mathbf{v}_n^\flat + \mathbf{u}_0^\flat\| dt$$

$$+ c \sup |\bar{\mathbf{u}}_0|_{L^2} \|\bar{\mathbf{u}}_0\| \int_0^T |\tilde{\mathcal{A}}_1 \bar{\mathbf{u}}_0|_{L^2} \|\bar{\mathbf{u}}_0\| dt \tag{73}$$

$$\leq c\|\mathbf{v}_n + \mathbf{u}_0^\flat\|_X^2 (R + \alpha_0)^2 \delta + c\delta^k \alpha_0^4$$

$$\leq c\delta(R^2 \|\mathbf{v}_n\|_X^2 + \alpha_0^4 + R^4) + c\delta^k \alpha_0^4.$$

We also have (note that $|NB(\mathbf{v}_n^\flat + \mathbf{u}_0^\flat, \mathbf{v}_n^\flat + \mathbf{u}_0^\flat)|_{L^2} \leq |B(\mathbf{v}_n^\flat + \mathbf{u}_0^\flat, \mathbf{v}_n^\flat + \mathbf{u}_0^\flat)|_{L^2}$)

$$|s_n|_{L^2}^2 \leq c\|\bar{\mathbf{v}}_n + \bar{\mathbf{u}}_0\|^2 |\mathcal{A}(\mathbf{v}_n^\flat + \mathbf{u}_0^\flat)|_{L^2} \|\mathbf{v}_n^\flat + \mathbf{u}_0^\flat\|$$

$$+ c\|\mathbf{v}_n^\flat + u_0^\flat\|^2 |\tilde{\mathcal{A}}_1(\bar{\mathbf{v}}_n + \bar{\mathbf{u}}_0)|_{L^2} \|\bar{\mathbf{v}}_n + \bar{\mathbf{u}}_0\|$$

$$+ c\|\mathbf{v}_n^\flat + \mathbf{u}_0^\flat\|_w^2 |\mathcal{A}(\mathbf{v}_n^\flat + \mathbf{u}_0^\flat)|_{L^2} \|\mathbf{v}_n^\flat + \mathbf{u}_0^\flat\|$$

$$+ c(\|\bar{\mathbf{v}}_{n+1}\|^2 + \|\bar{\mathbf{u}}_0\|^2 + \|\mathbf{u}_0^\flat\|^2) \tag{74}$$

and

$$\int_0^T |s_n|_{L^2}^2 dt \leq c \sup \|\bar{\mathbf{v}}_n + \bar{\mathbf{u}}_0\|^2 \int_0^T |\tilde{\mathcal{A}}_1(\bar{\mathbf{v}}_n + \bar{\mathbf{u}}_0)|_{L^2} \|\mathbf{v}_n^\flat + \mathbf{u}_0^\flat\| dt$$

$$+ c \sup \|\bar{\mathbf{v}}_n + \bar{\mathbf{u}}_0\| \sup \|\mathbf{v}_n^\flat + \mathbf{u}_0^\flat\| \int_0^T |\tilde{\mathcal{A}}_1(\bar{\mathbf{v}}_n + \bar{\mathbf{u}}_0)|_{L^2} \|\mathbf{v}_n^\flat + \mathbf{u}_0^\flat\| dt$$

$$+ c \sup \|\mathbf{v}_n^\flat + \mathbf{u}_0^\flat\|_w^2 \int_0^T |\mathcal{A}(\mathbf{v}_n^\flat + \mathbf{u}_0^\flat)|_{L^2} \|\mathbf{v}_n^\flat + \mathbf{u}_0^\flat\| dt + c\delta^k \alpha_0^2$$

$$+ c \int_0^T \|\bar{\mathbf{v}}_{n+1}\|^2 dt$$

$$\leq c(\|\mathbf{v}_n\|_X^2 + \alpha_0^2)(R + \alpha_0)^2 \delta + c\delta^k \alpha_0^2 + c \int_0^T \|\bar{\mathbf{v}}_{n+1}\|^2 dt$$

$$\leq c\delta(R^2 \|\mathbf{v}_n\|_X^2 + \alpha_0^4 + R^4) + c\delta^k \alpha_0^2 + c \int_0^T \|\bar{\mathbf{v}}_{n+1}\|^2 dt. \tag{75}$$

$\square$

Proposition 3.3. *We assume that (70) holds true. Then there exists a constant $\delta_0 > 0$ such that for $0 < \delta < \delta_0$, we have*

$$\|\mathbf{v}_{n+1}\|_X^2 \equiv \|\mathbf{v}_{n+1}^\flat\|_X^2 + \|\bar{\mathbf{v}}_{n+1}\|_{\bar{X}}^2 \le \delta^k R^2,$$

$$\int_0^T \|\mathbf{v}_{n+1}^\flat\|_w^2 dt \le \delta^2 R^2, \quad |\mathbf{v}_{n+1}^\flat(t)|_{L^2}^2 \le \delta^2 R^2, \tag{76}$$

where k is as in (27), $0 < k < 1$.

Proof. It clearly follows from Proposition 2.5 and (71) that

$$|\bar{\mathbf{v}}_{n+1}(t)|_{L^2}^2 \le c e^{N_0(T)}(\delta R^2 \|\mathbf{v}_n\|_X^2 + \delta \alpha_0^4 + \delta R^4 + \delta^k \alpha_0^4),$$

$$\|\bar{\mathbf{v}}_{n+1}(t)\|^2 \le e^{N_1(T)}(\delta R^2 \|\mathbf{v}_n\|_X^2 + \delta \alpha_0^4 + \delta R^4 + \delta^k \alpha_0^4),$$

$$\int_0^T |\tilde{A}_1 \bar{\mathbf{v}}_{n+1}|_{L^2}^2 dt \le e^{N_1(T)}(\delta R^2 \|\mathbf{v}_n\|_X^2 + \delta \alpha_0^4 + \delta R^4 + \delta^k \alpha_0^4), \tag{77}$$

where

$$N_0(t) = c \int_0^t \|\bar{\mathbf{u}}_0\|^2 dt \le c\alpha_0^2,$$

$$N_1(t) = c \int_0^t |\bar{u}_0|_{L^2} |\tilde{A}_1 \bar{u}_0|_{L^2} dt + c e^{2N_0(t)}(\delta R^4 + \delta \alpha_0^4 + \delta R^4 + \alpha_0^4)^2$$

$$+ \int_0^t \|\bar{\mathbf{u}}_0\|^4 dt$$

$$\le c\alpha_0^2 + c e^{2\alpha_0^2}(\delta R^4 + \delta \alpha_0^4 + \alpha_0^4)^2 + c\delta^k \alpha_0^4$$

$$\le c\alpha_0^2 + c e^{2\alpha_0^2}(\delta R^8 + \delta \alpha_0^8 + \alpha_0^8) + c\alpha_0^4 \equiv N_2. \tag{78}$$

It also follows from Proposition 2.6 and the estimates (75), (77) that

$$|\mathbf{v}_{n+1}^\flat(t)|_{L^2}^2 \le c(e^{N_2} + 1)(\delta R^2 \|\mathbf{v}_n\|_X^2 + \delta \alpha_0^4 + \delta R^4 + \delta^k \alpha_0^2 + \delta^k \alpha_0^4),$$

$$\|\mathbf{v}_{n+1}^\flat(t)\|_w^2 \le c(e^{N_2} + 1)(\delta R^2 \|\mathbf{v}_n\|_X^2 + \delta \alpha_0^4 + \delta R^4 + \delta^k \alpha_0^2 + \delta^k \alpha_0^4),$$

$$\int_0^T |A\mathbf{v}_{n+1}^\flat|_{L^2}^2 dt \le c(e^{N_2} + 1)(\delta R^2 \|\mathbf{v}_n\|_X^2 + \delta \alpha_0^4 + \delta R^4 + \delta^k \alpha_0^2 + \delta^k \alpha_0^4). \tag{79}$$

From (77), (79) we can write

$$\|\mathbf{v}_{n+1}\|_X^2 \equiv \|\bar{\mathbf{v}}_{n+1}\|_{\bar{X}}^2 + \|\mathbf{v}_{n+1}^\flat\|_X^2 \le \epsilon \|\mathbf{v}_n\|_X^2 + L_0, \tag{80}$$

where

$$\epsilon = c\delta(e^{N_2} + 1)R^2 = \delta^k \epsilon_0, \quad \epsilon_0 = c\delta^{1-k}(e^{N_2} + 1)R^2,$$

$$L_0 = c\delta(e^{N_2} + 1)(\alpha_0^4 + R^4) + c\delta^k(e^{N_2} + 1)(\alpha_0^4 + \alpha_0^2) = \delta^k \kappa_0, \tag{81}$$

$$\kappa_0 = c\delta^{1-k}(e^{N_2} + 1)(\alpha_0^4 + R^4) + c(e^{N_2} + 1)(\alpha_0^4 + \alpha_0^2).$$

Note that

$$ce^{N_2} = c\exp\left(c\alpha_0^2 + ce^{2\alpha_0^2}\alpha_0^8 + c\alpha_0^4\right)\exp\left(c\delta^2 e^{2\alpha_0^2}(R^8 + \alpha_0^8)\right)$$
$$= L_1 \cdot L_2, \tag{82}$$

where

$$L_1 = c\exp\left(c\alpha_0^2 + ce^{2\alpha_0^2}\alpha_0^8 + c\alpha_0^4\right),$$
$$L_2 = \exp\left(c\delta^2 e^{2\alpha_0^2}(R^8 + \alpha_0^8)\right), \tag{83}$$

and N_2 is given by (78). Therefore

$$\kappa_0 = c\delta^{1-k}e^{N_2}\alpha_0^4(\alpha_0^4 + R^4) + c(e^{N_2} + 1)(\alpha_0^4 + \alpha_0^2)$$
$$= \delta^{1-k}(L_1 L_2 + c)(\alpha_0^4 + R^4) + (L_1 L_2 + c)(\alpha_0^4 + \alpha_0^2). \tag{84}$$

Note that the constant c that appears in (81)–(83) is independent of R and δ (see Propositions 2.5, 2.6 and 3.1).

Using inequality (80) successively, we get

$$\|\mathbf{v}_{n+1}\|_X^2 \le \epsilon^n \|\mathbf{v}_0\|_X^2 + \frac{1 - \epsilon^n}{1 - \epsilon} L_0$$

$$\le \delta^k \epsilon_0^n \|\mathbf{v}_0\|_X^2 + \frac{1 - \epsilon^n}{1 - \epsilon}\delta^k \kappa_0 \tag{85}$$

$$\le \delta^k \left(\epsilon_0^n R^2 + \frac{1 - \epsilon^n}{1 - \epsilon}\kappa_0\right).$$

The goal is to choose R (independent of δ) such that

$$\|\mathbf{v}_{n+1}\|_X^2 \le \delta^k R^2, \tag{86}$$

for δ small enough (depending on R).

Now let us set

$$R^2 = 12(L_1 + c)(\alpha_0^4 + \alpha_0^2), \tag{87}$$

where L_1 is given by (83) and c is the constant that appears in (83). We then choose δ such that

$$\delta^{1-k}(L_1 L_2 + c)(\alpha_0^4 + R^4) \le L_1(\alpha_0^4 + \alpha_0^2),$$
$$L_2 \le 2, \tag{88}$$
$$\epsilon_0 \le \tfrac{1}{2}.$$

It follows that

$$\kappa_0 \le L_1(\alpha_0^4 + \alpha_0^2) + (2L_1 + c)(\alpha_0^4 + \alpha_0^2) \le \frac{R^2}{4}, \tag{89}$$

and (86) follows from (85) to (89).

It is clear that $(76)_2$ follows directly from Proposition 2.6 and the estimate (75). □

Proposition 3.4. *Let R be given by (87). We assume that δ is small enough so that (88) is satisfied. Let $\mathbf{v}_0 = \bar{\mathbf{v}}_0 + \mathbf{v}_0^\flat \in X$ such that (66) is satisfied. Then the sequences $(\bar{\mathbf{v}}_n)$, $(\mathbf{v}_n^\flat)$ given by (64)–(65) satisfy the estimates*

$$\|\mathbf{v}_n\|_X^2 \equiv \|\mathbf{v}_n^\flat\|_X^2 + \|\bar{\mathbf{v}}_n\|_{\hat{X}}^2 \le \delta^k R^2,$$

$$\int_0^T \|\mathbf{v}_n^\flat\|_w^2 dt \le \delta^2 R^2, \quad |\mathbf{v}_n^\flat(t)|_{L^2}^2 \le \delta^2 R^2, \tag{90}$$

where k is as in (27), $0 < k < 1$.

Proof. This result follows from the previous estimates. $\square$

Now, let us set $\mathbf{w}_{n+1} = \mathbf{v}_{n+1} - \mathbf{v}_n$. Then $\bar{\mathbf{w}}_{n+1}$ and $\mathbf{w}_{n+1}^\flat$ satisfy

$$\frac{d\bar{\mathbf{w}}_{n+1}}{dt} + \tilde{A}_1 \bar{\mathbf{w}}_{n+1} + E\bar{\mathbf{w}}_{n+1} + \tilde{B}(\bar{\mathbf{w}}_{n+1}, \bar{\mathbf{v}}_{n+1} + \bar{\mathbf{u}}_0) + \tilde{B}(\bar{\mathbf{v}}_n + \bar{\mathbf{u}}_0, \bar{\mathbf{w}}_{n+1})$$
$$+ h_n = 0 \text{ in } (MV)',$$
$$\bar{\mathbf{w}}_{n+1}(0) = 0, \tag{91}$$

$$\frac{d\mathbf{w}_{n+1}^\flat}{dt} + A_1 \mathbf{w}_{n+1}^\flat + \delta^{-2} A_2 \mathbf{w}_{n+1}^\flat + E\mathbf{w}_{n+1}^\flat + \Lambda \mathbf{w}_{n+1}^\flat + k_n = 0 \text{ in } (NV)',$$

$$\mathbf{w}_{n+1}^\flat(0) = 0, \tag{92}$$

where

$$h_n = MB(\mathbf{w}_n^\flat, \mathbf{v}_n^\flat + \mathbf{u}_0^\flat) + MB(\mathbf{v}_{n+1}^\flat + \mathbf{u}_0^\flat, \mathbf{w}_n^\flat), \tag{93}$$

and

$$
\begin{aligned}
k_n &= NB(\mathbf{w}_n, \mathbf{v}_n + \mathbf{u}_0) + NB(\mathbf{v}_{n-1} + \mathbf{u}_0, \mathbf{w}_n) + \Lambda \bar{\mathbf{w}}_{n+1} \\
&= B(\bar{\mathbf{w}}_n, \mathbf{v}_n^\flat + \mathbf{u}_0^\flat) + B(\mathbf{w}_n^\flat, \bar{\mathbf{v}}_n + \bar{\mathbf{u}}_0) \\
&\quad + NB(\mathbf{w}_n^\flat, \mathbf{v}_n^\flat + u_0^\flat) + B(\bar{\mathbf{v}}_{n-1} + \bar{\mathbf{u}}_0, \mathbf{w}_n^\flat) + B(\mathbf{v}_{n-1}^\flat + u_0^\flat, \bar{\mathbf{w}}_n) \\
&\quad + NB(\mathbf{v}_{n-1}^\flat + \mathbf{u}_0^\flat, \mathbf{w}_n^\flat) + \Lambda \bar{\mathbf{w}}_{n+1}.
\end{aligned}
\tag{94}
$$

Proposition 3.5. *We assume that δ is small enough so that (88) holds. Then the following estimates hold:*

$$\int_0^T |h_n|_{L^2}^2 dt \le c_4 \delta^k \|\mathbf{w}_n\|_X^2,$$

$$\int_0^T |k_n|_{L^2}^2 dt \le c_4 \delta^k \|\mathbf{w}_n\|_X^2 + c \int_0^T \|\bar{\mathbf{w}}_{n+1}\|^2 dt. \tag{95}$$

where $c_4 = c_4(\alpha_0)$ depends on α_0 for δ small enough.

Proof. From (14), we have

$$|h_n|_{L^2}^2 \le c\|\mathbf{w}_n^\flat\|_w^2 |A(\mathbf{v}_n^\flat + \mathbf{u}_0^\flat)|_{L^2} \|\mathbf{v}_n^\flat + \mathbf{u}_0^\flat\| + c\|\mathbf{v}_{n+1}^\flat + \mathbf{u}_0^\flat\|_w^2 |A\mathbf{w}_n^\flat|_{L^2} \|\mathbf{w}_n^\flat\| \tag{96}$$

and

$$\int_0^T |h_n|_{L^2}^2 dt \leq c \sup \|\mathbf{w}_n^\flat\|_w^2 \int_0^T |\mathcal{A}(\mathbf{v}_n^\flat + \mathbf{u}_0^\flat)|_{L^2} \|\mathbf{v}_n^\flat + \mathbf{u}_0^\flat\| dt$$

$$+ c \sup \|\mathbf{w}_n^\flat\| \|\mathbf{v}_{n+1}^\flat + \mathbf{u}_0^\flat\|_w \int_0^T |\mathcal{A}\mathbf{w}_n^\flat|_{L^2} \|\mathbf{v}_{n+1}^\flat + \mathbf{u}_0^\flat\|_w dt$$

$$\leq c_4 \delta \|\mathbf{w}_n\|_X^2. \tag{97}$$

We also have

$$|k_n|_{L^2}^2 \leq c \|\bar{\mathbf{w}}_n\|^2 |\mathcal{A}(\mathbf{v}_n^\flat + \mathbf{u}_0^\flat)|_{L^2} \|\mathbf{v}_n^\flat + \mathbf{u}_0^\flat\| + c \|\mathbf{w}_n^\flat\|_w^2 |\tilde{\mathcal{A}}_1(\bar{\mathbf{v}}_n + \bar{\mathbf{u}}_0)|_{L^2} \|\bar{\mathbf{v}}_n + \bar{\mathbf{u}}_0\|$$

$$+ c \|\mathbf{w}_n^\flat\|_w^2 |\mathcal{A}(\mathbf{v}_n^\flat + \mathbf{u}_0^\flat)|_{L^2} \|\mathbf{v}_n^\flat + \mathbf{u}_0^\flat\| + c \|\bar{\mathbf{v}}_{n-1} + \bar{\mathbf{u}}_0\|^2 |\mathcal{A}\mathbf{w}_n^\flat|_{L^2} \|\mathbf{w}_n^\flat\|$$

$$+ c \|\mathbf{v}_{n-1}^\flat + u_0^\flat\|_w^2 |\tilde{\mathcal{A}}_1 \bar{\mathbf{w}}_n|_{L^2} \|\bar{\mathbf{w}}_n\| + c \|\mathbf{v}_{n-1}^\flat + u_0^\flat\|_w^2 |\mathcal{A}\mathbf{w}_n^\flat|_{L^2} \|\mathbf{w}_n^\flat\|$$

$$+ c \|\bar{\mathbf{w}}_{n+1}\|^2, \tag{98}$$

which gives

$$\int_0^T |k_n|_{L^2}^2 dt \leq c \sup \|\bar{\mathbf{w}}_n\|^2 \int_0^T |\mathcal{A}(\mathbf{v}_n^\flat + \mathbf{u}_0^\flat)|_{L^2} \|\mathbf{v}_n^\flat + \mathbf{u}_0^\flat\| dt$$

$$+ c \sup \|\bar{\mathbf{v}}_n + \bar{\mathbf{u}}_0\| \|\mathbf{w}_n^\flat\|_w \int_0^T |\tilde{\mathcal{A}}_1(\bar{\mathbf{v}}_n + \bar{\mathbf{u}}_0)|_{L^2} \|\mathbf{w}_n^\flat\|_w dt$$

$$+ c \sup \|\bar{\mathbf{v}}_{n-1} + \bar{\mathbf{u}}_0\|^2 \int_0^T |\mathcal{A}\mathbf{w}_n^\flat|_{L^2} \|\mathbf{w}_n^\flat\| dt$$

$$+ c \sup \|\mathbf{v}_{n-1}^\flat + u_0^\flat\|_w \|\bar{\mathbf{w}}_n\| \int_0^T |\tilde{\mathcal{A}}_1 \bar{\mathbf{w}}_n|_{L^2} \|\mathbf{v}_{n-1}^\flat + \mathbf{u}_0^\flat\|_w dt$$

$$+ c \sup \|\mathbf{v}_{n-1}^\flat + \mathbf{u}_0^\flat\|_w \|\mathbf{w}_n^\flat\| \int_0^T |\mathcal{A}\mathbf{w}_n^\flat|_{L^2} \|_w \|\mathbf{w}_n^\flat\| dt$$

$$+ c \int_0^T \|\bar{\mathbf{w}}_{n+1}\|^2 dt$$

$$\leq c_4 \delta \|\bar{\mathbf{w}}_n\|_X^2 + c_4 \delta^k \|\mathbf{w}_n\|_X^2 + c_4 \delta^k \|\mathbf{w}_n^\flat\|_X^2 + c_4 \delta \|\mathbf{w}_n^\flat\|_X^2$$

$$+ c_4 \delta \|\mathbf{w}_n^\flat\|_X^2 + c \int_0^T \|\bar{\mathbf{w}}_{n+1}\|^2 dt$$

$$\leq c_4 \delta^k \|\mathbf{w}_n\|_X^2 + c \int_0^T \|\bar{\mathbf{w}}_{n+1}\|^2 dt, \tag{99}$$

and (95) follows. □

Proposition 3.6. *The assumptions are the same as in Proposition 3.4. Then the following estimate holds true:*

$$\|\mathbf{w}_{n+1}\|_X^2 \leq c_4 \delta^k \|\mathbf{w}_n\|_X^2. \tag{100}$$

Proof. From Proposition 2.5 and the estimates (95) on h_n, we derive that $\bar{\mathbf{w}}_{n+1}$ satisfies

$$|\bar{\mathbf{w}}_{n+1}(t)|_{L^2}^2 \leq c_4 \delta^k \|\mathbf{w}_n\|_X^2,$$

$$\|\bar{\mathbf{w}}_{n+1}(t)\|^2 \leq c_4 \delta^k \|\mathbf{w}_n\|_X^2, \tag{101}$$

$$\int_0^T |\tilde{\mathcal{A}}_1 \bar{\mathbf{w}}_{n+1}|_{L^2}^2 dt \leq c_4 \delta^k \|\mathbf{w}_n\|_X^2.$$

From Proposition 2.6 and the estimates (95) and (101), we derive that $\mathbf{w}_{n+1}^\flat$ satisfies

$$|\mathbf{w}_{n+1}^\flat(t)|_{L^2}^2 \leq c_4 \delta^k \|\mathbf{w}_n\|_X^2,$$

$$\|\mathbf{w}_{n+1}^\flat(t)\|_w^2 \leq c_4 \delta^k \|\mathbf{w}_n\|_X^2, \tag{102}$$

$$\int_0^T |\mathcal{A}\mathbf{w}_{n+1}^\flat|_{L^2}^2 dt \leq c_4 \delta^k \|\mathbf{w}_n\|_X^2.$$

We conclude that the flow $\mathbf{w}_{n+1} = \bar{\mathbf{w}}_{n+1} + \mathbf{w}_{n+1}^\flat$ satisfies

$$\|\mathbf{w}_{n+1}^\flat\|_X^2 \leq c_4 \delta^k \|\mathbf{w}_n\|_X^2. \tag{103}$$

Finally from (101) to (102), we get

$$\|\mathbf{w}_{n+1}\|_X^2 \leq c_4 \delta^k \|\mathbf{w}_n\|_X^2. \ \square \tag{104}$$

Hereafter, we also assume that δ is small enough such that (88) and

$$c_4 \delta^k < 1 \tag{105}$$

hold true.

Therefore (104) proves that the sequence $(\mathbf{v}_n = \bar{\mathbf{v}}_n + \mathbf{v}_n^\flat)$ is a Cauchy sequence. Moreover we have

$$\bar{\mathbf{v}}_n \longrightarrow \bar{\mathbf{v}}, \ \mathbf{v}_n^\flat \longrightarrow \mathbf{v}^\flat, \tag{106}$$

in $\tilde{X}$ and X, respectively and the following rate of convergence holds true

$$\|\mathbf{v}_n - \mathbf{v}\|_X^2 \equiv \|\bar{\mathbf{v}}_n - \bar{\mathbf{v}}\|_{\tilde{X}}^2 + \|\mathbf{v}_n^\flat - \mathbf{v}^\flat\|_X^2$$

$$\leq c \frac{(c_4 \delta^k)^n}{1 - c_4 \delta^k}. \tag{107}$$

Therefore, the following main result is proved.

Proposition 3.7. *We assume that δ is small enough so that (88) and (105) are satisfied. Then the sequence $(\mathbf{v}_n = \bar{\mathbf{v}}_n + \mathbf{v}_n^\flat)$ defined by (64),(65) converges to a solution $\mathbf{v} = \bar{\mathbf{v}} + \mathbf{v}^\flat$ of (59) in X, and $\mathbf{v}$ is the unique solution of (59) in X that satisfies $\|\mathbf{v}\|_X^2 \le R^2$. Furthermore the following convergence rate holds true:*

$$\|\mathbf{v}_n - \mathbf{v}\|_X^2 = \|\bar{\mathbf{v}}_n - \bar{\mathbf{v}}\|_{\tilde{X}}^2 + \|\mathbf{v}_n^\flat - \mathbf{v}^\flat\|_X^2$$
$$\le c \frac{(c_4 \delta^k)^n}{1 - c_4 \delta^k}. \tag{108}$$

Remark 3.8. The conditions (88) and (105) show that for any initial data and volume force satisfying (27), there exists δ_0 such that for $0 < \delta < \delta_0$, the PEs (5) have a unique strong solution $\mathbf{u} \in X$ that satisfies

$$\|\mathbf{u} - \mathbf{u}_0\|_X^2 \le R^2.$$

4 A Small Eddy Correction Method

Hereafter we set

$$F(\bar{\mathbf{u}}, \mathbf{u}^\flat) = \left(\frac{d\mathbf{u}^\flat}{dt} + \mathcal{A}_1 \mathbf{u}^\flat + \delta^{-2} \mathcal{A}_2 \mathbf{u}^\flat + E\mathbf{u}^\flat + NB(\bar{\mathbf{u}} + \mathbf{u}^\flat, \bar{\mathbf{u}} + \mathbf{u}^\flat) \right.$$
$$\left. + \Lambda(\bar{\mathbf{u}} + \mathbf{u}^\flat) - F^\flat, \ \mathbf{u}^\flat(0) - \mathbf{a}^\flat \right). \tag{109}$$

Then, (30) is equivalent to

$$F(\bar{\mathbf{u}}, \mathbf{u}^\flat) = (0, 0) \text{ in } (NV)' \times NV. \tag{110}$$

Supposing that the barotropic flow $\bar{\mathbf{u}}$ is know, formally applying the Newton iteration to (110), we get the following iterative procedure: assuming that the initial guess for the baroclinic (or the small eddy) component $\mathbf{u}_0^\flat = 0$ and the $(j-1)th$ approximation $\mathbf{u}_{j-1}^\flat \in NV$ is known for some integer j, find the jth approximation $\mathbf{u}_j^\flat \in NV$ such that

$$D_{\mathbf{u}^\flat} F(\bar{\mathbf{u}}, \mathbf{u}^\flat)(\mathbf{u}_j^\flat - \mathbf{u}_{j-1}^\flat) = -F(\bar{\mathbf{u}}, \mathbf{u}_{j-1}^\flat). \tag{111}$$

Simple calculations show that (111) reduces to

$$\frac{d\mathbf{u}_j^\flat}{dt} + \mathcal{A}_1 \mathbf{u}_j^\flat + \delta^{-2} \mathcal{A}_2 \mathbf{u}_j^\flat + E\mathbf{u}_j^\flat + \tilde{B}(\bar{\mathbf{u}}, \mathbf{u}_j^\flat) + \tilde{B}(\mathbf{u}_j^\flat, \bar{\mathbf{u}})$$
$$+ NB(\mathbf{u}_{j-1}^\flat, \mathbf{u}_j^\flat) + NB(\mathbf{u}_j^\flat, \mathbf{u}_{j-1}^\flat) - NB(\mathbf{u}_{j-1}^\flat, \mathbf{u}_{j-1}^\flat) + \Lambda(\bar{\mathbf{u}} + \mathbf{u}_j^\flat) \quad (112)$$
$$= F^\flat \text{ in } (NV)', \ \mathbf{u}_j^\flat(0) = \mathbf{a}^\flat.$$

Combining (112) with the barotropic equation (29) (with $\bar{\mathbf{u}}$ replaced by $\mathbf{v}$), we obtain the following small eddy correction method: let $\mathbf{w}_0 = 0$ and l be a fixed positive integer:

$$\frac{d\mathbf{v}}{dt} + \mathcal{A}_1\mathbf{v} + E\mathbf{v} + MB(\mathbf{v}+\mathbf{w}_l, \mathbf{v}+\mathbf{w}_l) = \bar{F} \text{ in } (MV)', \ \mathbf{v}(0) = \bar{\mathbf{a}}, \quad (113)$$

$$\frac{d\mathbf{w}_j}{dt} + \mathcal{A}_1\mathbf{w}_j + \delta^{-2}\mathcal{A}_2\mathbf{w}_j + E\mathbf{w}_j + \tilde{B}(\mathbf{v}, \mathbf{w}_j) + \tilde{B}(\mathbf{w}_j, \mathbf{v})$$
$$+NB(\mathbf{w}_{j-1}, \mathbf{w}_j) + NB(\mathbf{w}_j, \mathbf{w}_{j-1}) - NB(\mathbf{w}_{j-1}, \mathbf{w}_{j-1}) + \Lambda(\mathbf{v}+\mathbf{w}_j)$$
$$= F^\flat \text{ in } (NV)', \ \mathbf{w}_j(0) = \mathbf{a}^\flat,$$
$$\quad (114)$$

for $j = 1, 2, \ldots, l$.

Remark 4.1. For $l = 0$, (113)–(114) reduces to (since $\mathbf{w}_0 = 0$)

$$\frac{d\mathbf{v}}{dt} + \tilde{\mathcal{A}}_1\mathbf{v} + E\mathbf{v} + \tilde{B}(\mathbf{v}, \mathbf{v}) = \bar{F} \text{ in } (MV)', \ \mathbf{v}(0) = \bar{\mathbf{a}}, \quad (115)$$

which gives the well known QG model.

For $l = 1$, (113)–(114) become

$$\frac{d\mathbf{v}}{dt} + \tilde{\mathcal{A}}_1\mathbf{v} + E\mathbf{v} + MB(\mathbf{v}+\mathbf{w}_1, \mathbf{v}+\mathbf{w}_1) = \bar{F} \text{ in } (MV)', \ \mathbf{v}(0) = \bar{\mathbf{a}}, \quad (116)$$

$$\frac{d\mathbf{w}_1}{dt} + \mathcal{A}_1\mathbf{w}_1 + \delta^{-2}\mathcal{A}_2\mathbf{w}_1 + E\mathbf{w}_1 + \tilde{B}(\mathbf{v}, \mathbf{w}_1) + \tilde{B}(\mathbf{w}_1, \mathbf{v}) + \Lambda(\mathbf{v}+\mathbf{w}_1)$$
$$= F^\flat \text{ in } (NV)', \ \mathbf{w}_1(0) = \mathbf{a}^\flat,$$
$$\quad (117)$$

which has the form of the multilevel (nonlinear Galerkin) method studied in [27, 13, 19, 20] for the 2D Navier–Stokes equations.

4.1 Some A Priori Estimates

In this part, we prove the existence and uniqueness of strong solution to (113)–(114) when the aspect ratio δ is small enough.

Hereafter we set $\mathbf{X} = \tilde{X} \times X^l$. For $\mathbf{v} = (\bar{\mathbf{v}}, \mathbf{w}_1, \mathbf{w}_2, \ldots, \mathbf{w}_l) \in \mathbf{X}$, we set

$$\|\mathbf{v}\|_{\mathbf{X}}^2 = \|\bar{\mathbf{v}}\|_{\tilde{X}}^2 + \sup_i \|\mathbf{w}_i\|_X^2. \quad (118)$$

The existence and uniqueness of a strong solution is proved as in Sect. 2. Therefore, we will only sketch the proof.

4.1.1 Linear Problems

As previously, to (113)–(114) we associate the following system:

$$\frac{d\mathbf{v}^0}{dt} + \tilde{\mathcal{A}}_1\mathbf{v}^0 + E\mathbf{v}^0 = \bar{F} \text{ in } (MV)', \ \mathbf{v}^0(0) = \bar{\mathbf{a}}, \quad (119)$$

and for $j = 1, 2, \ldots, l$

$$\frac{d\mathbf{w}_j^0}{dt} + \mathcal{A}_1\mathbf{w}_j^0 + \delta^{-2}\mathcal{A}_2\mathbf{w}_j^0 + E\mathbf{w}_j^0 + \varLambda(\mathbf{v}^0 + \mathbf{w}_j^0) = F^\flat \text{ in } (NV)', \ \mathbf{w}_j^0(0) = \mathbf{a}^\flat.$$
$$(120)$$

Following Propositions 2.5 and 2.6, the unique strong solution $(\mathbf{v}^0, \mathbf{w}_1^0, \mathbf{w}_2^0, \ldots, \mathbf{w}_l^0) \in \mathbf{X}$ to (119)–(120) satisfied

$$\|\mathbf{v}^0\|_{\tilde{X}}^2 \le \delta^k \alpha_0^2,$$

$$\|\mathbf{v}^0\|_{\tilde{X}}^2 + \sup_j \|\mathbf{w}_j^0\|_X^2 \le \alpha_0^2,$$

$$(121)$$

$$\sup_j \int_0^T \|\mathbf{w}_j^0\|_w^2 dt \le \delta^2 \alpha_0^2, \ \sup_j |\mathbf{w}_j^0(t)|_{L^2}^2 \le \delta^2 \alpha_0^2,$$

where k is as in (27), $0 < k < 1$.

4.1.2 Nonlinear Problems

Now let us set $\vartheta = \mathbf{v} - \mathbf{v}^0$, $\eta_j = \mathbf{w}_j - \mathbf{w}_j^0$. Then ϑ and η_j satisfy

$$\frac{d\vartheta}{dt} + \tilde{\mathcal{A}}_1\vartheta + E\vartheta + \tilde{B}(\vartheta, \mathbf{v}^0) + \tilde{B}(\mathbf{v}^0, \vartheta) + \tilde{B}(\vartheta, \vartheta) + S_1 = 0 \text{ in } (MV)', \ \vartheta(0) = 0,$$
$$(122)$$

$$\frac{d\eta_j}{dt} + \mathcal{A}_1\eta_j + \delta^{-2}\mathcal{A}_2\eta_j + E\eta_j + \varLambda\eta_j + S_2 = 0 \ \text{ in } (NV)', \ \eta_j(0) = 0, \quad (123)$$

where

$$S_1 = \tilde{B}(\mathbf{v}^0, \mathbf{v}^0) + MB(\eta_l + \mathbf{w}_l^0, \eta_l + \mathbf{w}_l^0),$$

$$S_2 = \tilde{B}(\vartheta + \mathbf{v}^0, \eta_j + \mathbf{w}_j^0) + \tilde{B}(\eta_j + \mathbf{w}_j^0, \vartheta + \mathbf{v}^0) + NB(\eta_{j-1} + \mathbf{w}_{j-1}^0, \eta_j + \mathbf{w}_j^0)$$

$$+NB(\eta_j + \mathbf{w}_j^0, \eta_{j-1} + \mathbf{w}_{j-1}^0) - NB(\eta_{j-1} + \mathbf{w}_{j-1}^0, \eta_{j-1} + \mathbf{w}_{j-1}^0) + \varLambda\vartheta.$$
$$(124)$$

To solve (122)–(123), we consider the following iterative process

$$\frac{d\vartheta^{n+1}}{dt} + \tilde{\mathcal{A}}_1\vartheta^{n+1} + E\vartheta^{n+1} + \tilde{B}(\vartheta^{n+1}, \mathbf{v}^0) + \tilde{B}(\mathbf{v}^0, \vartheta^{n+1})$$

$$+\tilde{B}(\vartheta^{n+1}, \vartheta^{n+1}) + S_1^n = 0 \text{ in } (MV)',$$
$$(125)$$

$$\vartheta^{n+1}(0) = 0,$$

$$\frac{d\eta_j^{n+1}}{dt} + \mathcal{A}_1\eta_j^{n+1} + \delta^{-2}\mathcal{A}\eta_j^{n+1} + E\eta_j^{n+1} + \varLambda\eta_j^{n+1} + S_2^n = 0 \text{ in } (NV)', \ \eta_j^{n+1}(0) = 0,$$
$$(126)$$

where

$$S_1^n = \tilde{B}(\mathbf{v}^0, \mathbf{v}^0) + MB(\eta_l^n + \mathbf{w}_l^0, \eta_l^n + \mathbf{w}_l^0),$$

$$S_2 = \tilde{B}(\vartheta^n + \mathbf{v}^0, \eta_j^n + \mathbf{w}_j^0) + \tilde{B}(\eta_j^n + \mathbf{w}_j^0, \vartheta^n + \mathbf{v}^0)$$

$$+ NB(\eta_{j-1}^n + \mathbf{w}_{j-1}^0, \eta_j^n + \mathbf{w}_j^0) + NB(\eta_j^n + \mathbf{w}_j^0, \eta_{j-1}^n + \mathbf{w}_{j-1}^0)$$

$$- NB(\eta_{j-1}^n + \mathbf{w}_{j-1}^0, \eta_{j-1}^n + \mathbf{w}_{j-1}^0) + \Lambda \vartheta^{n+1}. \tag{127}$$

for $(\mathbf{v}_0, \eta_1^0, \eta_2^0, \dots, \eta_l^0) \in \mathbf{X}$ such that

$$\|\vartheta^0\|_{\tilde{X}}^2 \le \delta^k R^2,$$

$$\|\vartheta^0\|_{\tilde{X}}^2 + \sup_j \|\eta_j^0\|_X^2 \le R^2, \tag{128}$$

$$\sup_j \int_0^T \|\eta_j^0\|_w^2 dt \le \delta^2 R^2, \quad \sup_j |\eta_j^0(t)|_{L^2}^2 \le \delta^2 R^2,$$

where $R = R(\mathbf{a}, g, T, R_{e_1}, R_{e_2}) > 0$ *is given by* (87) *and* k *is as in* (27), $0 < k < 1$.

As in Sect. 3, the goal is to prove (using a fixed-point argument) that the sequence (ϑ^n, η_j^n) is convergent for δ small enough.

Proposition 4.2. *We assume that*

$$\|\vartheta^n\|_{\tilde{X}}^2 \le \delta^k R^2,$$

$$\|\vartheta^n\|_{X_1}^2 + \sup_j \|\eta_j^n\|_X^2 \le R^2, \tag{129}$$

$$\sup_j \int_0^T \|\eta_j^n\|_w^2 dt \le \delta^2 R^2,$$

where k *is as in (27),* $0 < k < 1$.

Then the following estimates hold true for S_1^n and S_2^n :

$$\int_0^T |S_1^n|_{L^2}^2 dt \le c(\delta R^2(\|\vartheta^n\|_{\tilde{X}}^2 + \sup_k \|\eta_k^n\|_X^2) + \delta R^4 + \delta \alpha_0^4 + \delta^k \alpha_0^4),$$

$$\int_0^T |S_2^n|_{L^2}^2 dt \le c(\delta R^2(\|\vartheta^n\|_{\tilde{X}}^2 + \sup_k \|\eta_k^n\|_X^2) + \delta R^4 + \delta \alpha_0^4) + c \int_0^T \|\vartheta^{n+1}\|^2 dt. \tag{130}$$

Proof. The proof, which is very similar to that of Proposition 3.1, follows from the inequalities

$$|S_1^n|_{L^2}^2 \le c\|\eta_l^n + \mathbf{w}_l^0\|_w^2 |A_2(\eta_l^n + \mathbf{w}_l^0)|_{L^2}\|\eta_l^n + \mathbf{w}_l^0\| + c|\mathbf{v}^0|_{L^2}|\tilde{A}_1\mathbf{v}^0|_{L^2}\|\mathbf{v}^0\|^2,$$
$$(131)$$

which gives

$$\int_0^T |S_1^n|_{L^2}^2 dt \le c\delta(R^2 \sup_j \|\eta_j^n\|_X^2 + \alpha_0^4 + R^4) + c\delta^k\alpha_0^4. \tag{132}$$

We also have

$$|S_2^n|_{L^2}^2 \le c\|\vartheta^n + \mathbf{v}^0\|^2 |A(\eta_j^n + \mathbf{w}_j^0)|_{L^2}\|\eta_j^n + \mathbf{w}_j^0\| + c\|\eta_j^n$$

$$+\mathbf{w}_j^0\|^2|\tilde{A}_1(\vartheta^n + \mathbf{v}^0)|_{L^2}\|\vartheta^n + \mathbf{v}^0\| + c\|\eta_{j-1}^n$$

$$+\mathbf{w}_{j-1}^0\|_w^2|A(\eta_j^n + \mathbf{w}_j^0)|_{L^2}\|\eta_j^n$$

$$+\mathbf{w}_j^0\| + c\|\eta_j^n + \mathbf{w}_j^0\|_w^2|A(\eta_{j-1}^n + \mathbf{w}_{j-1}^0)|_{L^2}\|\eta_{j-1}^n + \mathbf{w}_{j-1}^0\|$$

$$+c\|\eta_{j-1}^n + \mathbf{w}_{j-1}^0\|_w^2|A(\eta_{j-1}^n + \mathbf{w}_{j-1}^0)|_{L^2}\|\eta_{j-1}^n + \mathbf{w}_{j-1}^0\| + c\|\vartheta^{n+1}\|^2, \quad (133)$$

which gives

$$\int_0^T |S_2^n|_{L^2}^2 dt \le c\delta(R^2(\|\vartheta^n\|_{\tilde{X}}^2 + \sup_j \|\eta_j^n\|_X^2) + \alpha_0^4 + R^4) + c\int_0^T \|\vartheta^{n+1}\|^2 dt.$$
$$(134)$$

□

Proposition 4.3. *Let R be given by (87). We assume that (129) holds true. Then for $\delta < \delta_0$, we have*

$$\|\vartheta^{n+1}\|_{\tilde{X}}^2 + \sup_j \|\eta_j^{n+1}\|_X^2 \le \delta^k R^2,$$

$$(135)$$

$$\sup_j \int_0^T \|\eta_j^{n+1}\|_w^2 dt \le \delta^2 R^2, \; \sup_j |\eta_j^{n+1}(t)|_{L^2}^2 \le \delta^2 R^2,$$

where k is as in (27), $0 < k < 1$.

Proof. It clearly follows from Propositions 2.5 and 4.1 that

$$|\vartheta^{n+1}(t)|_{L^2}^2 \le ce^{N_0(T)}(\delta R^2(\|\vartheta^n\|_{\tilde{X}}^2 + \sup_j \|\eta_j^n\|_X^2) + \delta\alpha_0^4 + \delta R^4 + \delta^k\alpha_0^4),$$

$$\|\vartheta^{n+1}(t)\|^2 \le e^{N_2}(\delta R^2(\|\vartheta^n\|_{\tilde{X}}^2 + \sup_j \|\eta_j^n\|_X^2) + \delta\alpha_0^4 + \delta R^4 + \delta^k\alpha_0^4),$$

$$\int_0^T |\tilde{A}_1\vartheta^{n+1}|_{L^2}^2 dt \le e^{N_2}(\delta R^2(\|\vartheta^n\|_{\tilde{X}}^2 + \sup_j \|\eta_j^n\|_X^2) + \delta\alpha_0^4 + \delta R^4 + \delta^k\alpha_0^4).$$
$$(136)$$

It also follows from Propositions 2.6 and 4.1, the estimates (134) and (136), that

$$\sup_j |\eta_j^{n+1}(t)|_{L^2}^2 \le c(e^{N_2}+1)(\delta R^2(\|\vartheta^n\|_{\tilde{X}}^2 + \sup_j \|\eta_j^n\|_X^2) + \delta\alpha_0^4 + \delta R^4 + \delta^k\alpha_0^4),$$

$$\sup_j \|\eta_j^{n+1}(t)\|_w^2 \le c(e^{N_2}+1)(\delta R^2(\|\vartheta^n\|_{\tilde{X}}^2 + \sup_j \|\eta_j^n\|_X^2) + \delta\alpha_0^4 + \delta R^4 + \delta^k\alpha_0^4),$$

$$\sup_j \int_0^T |A\eta_j^{n+1}|_{L^2}^2 dt$$

$$\le c(e^{N_2}+1)(\delta R^2(\|\vartheta^n\|_{\tilde{X}}^2 + \sup_j \|\eta_j^n\|_X^2) + \delta\alpha_0^4 + \delta R^4 + \delta^k\alpha_0^4).$$

$$(137)$$

It is clear that $(135)_2$ follows from Proposition 2.6 and (134) (see Proof of Proposition 3.3).   □

Proposition 4.4. *Let R be given by (87). We assume that δ is small enough so that (88) is satisfied. Let $(\vartheta^0, \eta_1^0, \ldots, \eta_l^0) \in \mathbf{X}$ such that (128) is satisfied. Then the sequence $(\vartheta^n, \eta_1^n, \eta_2^n, \ldots, \eta_l^n) \in \mathbf{X}$ given by (125)–(126) satisfy the estimates*

$$\|\vartheta^n\|_{\tilde{X}}^2 + \sup_j \|\eta_j^n\|_X^2 \le \delta^k R^2,$$

$$(138)$$

$$\sup_j \int_0^T \|\eta_j^n\|_w^2 dt \le \delta^2 R^2, \quad \sup_j |\eta_j^n(t)|_{L^2}^2 \le \delta^2 R^2,$$

where k is as in (27), $0 < k < 1$.

Proof. It follows from the previous estimates.   □

Now, let us set $\theta^{n+1} = \vartheta^{n+1} - \vartheta^n$, $q_j^{n+1} = \eta_j^{n+1} - \eta_j^n$. Then $(\theta^{n+1}, q_j^{n+1})$ satisfy

$$\frac{d\theta^{n+1}}{dt} + \tilde{A}_1\theta^{n+1} + E\theta^{n+1} + \tilde{B}(\theta^{n+1}, \vartheta^{n+1} + \mathbf{v}^0)$$

$$+\tilde{B}(\vartheta^n + \mathbf{v}^0, \theta^{n+1}) + K_1^n = 0 \text{ in } (MV)', \quad \theta^{n+1}(0) = 0, \quad (139)$$

$$\frac{dq_j^{n+1}}{dt} + \mathcal{A}_1 q_j^{n+1} + \delta^{-2}\mathcal{A}_2 q_j^{n+1} + E q_j^{n+1} + \Lambda q_j^{n+1} + K_2^n = 0 \text{ in } (NV)',$$

$$q_j^{n+1}(0) = 0, \quad (140)$$

where

$$K_n^1 = MB(q_l^n, \eta_l^{n-1} + \mathbf{w}_l^0) + MB_2(\eta_l^n + \mathbf{w}_l^0, q_l^n), \quad (141)$$

and

$$K_n^2 = \tilde{B}(\theta^n, \eta_j^n + \mathbf{w}_j^0) + \tilde{B}(q_j^n, \vartheta^n + \mathbf{v}^0) + NB(q_j^n, \eta_j^n + \mathbf{w}_j^0)$$

$$+\tilde{B}(\vartheta^{n-1} + \mathbf{v}^0, q_j^n) + \tilde{B}(\eta_j^{n-1} + \mathbf{w}_j^0, \theta^n) + NB(\eta_j^{n-1} + \mathbf{w}_j^0, q_j^n) + \Lambda\theta^{n+1}.$$

$$(142)$$

Proposition 4.5. *We assume that δ is small enough so that (88) holds. Then the following estimates hold:*

$$\int_0^T |K_n^1|_{L^2}^2 dt \leq c_4 \delta^k (\|\theta^n\|_{\tilde{X}}^2 + \sup_j \|q_j^n\|_X^2),$$

$$\int_0^T |K_n^2|_{L^2}^2 dt \leq c_4 \delta^k (\|\theta^n\|_{\tilde{X}}^2 + \sup_j \|q_j^n\|_X^2) + c \int_0^T \|\theta^{n+1}\|^2 dt. \tag{143}$$

Proof. The proof is very similar to that of Proposition 3.4.   □

Proposition 4.6. *We assume that δ is small enough so that (88) holds. Then the following estimates hold:*

$$\|\theta^{n+1}\|_{\tilde{X}}^2 + \sup_j \|q_j^{n+1}\|_X^2 \leq c_4 \delta^k \left(\|\theta^n\|_{\tilde{X}}^2 + \sup_j \|q_j^n\|_X^2 \right). \tag{144}$$

Proof. The proof, which is similar to that of Proposition 3.5, follows from (143). Moreover, the following results is proved.   □

Proposition 4.7. *We assume that δ is small enough so that (88) and (105) are satisfied. Then the sequence $(\vartheta^n, \eta_1^n, \eta_2^n, \ldots, \eta_l^n) \in \mathbf{X}$ defined by (125),(126) converges to a solution $(\vartheta, \eta_1, \eta_2, \ldots, \eta_l)$ to (125)–(126) in $\mathbf{X}$. Moreover, $(\vartheta, \eta_1, \eta_2, \ldots, \eta_l)$ is the unique solution to (122)–(123) in $\mathbf{X}$ that satisfies*

$$\|\vartheta^n - \vartheta\|_{\tilde{X}}^2 + \sup_j \|\eta_j^n - \eta_j\|_X^2 \leq R^2.$$

Furthermore the following convergence rate holds true:

$$\|\vartheta^n - \vartheta\|_{\tilde{X}}^2 + \sup_j \|\eta_j^n - \eta_j\|_X^2 \leq c \frac{(c_4 \delta^k)^n}{1 - c_4 \delta^k}. \tag{145}$$

5 Convergence of the Method

In this part, we study the convergence of the small eddy correction method presented in Sect. 4. We prove that the method converges and we estimate the rate of convergence with respect to the aspect ratio δ.

Hereafter, we set $\mathbf{u}_l = \mathbf{v} + \mathbf{w}_l$, $\zeta = \mathbf{u} - \mathbf{u}_l$, $\varepsilon_j = \mathbf{w}_j - \mathbf{w}_{j-1}$. In particular, $\varepsilon_1 = \mathbf{w}_1$.

Using (28) and (113)–(114), it is clear that $\mathbf{u}_l$ and ζ satisfy:

$$\frac{d\mathbf{u}_l}{dt} + \mathcal{A}_1 \mathbf{u}_l + \delta^{-2} \mathcal{A}_2 \mathbf{u}_l + E\mathbf{u}_l + B(\mathbf{u}_l, \mathbf{u}_l) - NB(\varepsilon_l, \varepsilon_l) + \Lambda \mathbf{u}_l = F \text{ in } V', \ \mathbf{u}_l(0) = \mathbf{a}, \tag{146}$$

$$\frac{d\zeta}{dt} + \mathcal{A}_1 \zeta + \delta^{-2} \mathcal{A}_2 \zeta + E\zeta + B(\zeta, \mathbf{u}) + B(\mathbf{u}_l, \zeta) + NB(\varepsilon_l, \varepsilon_l) + \Lambda \zeta = 0 \text{ in } V', \ \zeta(0) = 0. \tag{147}$$

Taking the vertical average of (147), we derive that the barotropic and baro-clinic flows $\bar{\zeta}$ and ζ^b satisfy

$$\frac{d\bar{\zeta}}{dt} + \tilde{A}_1\bar{\zeta} + E\bar{\zeta} + \tilde{B}(\bar{\zeta}, \bar{\mathbf{u}}) + \tilde{B}(\mathbf{v}, \bar{\zeta}) + MB(\zeta^b, \mathbf{u}^b) + MB(\mathbf{w}_l, \zeta^b) = 0 \text{ in } (MV)',$$

$$\bar{\zeta}(0) = 0, \tag{148}$$

$$\frac{d\zeta^b}{dt} + A_1\zeta^b + \delta^{-2}A_2\zeta^b + E\zeta^b + \tilde{B}(\bar{\zeta}, \mathbf{u}^b) + \tilde{B}(\zeta^b, \bar{\mathbf{u}}) + \tilde{B}(\mathbf{v}, \zeta^b) + \tilde{B}(\mathbf{w}_l, \bar{\zeta})$$

$$+ NB(\zeta^b, \mathbf{u}^b) + NB(\mathbf{w}_l, \zeta^b) + NB(\varepsilon_l, \varepsilon_l) + \Lambda(\bar{\zeta} + \zeta^b) = 0 \text{ in } (NV)', \ \zeta^b(0) = 0. \tag{149}$$

Hereafter, we set $\alpha = \min(\alpha_1, \alpha_2)$, where α_1, α_2 are given by (14), (20).

Note that

$$|\langle \tilde{B}(\bar{\zeta}, \bar{\mathbf{u}}), \bar{\zeta}\rangle| \le c\|\bar{\zeta}\| |\bar{\zeta}|_{L^2} \|\bar{\mathbf{u}}\| \le \frac{\alpha}{8}\|\bar{\zeta}\|^2 + c\|\bar{\mathbf{u}}\|^2 |\bar{\zeta}|_{L^2}^2, \tag{150}$$

$$|\langle MB(\zeta^b, \mathbf{u}^b), \bar{\zeta}\rangle\rangle| \le |\langle B(\zeta^b, \bar{\zeta}), \mathbf{u}^b\rangle|$$

$$\le c|\zeta^b|_{L^2}^{\frac{1}{2}}\|\zeta^b\|^{\frac{1}{2}}\|\bar{\zeta}\| |\mathbf{u}^b|_{L^2}^{\frac{1}{2}}\|\mathbf{u}^b\|^{\frac{1}{2}} \tag{151}$$

$$\le \frac{\alpha}{8}\|\bar{\zeta}\|^2 + c|\zeta^b|_{L^2}\|\zeta^b\| |\mathbf{u}^b|_{L^2}\|\mathbf{u}^b\|,$$

$$|\langle MB(\mathbf{w}_l, \zeta^b), \bar{\zeta}\rangle\rangle| \le |\langle B(\mathbf{w}_l, \bar{\zeta}), \zeta^b\rangle|$$

$$\le c|\mathbf{w}_l|_{L^2}^{\frac{1}{2}}\|\mathbf{w}_l\|^{\frac{1}{2}}\|\bar{\zeta}\| |\zeta^b|_{L^2}^{\frac{1}{2}}\|\zeta^b\|^{\frac{1}{2}} \tag{152}$$

$$\le \frac{\alpha}{8}\|\bar{\zeta}\|^2 + c|\mathbf{w}_l|_{L^2}\|\mathbf{w}_l\| |\zeta^b|_{L^2}\|\zeta^b\|,$$

Therefore, multiplying (148) by $\bar{\zeta}$ and using (150)–(152) yields

$$|\bar{\zeta}(t)|_{L^2}^2 + c\int_0^t \|\bar{\zeta}\|^2 ds \le c\sup_s(|\mathbf{u}^b(s)|_{L^2}\|\mathbf{u}^b(s)\| + |\mathbf{w}_l(s)|_{L^2}\|\mathbf{w}_l(s)\|) \tag{153}$$

$$\times \int_0^T \|\zeta^b\|_w^2 ds.$$

We also have

$$|\langle \tilde{B}(\bar{\zeta}, \mathbf{u}^b), \zeta^b)\rangle| = |\langle \tilde{B}(\bar{\zeta}, \zeta^b), \mathbf{u}^b\rangle|$$

$$\le c|\bar{\zeta}|_{L^2}^{\frac{1}{2}}\|\bar{\zeta}\|^{\frac{1}{2}}\|\zeta^b\| |\mathbf{u}^b|_{L^2}^{\frac{1}{2}}\|\mathbf{u}^b\|^{\frac{1}{2}} \tag{154}$$

$$\le \frac{\alpha}{8}\|\zeta^b\|^2 + c\|\bar{\zeta}\|^2 |\mathbf{u}^b|_{L^2}\|\mathbf{u}^b\|,$$

$$|\langle \tilde{B}(\zeta^b, \bar{\mathbf{u}}), \zeta^b)\rangle| \le c\|\zeta^b\|_w\|\bar{u}\| |\zeta^b|_{L^2}^{\frac{1}{4}}\|\zeta^b\|^{\frac{3}{4}} \le \frac{\alpha}{8}\|\zeta^b\|_w^2 + c\|\bar{u}\|^8 |\zeta^b|_{L^2}^2, \tag{155}$$

$$|\langle NB(\zeta^\flat, \mathbf{u}^\flat), \zeta^\flat)\rangle| \leq c\|\zeta^\flat\|_w\|\mathbf{u}^\flat\|\|\zeta^\flat\|_{L^2}^{\frac{1}{4}}\|\zeta^\flat\|^{\frac{3}{4}} \leq \frac{\alpha}{8}\|\zeta^\flat\|_w^2 + c\|\mathbf{u}^\flat\|^8|\zeta^\flat|_{L^2}^2,$$
$$(156)$$

$$|\langle B(\mathbf{w}_l, \bar{\zeta}), \zeta^\flat)\rangle| \leq c|\mathbf{w}_l|_{L^2}^{\frac{1}{2}}\|\mathbf{w}_l\|^{\frac{1}{2}}\|\bar{\zeta}\|\|\zeta^\flat\|_{L^2}^{\frac{1}{2}}\|\zeta^\flat\|^{\frac{1}{2}}$$
$$\leq \frac{\alpha}{8}\|\zeta^\flat\|^2 + c|\mathbf{w}_l|_{L^2}\|\mathbf{w}_l\|\|\bar{\zeta}\|^2,$$
$$(157)$$

$$|\langle NB(\varepsilon_l, \varepsilon_l), \zeta^\flat)\rangle| \leq c\|\varepsilon_l\|_w\|\varepsilon_l\|\|\zeta^\flat\|_{L^2}^{\frac{1}{4}}\|\zeta^\flat\|^{\frac{3}{4}} \leq \frac{\alpha}{8}\|\zeta^\flat\|_w^2 + c\|\varepsilon_l\|_w^4, \qquad (158)$$

$$|\Lambda\bar{\zeta}|_{L^2} \leq c\|\bar{\zeta}\|. \qquad (159)$$

Therefore, multiplying (149) by $\zeta^\flat$ and using (154)–(159) yields

$$|\zeta^\flat(t)|_{L^2}^2 + c\int_0^t \left(\|\zeta^\flat\|_w^2 + \delta^{-2}\left|\frac{\partial\zeta^\flat}{\partial z}\right|^2\right) ds \leq c\sup_s(|\mathbf{u}^\flat(s)|_{L^2}\|\mathbf{u}^\flat(s)\|$$

$$+|\mathbf{w}_l(s)|_{L^2}\|\mathbf{w}_l(s)\|)\int_0^t \|\bar{\zeta}\|^2 ds + c\sup_s\|\varepsilon_l(s)\|_w^2\int_0^t \|\varepsilon_l\|_w^2 ds + c\int_0^t \|\bar{\zeta}\|^2 ds,$$

$$\leq c\sup_s(|\mathbf{u}^\flat(s)|_{L^2}\|\mathbf{u}^\flat(s)\| + |\mathbf{w}_l(s)|_{L^2}\|\mathbf{w}_l(s)\|)^2\int_0^t \|\zeta^\flat\|_w^2 dt + c\delta\|\varepsilon_l\|_X^3$$

$$+\sup_s(|\mathbf{u}^\flat(s)|_{L^2}\|\mathbf{u}^\flat(s)\| + |\mathbf{w}_l(s)|_{L^2}\|\mathbf{w}_l(s)\|)\int_0^t \|\zeta^\flat\|_w^2 dt.$$
$$(160)$$

From Remark 3.1, Propositions 3.2 and 4.2, we have

$$|\mathbf{u}^\flat(s)|_{L^2} \leq c_4\delta, \quad |\mathbf{w}_l(s)|_{L^2} \leq c_4\delta. \qquad (161)$$

Therefore, (121) and (138) yield

$$|\zeta^\flat(t)|_{L^2}^2 + (c - c_4\delta)\int_0^T \|\zeta^\flat\|_w^2 ds + \frac{c}{\delta^2}\int_0^T \left|\frac{\partial\zeta^\flat}{\partial z}\right|^2 ds \leq c\delta\|\varepsilon_l\|_X^3. \qquad (162)$$

Finally, for δ small enough such that

$$c - c_4\delta > 0, \qquad (163)$$

we have (see (153))

$$|\zeta^\flat(t)|_{L^2}^2 \leq c_4\delta\|\varepsilon_l\|_X^3, \quad \int_0^T |\zeta^\flat(t)|_{L^2}^2 dt \leq c_4\delta^3\|\varepsilon_l\|_X^3, \quad \int_0^T \|\zeta^\flat\|_w^2 ds \leq c_4\delta\|\varepsilon_l\|_X^3,$$

$$|\bar{\zeta}(t)|_{L^2}^2 \leq c_4\delta\|\varepsilon_l\|_X^3, \quad \int_0^T |\bar{\zeta}(t)|_{L^2}^2 dt \leq c_4\delta^3\|\varepsilon_l\|_X^3.$$
$$(164)$$

The next step is to derive some a priori estimates on ε_k.

Note that ε_j satisfies (for $2 \le j \le l$)

$$\frac{d\varepsilon_j}{dt} + \mathcal{A}_1\varepsilon_j + \delta^{-2}\mathcal{A}_2\varepsilon_j + E\varepsilon_j + \tilde{B}(\mathbf{v},\varepsilon_j) + \tilde{B}(\varepsilon_j,\mathbf{v})$$

$$+NB(\mathbf{w}_{j-1},\varepsilon_j) + NB(\varepsilon_j,\mathbf{w}_{j-1}) + NB(\varepsilon_{j-1},\varepsilon_{j-1}) + \Lambda\varepsilon_j = 0 \text{ in } (NV)',$$

$$\varepsilon_j(0) = 0.$$

$$(165)$$

We also have

$$|\langle\tilde{B}(\varepsilon_j,\mathbf{v}),\varepsilon_j\rangle| \le c\|\varepsilon_j\|_w\|\mathbf{v}\||\varepsilon_j|_{L^2}^{\frac{1}{4}}\|\varepsilon_j\|^{\frac{3}{4}} \le \frac{\alpha}{8}\|\varepsilon_j\|_w^2 + c\|\mathbf{v}\|^8|\varepsilon_j|_{L^2}^2, \quad (166)$$

$$\langle NB(\mathbf{w}_{j-1},\varepsilon_j),\varepsilon_j\rangle = 0, \quad (167)$$

$$|\langle NB(\varepsilon_j,\mathbf{w}_{j-1}),\varepsilon_j\rangle| \le c\|\varepsilon_j\|_w\|\mathbf{w}_{j-1}\||\varepsilon_j|_{L^2}^{\frac{1}{4}}\|\varepsilon_j\|^{\frac{3}{4}}$$

$$\le \frac{\alpha}{8}\|\varepsilon_j\|_w^2 + c\|\mathbf{w}_{j-1}\|^8|\varepsilon_j|_{L^2}^2, \quad (168)$$

$$|\langle NB(\varepsilon_{j-1},\varepsilon_{j-1}),\varepsilon_j\rangle| \le c\|\varepsilon_{j-1}\|_w\|\varepsilon_{j-1}\||\varepsilon_j|_{L^2}^{\frac{1}{4}}\|\varepsilon_j\|^{\frac{3}{4}}$$

$$\le \frac{\alpha}{8}\|\varepsilon_j\|_w^2 + c\|\varepsilon_{j-1}\|_w^2\|\varepsilon_{j-1}\|^2, \quad (169)$$

Therefore, multiplying (165) by ε_j and using (166)–(169) yields

$$|\varepsilon_j(t)|_{L^2}^2 + c\int_0^t \|\varepsilon_j\|_w^2 ds \le c_4\delta\|\varepsilon_{j-1}\|_X^3 \quad (170)$$

We also have

$$|\langle\tilde{B}(\varepsilon_j,\mathbf{v}),\mathcal{A}\varepsilon_j\rangle| \le c\|\varepsilon_j\|_w\|\mathbf{v}\|^{\frac{1}{2}}|\tilde{\mathcal{A}}_1\mathbf{v}|_{L^2}^{\frac{1}{2}}|\mathcal{A}\varepsilon_j|_{L^2}$$

$$\le \frac{\alpha}{8}|\mathcal{A}\varepsilon_j|_{L^2}^2 + c\|\varepsilon_j\|_w^2\|\mathbf{v}\||\tilde{\mathcal{A}}_1\mathbf{v}|_{L^2}, \quad (171)$$

$$|\langle\tilde{B}(\mathbf{v},\varepsilon_j),\mathcal{A}\varepsilon_j\rangle| \le c\|\mathbf{v}\|_w\|\varepsilon_j\|^{\frac{1}{2}}|\mathcal{A}\varepsilon_j|_{L^2}^{\frac{1}{2}}|\mathcal{A}\varepsilon_j|_{L^2} \le \frac{\alpha}{8}|\mathcal{A}\varepsilon_j|_{L^2}^2 + c\|\mathbf{v}\|_w^4\|\varepsilon_j\|^2, \quad (172)$$

$$|\langle NB(\mathbf{w}_{j-1},\varepsilon_j),\mathcal{A}\varepsilon_j\rangle| \le c\|\mathbf{w}_{j-1}\|_w\|\varepsilon_j\|^{\frac{1}{2}}|\mathcal{A}\varepsilon_j|_{L^2}^{\frac{1}{2}}|\mathcal{A}\varepsilon_j|_{L^2}$$

$$\le \frac{\alpha}{8}|\mathcal{A}\varepsilon_j|_{L^2}^2 + c\|\mathbf{w}_{j-1}\|_w^4\|\varepsilon_j\|^2, \quad (173)$$

$$|\langle NB(\varepsilon_j,\mathbf{w}_{j-1}),\mathcal{A}\varepsilon_j\rangle| \le c\|\varepsilon_j\|_w\|\mathbf{w}_{j-1}\|^{\frac{1}{2}}|\mathcal{A}\mathbf{w}_{j-1}|_{L^2}^{\frac{1}{2}}|\mathcal{A}\varepsilon_j|_{L^2}$$

$$\le \frac{\alpha}{8}|\mathcal{A}\varepsilon_j|_{L^2}^2 + c\|\mathbf{w}_{j-1}\||\mathcal{A}\mathbf{w}_{j-1}|_{L^2}\|\varepsilon_j\|^2, \quad (174)$$

$$|\langle NB(\varepsilon_{j-1},\varepsilon_{j-1}),\mathcal{A}\varepsilon_j)\rangle| \le c\|\varepsilon_{j-1}\|_w\|\varepsilon_{j-1}\|^{\frac{1}{2}}|\mathcal{A}\varepsilon_{j-1}|_{L^2}^{\frac{1}{2}}|\mathcal{A}\varepsilon_j|_{L^2}$$

$$\le \frac{\alpha}{8}|\mathcal{A}\varepsilon_j|_{L^2}^2 + c\|\varepsilon_{j-1}\|_w^2|\mathcal{A}\mathbf{w}_{j-1}|_{L^2}\|\varepsilon_{j-1}\|. \quad (175)$$

Therefore, multiplying (165) by $\mathcal{A}\varepsilon_j$ and using (171)–(175) yields

$$\|\varepsilon_j(t)\|_w^2 + c\int_0^t |\mathcal{A}\varepsilon_j|_{L^2}^2 ds \leq c \sup_s \|\varepsilon_{j-1}(s)\| \|\varepsilon_{j-1}(s)\|_w \int_0^T \|\varepsilon_{j-1}\|_w |\mathcal{A}\varepsilon_{j-1}|_{L^2} dt$$

$$\leq c_4 \delta \|\varepsilon_{j-1}\|_X^3.$$

$$(176)$$

Finally we have (for $j \geq 2$)

$$\|\varepsilon_j\|_X^2 \leq c_4 \delta \|\varepsilon_{j-1}\|_X^3, \tag{177}$$

which gives

$$\|\varepsilon_l\|_X \leq \delta^{-1} \left(\delta^{1/2}\right)^{(3/2)^{l-1}} \left(c_4 \delta^{\frac{1}{2}} \|\varepsilon_1\|_X\right)^{(3/2)^{l-1}}. \tag{178}$$

Hereafter, we assume that δ is small enough such that

$$c_4 \delta^{\frac{1}{2}} \|\varepsilon_1\|_X < 1. \tag{179}$$

Note that $\|\varepsilon_1\|_X = \|\mathbf{w}_1\|_X$ is bounded (although not necessary small).

Therefore, we have (see (178))

$$\|\varepsilon_l\|_X^3 \leq \delta^{-3} \delta^{(3/2)^l}, \quad \|\varepsilon_l\|_X^2 \leq \delta^{-2} \delta^{(3/2)^{l-1}}, \tag{180}$$

and (164) yields

$$|\zeta(t)|_{L^2}^2 = |\bar{\zeta}(t)|_{L^2}^2 + |\zeta^\flat(t)|_{L^2}^2 \leq c_4 \delta^2 \|\varepsilon_l\|_X^2 \leq c_4 \delta^{(3/2)^{l-1}},$$

$$\int_0^T |\zeta(t)|_{L^2}^2 dt \leq c_4 \|\varepsilon_l\|_X^3 \leq c_4 \delta^{(3/2)^l}. \tag{181}$$

Finally, the following convergence result is proved.

Theorem 5.1. *We assume that the data satisfy (27). Then there exits a constant $\delta_1 > 0$ independent of l such that for $0 < \delta < \delta_1$, the error $\zeta(t) = u(t) - u_l(t)$ satisfies*

$$|\zeta(t)|_{L^2}^2 \leq c_4 \delta^{(3/2)^{l-1}}, \quad \int_0^T |\zeta(t)|_{L^2}^2 dt \leq c_4 \delta^{(3/2)^l}. \tag{182}$$

6 Numerical Results

In this section, we present some numerical simulations of the PEs of the ocean with continuous density stratification. The goal is to compare the solution obtained for the PEs (29)–(30) to that of the small eddy correction algorithms presented above. Hereafter we restrict ourselves to $l = 1$. More simulations for larger values of l will be presented elsewhere. To avoid dealing with the divergence-free condition, we first rewrote the barotropic equations in the vorticity stream-function formulation.

In our experiments, the basin configuration is the (nondimensional) cube $[0,1] \times [0,1] \times [-1,0]$. Let us simply recall that this is a two-gyre, wind-driven ocean problem with a steady sinusoidal wind stress (maximum $\tau_0 = 1$ dyne cm^{-4}) in a basin that is $L_1 \times L_1 \times H_1$ km (east–west $\times$ north–south $\times$ bottom-surface extent). The Coriolis parameter is given by $f = f_0 + \beta y$, $f_0 = 9.3 \times 10^{-5}s^{-1}$, $\beta = 2. \times 10^{-11}$ m^{-1}s^{-1}. The model does not include bottom topography. The ocean is forced by a steady wind stress $\tau_0 = (\tau_0^x, \tau_0^y) = (-10^{-4}\cos(2\pi y/L_1), 0)$ and a density variation. Others dimensional quantities are given by $U_1 = 10^{-1}$m s^{-1}, $g = 9.8$ m s^{-2}, $L_1 = 2.10^6$ m and $H_1 = 4,000$ m, which gives $\delta = H_1/L_1 = 2.10^{-3}$. The initial condition is given by $v = \rho = 0$ at $t = 0$. The details on the numerical method is given in [24]. Let us simply recall that all the operators in (29)–(30) are discretized using a second order central difference scheme. The Jacobian operator is approximated using Arakawa's method [29]. For the time integration of the model, we use a fourth-order Adams–Bashforth method. In all the computations presented in this article, the (nondimensional) time step is $\triangle t = 10^{-4}$. For the space discretization, we take 100×100 points in the x–y plane and 10 points on the vertical direction. For the boundary condition (6), we take $\tau_v = \rho^* = 0$ and we compensate with a forcing term F_1 in $(5)_1$ defined by

$$F_1 = c(g(z)\tau_0, 0), \tag{183}$$

where $g(z)$ is defined by

$$g(z) = 0.5(1 + \tanh((z/H_1 + z_1)/\epsilon_1)), \tag{184}$$

where z_1 and ϵ_1 are very small constants chosen such that the forcing F_1 is nonzero only on the first couple layers from the surface of the ocean.

The forcing term F_2 in $(5)_3$ has the form

$$F_2(z) = c\frac{\partial^2 \rho_s}{\partial z^2}/\rho_0, \tag{185}$$

where ρ_s is given by

$$\rho_s(z) = 1028 - 3\exp(10z/H_1), \tag{186}$$

and c is a constant.

Simulations 1 (steady state solutions). In this simulation, we compare the two models (29)–(30) and (116)–(117) when the solution converges to a steady state. For $R_{e_1} = R_{e_2} = R_{t_1} = R_{t_2} = 100$, the solutions obtained with the two models converge to a steady state characterized by two gyres, one subpolar cyclonic and one subtropical anticyclonic. The two gyres are separated by a meandering jet, [9, 10]. Figure 1 shows the barotropic streamfunctions obtained with the two models. As one can see, model (116)–(117) approximates very well the PEs (29)–(30).

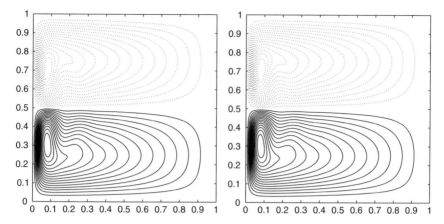

Fig. 1. Barotropic stream-function at the steady state, $\delta = 1.10^{-3}$. PEs on the left and reduced model (116)–(117) on the right

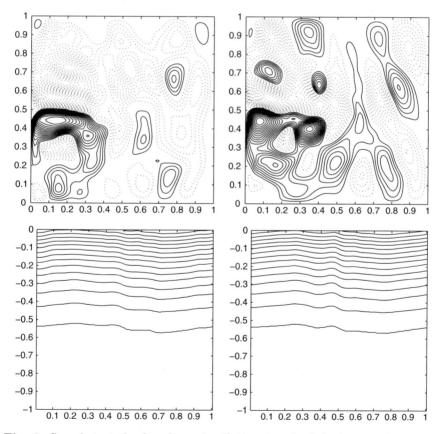

Fig. 2. Snapshot at the (nondimensional) time $t = 5$ of the barotropic stream-function and the total density at $x = 0.25$. PEs on the left and reduced model (116)–(117) on the right

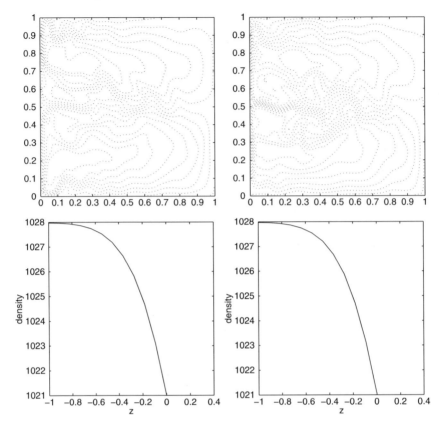

Fig. 3. Snapshot at the (nondimensional) time $t = 5$ of the surface density deviation and the total density at $(x, y) = (0.5, 0.5)$. PEs on the left and reduced model (116)–(117) on the right

Simulations 2 (time-dependent solutions). For the two models (29)–(30) and (116)–(117), the following Figs. 2–5 show a time-sequence of the surface density deviation (that is $\rho(x, y, 0)$), the barotropic stream-functions, the total density at $x = 0.25$ (that is $(\rho_s + \rho)(0.25, y, z)$) and the total density at $(x, y) = (0.5, 0.5)$, (that is $(\rho_s + \rho)(0.5, 0.5, z)$) for the Reynolds numbers $R_{e_1} = 2.10^3$, $R_{e_2} = 10^2$, $R_{t_1} = 10^3$, $R_{t_2} = 10^2$. For these values of the Reynolds number, the flow remains time-dependent. From these figures, it is clear that the density fields obtained with the two models are similar. The situation is a little different for the barotropic stream-functions as the solutions obtained with the two models present more differences. However, Fig. 6 shows that the time-average of the two flows are very similar. This seems to confirm what is already believed in oceanography that, from the climate point of view (where the main focus is on the time-average of the flow) the interactions between the baroclinic and the barotropic mode do not need to be accurately represented [6].

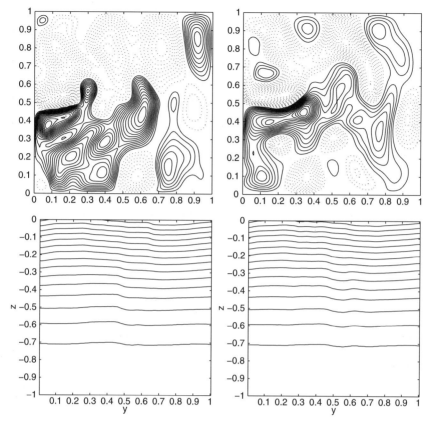

Fig. 4. Snapshot at the (nondimensional) time $t = 10$ of the barotropic stream-function and the total density at $x = 0.25$. PEs on the left and reduced model (116)–(117) on the right

7 Conclusion

The purpose of this article was to present a small eddy correction method for the numerical solution of the PEs of the ocean. Considering the interaction between the baroclinic and barotropic flows and using the idea of the Newton iteration, a small eddy correction method was proposed for the PEs of the ocean. We assume that the barotropic approximation to the solution is known. Formally applying the Newton iterative procedure to the baroclinic flow equation, we then generate approximate systems. It was shown that the initial step ($l = 0$) leads to the well known quasi-geostrophic equations and the next step ($l = 1$) yields a nonlinear Galerkin type approximation. Some numerical simulations for $l = 1$ show that the method can accurately approximate the PEs of the ocean. For the simulations presented in this article, we did not observed any CPU gain for the reduced model (116)–(117) compared

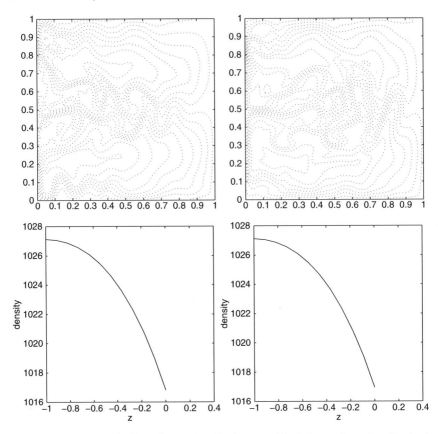

Fig. 5. Snapshot at the (nondimensional) time $t = 10$ of the surface density deviation and the total density at $(x, y) = (0.5, 0.5)$. PEs on the left and reduced model (116)–(117) on the right

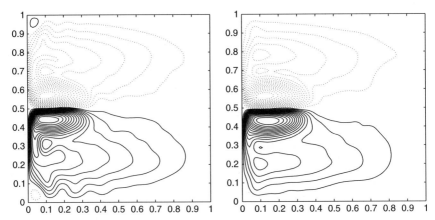

Fig. 6. Time-average (over $[0, 10]$) of the barotropic stream-function. PEs on the left and reduced model (116)–(117) on the right

to the PEs (29)–(30), the primary reason being the simplicity of the time discretization used in this article (a fourth-order Adams–Bashforth method for both the barotropic and baroclinic flows). An efficient time discretization of the model presented in this article may lead to considerable savings in calculation costs. In fact, such method should take advantage of the time scale differences between the barotropic and baroclinic modes. For instance:

- One can use different time steps for the small and large scales.
- One can use different schemes for the small and large scales.
- One can freeze the small scales over an interval of time.

These approaches, already used with success (significant reduction in CPU time cost) for the Navier–Stokes equations (see [4, 13]) are currently under development by the authors.

Acknowledgments

This work was supported in part by NSF Grant DMS0604235, DOE Grant DE-FG02-01ER63251:A000, and by the Research Fund of Indiana University.

References

1. A. Belmiloudi. Asymptotic behavior for the perturbation of the primitive equations of the ocean with vertical viscosity. *Canad. Appl. Math. Quart.*, 8(2):97–139, 2000.
2. C. Cao and E. S. Titi. Global well-posedness of the three-dimensional viscous primitive equations of large scale ocean andatmosphere dynamics. *arXiv: Math. AP/0503028*, 2, 2005.
3. J. G. Charney. Geostrophic turbulence. *J. Atmos. Sci.*, 28:1087–1095, 1971.
4. T. Dubois, F. Jauberteau, and R. Temam. *Dynamic multilevel methods and the numerical simulation of turbulence.* Cambridge University Press, Cambridge, 1999.
5. T. Dubois, F. Jauberteau, R. M. Temam, and J. Tribbia. Different time schemes to compute the large and small scales for the shallow water problem. *J. Comp. Phys.*, 207:660–694, 2005.
6. S. M. Griffies, C. Boening, F. O. Bryan, E. P. Chassignet, R. Gerdes, H. Hasumi, A. Hirst, A. M. Treguier, and D. Webb. Developments in ocean climate modelling. *Ocean Modell.*, 2:123–192, 2000.
7. Jack K. Hale. *Asymptotic behavior of dissipative systems*, Vol. 25. American Mathematical Society, Providence, RI, mathematical surveys and monographs edition, 1988.
8. G. J. Haltiner and R. T. Williams. *Numerical prediction and dynamic meteorology.* Wiley, New York, 1980.
9. W. R. Holland. The role of mesoscale eddies in the general circulation of the ocean. *J. Phys. Oceanogr.*, 8:363–392, 1978.
10. W. R. Holland and P. B. Rhines. An example of eddy-induced ocean circulation. *J. Phys. Oceanogr.*, 10:1010–1031, 1980.

11. Y. Hou and K. Li. A small eddy correction method for nonlinear dissipative evolutionary equations. *SIAM J. Numer. Anal.*, 41(3):1101–1130, 2003.
12. C. Hu, R. Temam, and M. Ziane. The primitive equations of the large scale ocean under the small depth hypothesis. *Discrete Contin. Dyn. Syst.*, 9(1):97–131, 2003.
13. F. Jauberteau, C. Rosier, and R. Temam. The nonlinear Galerkin method in computational fluid dynamic. *Appl. Numer. Math.*, 6(5):361–370, 1990.
14. G. M. Kobelkov. Existence of a solution 'in the large' for the 3*d* large-scale ocean dynamics equations. *C. R. Math. Acad. Sci. Paris*, 343:283–286, 2006.
15. I. Kukavica and M. Ziane. On the regularity of the primitive equations of the ocean. To appear.
16. J. L. Lions, R. Temam, and S. Wang. On the equations of large-scale ocean. *Nonlinearity*, 5:1007–1053, 1992.
17. J. L. Lions, R. Temam, and S. Wang. Models of the coupled atmosphere and ocean (CAO I). *Comput. Mech. Adv.*, 1:3–54, 1993.
18. J. L. Lions, R. Temam, and S. Wang. Numerical analysis of the coupled atmosphere and ocean models (CAOII). *Comput. Mech. Adv.*, 1:55–120, 1993.
19. M. Marion and R. Temam. Nonlinear Galerkin methods. *SIAM J. Numer. Anal.*, 26(5):1139–1157, 1989.
20. M. Marion and R. Temam. Nonlinear Galerkin methods: the finite elements case. *Numer. Math.*, 57(3):205–226, 1990.
21. B. Di Martino and P. Orenga. Resolution to a three-dimensional physical oceanographic problem using the non-linear Galerkin method. *Int. J. Num. Meth. Fluids*, 30:577–606, 1999.
22. M. Petcu, R. Temam, and M. Ziane. *Some Mathematical Problems in Geophysical Fluid Dynamics*, 2008.
23. G. R. Sell and Y. You. *Dynamics of evolutionary equations*, Vol. 143. Springer, New York, Applied Mathematical Sciences edition, 2002.
24. E. Simmonet, T. Tachim Medjo, and R. Temam. Barotropic–baroclinic formulation of the primitive equations of the ocean. *Appl. Anal.*, 82(5):439–456, 2003.
25. E. Simmonet, T. Tachim Medjo, and R. Temam. On the order of magnitude of the baroclinic flow in the primitive equations of the ocean. *Ann. Mat. Pura Appl.*, 185, 2006, S293–S313.
26. R. Temam. *Infinite dimensional dynamical systems in mechanics and physics*, Vol. 68. Appl. Math. Sci., Springer-Verlag, New York, second edition, 1988.
27. R. Temam. Stability analysis of the nonlinear Galerkin method. *Math. Comp.*, 57(196):477–505, 1991.
28. R. Temam and M. Ziane. Some mathematical problems in geophysical fluid dynamics. In S. Friedlander and D. Serre, editors, *Handbook of Mathematical Fluid Dynamics, Vol. 3*, pages 535–658. Elsevier, 2004.
29. W. M. Washington and C. L. Parkinson. *An Introduction to Three-Dimensional Climate Modeling*. Oxford University Press, Oxford, 1986.

Part II

Modeling and Control of Phenomena

Aspects of Modeling Transport in Small Systems with a Look at Motor Proteins

D. Kinderlehrer

1 Introduction

Diffusion-mediated transport is a phenomenon in which directed motion is achieved as a result of two opposing tendencies: diffusion, which spreads the particles uniformly through the medium, and transport, which concentrates the particles at some special sites. It is implicated in the operation of many molecular level systems. These include some liquid crystal and lipid bilayer systems, and, especially, the motor proteins responsible for eukaryotic cellular traffic. All of these systems are extremely complex and involve subtle interactions on widely varying scales. The chemical/mechanical transduction in motor proteins is, by contrast to many materials microstructure situations, quite distant from equilibrium. These bio-systems function in a dynamically metastable range. There is an enormous biological literature about this and a considerable math-biology and biophysics literature, e.g., [1, 2, 8, 9, 14, 15, 19, 20, 23, 25–28].

Our approach is to look at a dissipation principle for such situations and its relationship to the Monge–Kantorovich mass transfer problem, e.g., [16]. In effect, we begin with simple – but not too simple – assumptions of motion along a track followed by statistical assumptions which provide us an ensemble. The procedure permits us to establish consistent thermodynamical dissipation principles from which evolution equations follow. In a given instance, the dissipation principle identifies the thermodynamic free energy, the conformational changes that result, for example, from ATP hydrolysis reactions, and dissipation.

The simplest diffusional transport equation in this context is the Fokker–Planck Equation with a periodic potential. Diffusion spreads density and potential attracts density to specific cites but this typical system does not exhibit biased transport. Something else has to be present in system for transport to take place. We investigate possible relationships between the potentials in the system and the conformational changes that lead to the accumulation of mass at one end of the track.

2 A Dissipation Principle

Our dissipation principle, based on extending the simple spring-mass-dashpot to an ensemble, provides natural weak topology kinetics for the motion of molecular motors. The dissipation, we shall see, is related to the Wasserstein metric, a Monge–Kantorovich metric on probability distributions. The motion of a spring-mass-dashpot in a highly viscous environment may be expressed by the ordinary differential equation

$$\gamma \frac{d\xi}{dt} + \psi'(\xi) = 0, \ 0 < t < \tau$$
$$\xi(0) = x \ . \tag{1}$$

We think of $x \in \Omega = (0,1)$ and $\tau > 0$ a relaxation time. ψ is a given potential. Multiplying the equation by (1) by $d\xi/dt$ and integrating gives

$$\gamma \int_0^\tau (\frac{d\xi}{dt})^2 + \psi(\xi(\tau)) = \psi(x), \tag{2}$$

expressing that, in this simple system, the dissipation plus energy at the terminal state is the initially supplied energy. We have a trajectory like as pictured in Fig. 1 left. Consider now an ensemble $\{\xi(t,x)\}$ of spring-mass-dashpots distributed by a density $\rho^*(x)$. We average (2) over Ω, which renders the individual spring-mass-dashpots indistinguishable, Fig. 1 right. We must, thus, contribute entropy to the resulting configuration. The most convenient such entropy is combinatorial indeterminacy leading us to

$$\gamma \int_0^\tau \int_\Omega (\frac{d\xi}{dt})^2 \rho^*(\xi(\tau,x)) dx dt + \int_\Omega \{\psi(\xi(\tau,x))\rho^*(\xi(\tau,x))$$
$$+\sigma\rho^*(\xi(\tau,x)) \log \rho^*(\xi(\tau,x))\} dx \tag{3}$$
$$\leqq \int_\Omega \{\psi(x)\rho^*(x) + \sigma\rho^*(x) \log \rho^*(x)\} dx.$$

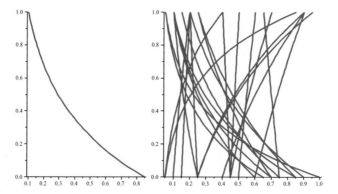

Fig. 1. A trajectory (*left*) and an ensemble of trajectories (*right*)

This Lagrangian formulation (3) may be written in Eulerian terms by setting

$$\phi(x,t) = \xi(t,x), \text{ family of transfer mappings}$$
$$v(\xi,t) = \phi_t(x,t), \text{ with } \xi = \phi(x,t) \text{ velocity}$$
$$\rho(\xi,t) \text{ family of transformations given by} \tag{4}$$

$$\int_\Omega \zeta\rho d\xi = \int_\Omega \zeta(\phi(x,t))\rho^*(x)dx.$$

The descriptions are linked by the continuity equation

$$\frac{\partial}{\partial t}\rho + \frac{\partial}{\partial x}(v\rho) = 0. \tag{5}$$

The Eulerian expression of (3) is

$$\gamma \int_0^\tau \int_\Omega v^2\rho dx dt + \int_\Omega \{\psi\rho + \sigma\rho\log\rho\}dx \leqq \int_\Omega \{\psi\rho^* + \sigma\rho^*\log\rho^*\}dx. \tag{6}$$

The dissipation principle is to choose the most likely configuration satisfying (6) among all probability densities on Ω, which means minimizing its left hand side. In particular, we can begin by minimizing the first term for given initial state ρ^* and terminal state ρ. This gives rise to a Monge–Kantorovich metric or Wasserstein metric $d(\rho, \rho^*)$, namely

$$\frac{1}{\tau}d(\rho, \rho^*)^2 = \min_A \int_0^\tau \int_\Omega v^2\rho dx dt, \tag{7}$$

where the minimum is taken over the set A of deformations and velocities satisfying the continuity condition (5) and initial state ρ^* and terminal state ρ. The Wasserstein metric may be defined in several ways, in particular,

$$d(\rho, \rho^*)^2 = \min_P \int_{\Omega\times\Omega} |x - \xi|^2 dp(x,\xi),$$

where P denotes the set of joint distributions whose marginal densities are ρ^* and ρ. Obviously it can be extended to probability measures and the resulting metric induces the weak-* topology on bounded subsets of $C(\bar{\Omega})'$, cf. Villani, [29]. The equivalence (7) is due to Benamou and Brenier [3], and is also found in work of Otto [22].

Thus we are led to the dissipation interpretation of the variational principle: given ρ^* find ρ subject to

$$\frac{1}{2\tau}d(\rho, \rho^*)^2 + \int_\Omega \{\psi\rho + \sigma\rho\log\rho\}dx = \min. \tag{8}$$

Metrics and variational principles give rise to implicit schemes. Given $\rho^{(k-1)}$, determine $\rho^{(k)}$ by setting $\rho^{(k-1)} = \rho^*$ and $\rho^{(k)} = \rho$, the solution, in (8). For fixed τ we set

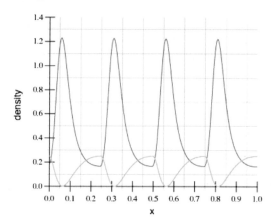

Fig. 2. The stationary solution of a Fokker–Planck Equation with its potential illustrating diffusion and transport but not biased transport

$$\rho_\tau(x,t) = \rho^{(k)}(x), \ k\tau \leqq t < (k+1)\tau. \tag{9}$$

According to Jordan et al. [16] and Otto [21],

$$\rho_\tau \to \rho,$$

where ρ is the solution of the Fokker–Planck Equation

$$\frac{\partial \rho}{\partial t} = \frac{\partial}{\partial x}\{\sigma \frac{\partial \rho}{\partial x} + \psi' \rho\}, \text{ in } \Omega, \ t > 0$$

$$\sigma \frac{\partial}{\partial x}\rho + \psi' \rho = 0 \text{ on } \partial\Omega, \ t > 0. \tag{10}$$

The iteration scheme (8) expresses the Fokker–Planck Equation as a form of gradient flow in the weak* topology for a natural free energy. Equation (10) itself illustrates diffusion and transport as described in the introduction. Diffusion spreads density and the potential attracts density to specific sites. It does not show biased transport, cf. Fig. 2. For this, something additional must be present in the system. We shall investigate this by looking at a description of kinesin 1 molecular motors.

3 A Look at Multiple State Motors

Conventional kinesin has two identical head domains (heavy chains) which walk in a hand over hand fashion along a rigid microtubule. This is an intricate process with a complicated transformation path comprising both the ATP hydrolysis (chemical states) and the motion (mechanical states), [14, 28]. For a crude reckoning, at a gross combinatorial level, each head is attached or in motion and is nucleotide bound or not. Assuming that a given motor has one head bound and one free at any instant leads to eight possible pathways for each cycle. We shall give a simplified description by considering the nucleotide binding and then the subsequent motion. Our dissipation/variational principle is flexible enough to accommodate this process. We shall consider an ensemble of motors with n states, which in the earlier mechanical language may be described as an ensemble of conformation changing spring-mass-dashpots. Conformation changes occur, typically, at specific places and are represented by changes in population among the various states. More complicated types of conformation changes are possible and we reserve this for a future discussion. For details of how this scheme is applied to kinesin 1, we refer to [5, 10, 12].

Let $\sigma > 0$, $\psi_1, \ldots, \psi_n$ be smooth non-negative functions of period $1/N$, and $A = (a_{ij})$ a smooth rate matrix of period $1/N$, that is, a_{ij} are $1/N$−periodic functions with

$$a_{ii} \leqq 0, \ a_{ij} \geqq 0 \text{ for } i \neq j \text{ and}$$

$$\sum_{i=1,\ldots,n} a_{ij} = 0, \ j = 1, \ldots, n. \tag{11}$$

Note that for $\tau > 0$ small, the matrix $P = \mathbf{1} + \tau A$ is a probability matrix. Let

$$F(\eta) = \int_\Omega \sum_{i=1,\ldots,n} \{\psi_i \eta_i + \sigma \eta_i \log \eta_i\} dx \tag{12}$$

$$\eta_i \geqq 0 \text{ and } \int_\Omega \sum_{i=1,\ldots,n} \eta_i dx = 1$$

denote the free energy. Now given a state ρ^*, determine its successor state ρ by resolving the variational principle

$$\frac{1}{2\tau} \sum_{i=1,\ldots,n} d(\rho_i, (P\rho^*)_i)^2 + F(\rho) = \min,$$

$$\int_\Omega \rho_i dx = \int_\Omega (P\rho^*)_i dx, \tag{13}$$

where P is the probability matrix above. For this functional we can perform the same iterative procedure that we indicated for the original one (8): determine $\rho^{(k)}$ by setting $\rho^{(k-1)} = \rho^*$ and $\rho^{(k)} = \rho$, the solution, in (13). Again we may set

$$\rho_\tau(x, t) = \rho^{(k)}(x), \ k\tau \leqq t < (k+1)\tau \tag{14}$$

and ask about $\rho = (\rho_1, \ldots, \rho_n)$ which arises as

$$\rho = \lim_{\tau \to 0} \rho_\tau.$$

It is the solution of the weakly coupled parabolic system

$$\frac{\partial \rho_i}{\partial t} = \frac{\partial}{\partial x}\left(\sigma \frac{\partial \rho_i}{\partial x} + \psi_i' \rho_i\right) + \sum_{j=1,\ldots,n} a_{ij}\rho_j = 0 \text{ in } \Omega, \ t > 0,$$

$$\sigma \frac{\partial \rho_i}{\partial x} + \psi_i'\rho_i = 0 \text{ on } \partial\Omega, \ t > 0, \ i = 1,\ldots,n,$$

$$\rho_i \geqq 0 \text{ in } \Omega, \ \int_\Omega (\rho_1 + \cdots + \rho_n)dx = 1, \ t > 0. \tag{15}$$

The possibility of including terms of inferior order in this type of implicit scheme seems to have been first observed in [18].

Before proceeding further, let us review what we intend by transport. In a chemical or conformational change process, a reaction coordinate (or coordinates) must be specified. This is the independent variable. In a mechanical system, it is usually evident what this coordinate must be. In our situation, even though both conformational change and mechanical effects are present, it is natural to specify the distance along the motor track, the microtubule, here the interval Ω, as the independent variable. We interpret the migration of density during the evolution to one end of the track as evidence of transport. This leads us to the stationary state of the system $\rho^\sharp$ where

$$\rho(x,t) \to \rho^\sharp(x) \text{ as } t \to \infty. \tag{16}$$

We shall omit the superscript, writing $\rho(x) = \rho^\sharp(x)$. So, obviously, ρ is the solution of the stationary system of ordinary differential equations

$$\frac{d}{dx}\left(\sigma \frac{d\rho_i}{dx} + \psi_i'\rho_i\right) + \sum_{j=1,\ldots,n} a_{ij}\rho_j = 0 \text{ in } \Omega$$

$$\sigma \frac{d\rho_i}{dx} + \psi_i'\rho_i = 0 \text{ on } \partial\Omega, \ i = 1,\ldots,n, \tag{17}$$

$$\rho_i \geqq 0 \text{ in } \Omega, \ \int_\Omega (\rho_1 + \cdots + \rho_n)dx = 1.$$

Directed transport results from functional relationships in the system (15) or (17). First note that detailed balance for the conformational changes must be broken. In this situation, where the solution ρ of (17) satisfies

$$A\rho = 0 \text{ in } \Omega, \tag{18}$$

the ordinary differential equations decouple leading to a collection of functions $\rho_1,\ldots,\rho_n$ any of which has a plot like that depicted in Fig. 2. Let us first direct our attention to the potentials $\psi_1,\ldots,\psi_n$. In Fig. 3 two configurations with symmetric wells are depicted. Detailed balance, (18), is not satisfied and there is lack of transport. Asymmetric potentials are thought to play a role in transport. In Fig. 4, we see that this is not necessarily the case.

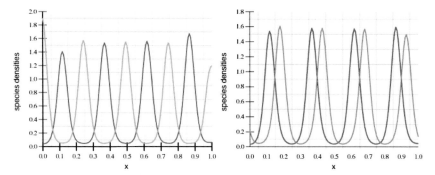

Fig. 3. Solutions of (17) for two state systems with symmetric potentials placed symmetrically (*left*) and symmetric potentials placed asymmetrically (*right*) showing lack of transport of density. A was chosen constant to optimize the possibility of transport. Detailed balance is not satisfied by the solutions

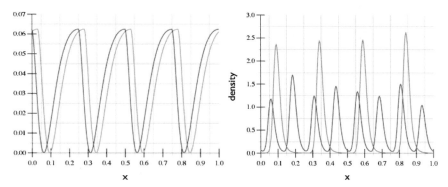

Fig. 4. Slightly shifted asymmetric wells, *left*, and the solutions of (17), *right*, illustrating lack of transport. The matrix A was chosen to optimize transport possibilities

If we adopt the pragmatic notion that in a two species system, the two species function in the same way, we are led to interdigitated potentials ψ_j of the form in Fig. 5. This is not a reason, of course. We discuss this further below. We are led to the intriguing question of the relationship between the ψ_j and A. Even under the most propitious circumstances, one may always add to the system independent uncoupled equations. So it is necessary, in view of (11), that

$$a_{ii} \not\equiv 0 \text{ in } \Omega.$$

The basic mechanism of diffusional transport is that mass is transported to specific sites determined by minima and local minima of the potential. For directed transport, to the left toward $x = 0$, for example, in any subinterval of a period interval, there should be some ψ_i which is increasing. This explains the result shown in Fig. 4, where the potentials are asymmetric and transport is not present. Moreover, some interchange must take place: mass in states associated to each of the ψ_j which is decreasing should have the opportunity

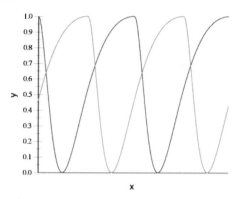

Fig. 5. Asymmetric periodic potentials symmetrically interdigitated, the configuration which promotes transport

to change to the ith-state. This is reminiscent of an ergodic hypothesis. It does not say that all states are connected, but it will be a very strong condition since it will be required to hold near all the minima of all of the potentials. In the neck linker example we have mentioned, the condition fails and so does the conclusion of our theorem. We give a more precise statement with this theorem, [11], and see [4] for a different point of view.

Theorem 3.1. *Suppose that ρ is a positive solution of* (17), *where the coefficients a_{ij}, $i,j = 1,\ldots,n$ and the ψ_i, $i = 1,\ldots,n$ are smooth and $1/N$-periodic in $\overline{\Omega}$. Suppose that (11) holds and also that the following conditions are satisfied.*

(1) Each ψ_i' has only a finite number of zeros in $\overline{\Omega}$.
(2) There is some interval in which $\psi_i' > 0$ for all $i = 1,\ldots,n$.
(3) In any interval in which no ψ_i' vanishes, $\psi_j' > 0$ in this interval for at least one j.
(4) If $I, |I| < 1/N$, is an interval in which $\psi_i' > 0$ for $i = 1,\ldots,p$ and $\psi_i' < 0$ for $i = p+1,\ldots,n$, and a is a zero of at least one of the ψ_k' which lies within ϵ of the right-hand end of I, then for ϵ sufficiently small, there is at least one index i, $i = 1,\ldots,p$, with $a_{ij} > 0$ in $(a-\eta, a)$ for some $\eta > 0$, all $j = p+1,\ldots,n$.

Then, there exist positive constants K, M independent of σ such that

$$\sum_{i=1}^{n} \rho_i\left(x + \frac{1}{N}\right) \leqq Ke^{-\frac{M}{\sigma}} \sum_{i=1}^{n} \rho_i(x), \quad x \in \Omega, \ x < 1 - \frac{1}{N} \qquad (19)$$

for sufficiently small σ.

An example is given in Fig. 6, where the potentials from Fig. 5 were employed and the 2×2 matrix A had support in a neighborhood of the well minima. This situation is consistent with the experimental results reported some time ago by

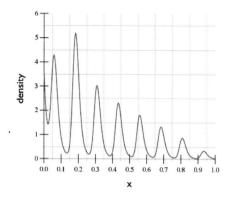

Fig. 6. Computed solution for two state rachet for interdigitated asymmetric potentials, period 4. Plot shows summed density $\rho_1 + \rho_2$

Hackney [8]. The methods described here may be elaborated to include some additional features, even without incorporating multiple reaction coordinates, for example. In [12] we describe how the kinesin 1 necklinker, described in [28] and analyzed in [9] falls into this framework. The central feature of additional conformational changes, or more generally rapid localized changes, is that they act by altering the populations of the species but without potentials. Thus, one is led to a dissipation principle

$$\frac{1}{2\tau} \sum_{i=1,\ldots,m} d(\rho_i, (P\rho^*)_i)^2 + F(\rho) = \min,$$
$$\int_\Omega \rho_i dx = \int_\Omega (P\rho^*)_i dx, \tag{20}$$

where the number m of species is larger than the number n of potentials. The resulting equations are a mixed system of partial differential equations and ordinary differential equations.

Finally we note that there are other mechanisms of diffusion mediated transport, in particular, the flashing rachet. This is discussed in [2] and from the analysis point of view in [7] and [17]. Not all apparent rachet type mechanisms are in fact such. For a discussion of this in relation to the Parrondo coin toss, see [13].

Acknowledgement

This work is partially supported by the National Science Foundation Grants DMS 0305794 and DMS 0405343.

References

1. AJDARI, A. AND PROST, J. (1992) Mouvement induit par un potentiel périodique de basse symétrie: dielectrophorese pulse, C. R. Acad. Sci. Paris t. **315**, Série II, 1653

2. ASTUMIAN, R.D. (1997) Thermodynamics and kinetics of a Brownian motor, Science **276**, 917–922

3. BENAMOU, J.-D. AND BRENIER, Y. (2000) A computational fluid mechanics solution to the Monge-Kantorovich mass transfer problem, Numer. Math. **84**, 375–393

4. CHIPOT, M., HASTINGS, S., AND KINDERLEHRER, D. (2004) Transport in a molecular motor system, Math. Model. Numer. Anal. (M2AN) **38**, no. 6, 1011–1034

5. CHIPOT, M., KINDERLEHRER, D., AND KOWALCZYK, M. (2003) A variational principle for molecular motors, Meccanica **38**, 505–518

6. DOERING, C., ERMENTROUT, B., AND OSTER, G. (1995) Rotary DNA motors. Biophys. J. **69**, no. 6, 2256–2267

7. DOLBEAULT, J., KINDERLEHRER, D., AND KOWALCZYK, M. (2004) Remarks about the flashing rachet, Partial Differential Equations and Inverse Problems (Conca, C. et al. eds), Cont. Math. **362** AMS, 167–176

8. HACKNEY, D.D. (1996) The kinetic cycles of myosin, kinesin, and dynein, Ann. Rev. Physiol. **58**, 731–750

9. HACKNEY, D.D., STOCK, M.F., MOORE, J., AND PATTERSON, R. (2003) Modulation of kinesin half-site ADP release and kinetic processvity by a spacer between the head grounps, Biochemistry **42**, 12011–12018

10. HASTINGS, S. AND KINDERLEHRER, D. (2005) Remarks about diffusion mediated transport: thinking about motion in small systems, Nonconvex Optim. Appl. **79**, 497–511

11. HASTINGS, S., KINDERLEHRER, D., AND MCLEOD, J.B. Diffusion mediated transport in multiple state systems, to appear SIAM J. Math. Anal.

12. HASTINGS, S., KINDERLEHRER, D., AND MCLEOD, J.B. Diffusion mediated transport with a look at motor proteins, to appear

13. HEATH, D., KINDERLEHRER, D., AND KOWALCZYK, M. (2002) Discrete and continuous ratchets: from coin toss to molecular motor, Discrete Contin. Dyn. Syst. Ser. B **2**, no. 2, 153–167

14. HOWARD, J. (2001) Mechanics of Motor Proteins and the Cytoskeleton, Sinauer Associates, Sunderland, MA

15. HUXLEY, A.F. (1957) Muscle structure and theories of contraction, Prog. Biophys. Biophys. Chem. **7**, 255–318

16. JORDAN, R., KINDERLEHRER, D., AND OTTO, F. (1998) The variational formulation of the Fokker-Planck equation, SIAM J. Math. Anal. **29**, no. 1, 1–17

17. KINDERLEHRER, D. AND KOWALCZYK, M (2002) Diffusion-mediated transport and the flashing ratchet, Arch. Rat. Mech. Anal. **161**, 149–179

18. KINDERLEHRER, D. AND WALKINGTON, N. (1999) Approximation of parabolic equations based upon Wasserstein's variational principle, Math. Model. Numer. Anal. (M2AN) **33**, no. 4, 837–852

19. OKADA, Y. AND HIROKAWA, N. (1999) A processive single-headed motor: kinesin superfamily protein KIF1A, Science **283**, 1152–1157

20. OKADA, Y. AND HIROKAWA, N. (2000) Mechanism of the single headed processivity: diffusional anchoring between the K-loop of kinesin and the C terminus of tubulin, Proc. Nat. Acad. Sci. **7**, no. 2, 640–645

21. OTTO, F. (1998) Dynamics of labyrinthine pattern formation: a mean field theory, Arch. Rat. Mech. Anal. **141**, 63–103

22. OTTO, F. (2001) The geometry of dissipative evolution equations: the porous medium equation, Comm. PDE **26**, 101–174

23. PARMEGGIANI, A., JÜLICHER, F., AJDARI, A., AND PROST, J. (1999) Energy transduction of isothermal ratchets: generic aspects and specific examples close and far from equilibrium, Phys. Rev. E **60**, no. 2, 2127–2140

24. PERTHAME, B. The general relative entropy principle

25. PESKIN, C.S., ERMENTROUT, G.B., AND OSTER, G.F. (1995) *The correlation ratchet: a novel mechanism for generating directed motion by ATP hydrolysis*, in Cell Mechanics and Cellular Engineering (V.C. Mow et al. eds.), Springer, New York

26. REIMANN, P. (2002) Brownian motors: noisy transport far from equilibrium, Phys. Rep. **361**, nos. 2–4, 57–265

27. SCHLIWA, M., ed. (2003) Molecular Motors, Wiley-VCH, Wennheim

28. VALE, R.D. AND MILLIGAN, R.A. (2000) The way things move: looking under the hood of motor proteins, Science **288**, 88–95

29. VILLANI, C. (2003) Topics in optimal transportation, AMS Graduate Studies in Mathematics vol. 58, Providence

Optimal Control of Ill-Posed Parabolic Distributed Systems

A. Omrane

Summary. We show that the low-regret notion of Lions [C. R. Acad. Sci. Paris Ser. I Math. 315:1253–1257, 1992] is well adapted for the control of the ill-posed heat problem. Passing to the limit, we give a characterization of the no-regret control by a singular optimality system. No Slater hypothesis on the admissible set of controls $\mathcal{U}_{\mathrm{ad}}$ is necessary, since we use a corrector of order zero argument. The result is a generalization of the low-regret control [Dorville et al., Appl. Math. Lett. 17:549–552, 2004; Dorville et al., C. R. Acad. Sci. Paris Ser. I Math. 338:921–924, 2004] to the no-regret control optimality system.

1 Introduction

The question of existence and characterization of the optimal control for singular problems, and thus for ill-posed problems has not been considered in detail, since there is a lack of regularity of the state solution. Moreover, the assumptions on the control problem are mostly chosen in such a way that some standard methods can be applied to derive the existence of solutions. Most of the work is then to derive necessary or sufficient optimality conditions.

We consider in this paper the prototype of ill-posed problems: the ill-posed backward heat problem, with controls v in a non-empty closed convex subset $\mathcal{U}_{\mathrm{ad}}$ of the Hilbert space L^2, and a cost functional.

It is well known that if we add the Slater type hypothesis as: $\mathcal{U}_{\mathrm{ad}}$ *having a non-empty interior*, then we ensure the existence of an optimal control, using standard methods (see [4]). But, we do not know if this hypothesis is really necessary!

One way to deal with this problem – instead of answering to the question – is to propose another approach, where the Slater hypothesis on $\mathcal{U}_{\mathrm{ad}}$ is not needed. Convex cones as $(L^2)^+$ which are of empty interior may be used as set of admissible controls v. We here use the regularization approach in a first part, and the null-controllability approach in a second part. In both methods, a new data is introduced. This data is supposed to be chosen as largely as

possible, say in a vector space. That is the data is *incomplete*, and in this
case, the standard methods as the penalization method are not adapted (see
[6, 9, 10]). We then seek for the low-regret control of the distributed system
of incomplete data obtained. And thus, the no-regret control of the original
problem appears naturally by passage to the limit, which is possible without
the Slater hypothesis.

Roughly speaking, the low-regret control u^γ satisfies to the following inequality:

$$J(u^\gamma, g) \leq J(0, g) + \gamma \|g\|_Y^2 \qquad \forall\, g \text{ in a Hilbert space } Y,$$

where γ is a small positive parameter (g being the pollution or the incomplete
data). With the low-regret control we admit the possibility of making a choice
of controls v 'slightly worse' than by doing better than $v = 0$ – but 'not much'
if we choose γ small enough – compared to the worst things that could happen
with the 'pollution' g.

In the no-regret concept, we search for the control u, if it exists, which
makes things better than $v = 0$, for any given perturbation parameter. It is
the limit when $\gamma \to 0$, of the family of low-regret controls u^γ.

This concept is previously introduced by Savage [11] in statistics. Lions was
the first to use it to control distributed systems of incomplete data, motivated
by a number of applications in economics, and ecology as well (see for instance
[7, 8]).

In [9] (see also [10]), Nakoulima et al. give a precise optimality system
(which is a singular optimality system). In [10], the no-regret control for
problems of incomplete data, in both the stationary and evolution cases is
characterized. A number of applications is given too.

In the literature mentioned above, the regular problems are considered
only. Moreover, the set of controls was a Hilbert space. In this article, we
generalize the study to the control of ill-posed problems, where the controls
are in a closed convex subset of a Hilbert space only. Without loss of generality,
we consider the prototype of ill-posed problems: *the ill-posed backward heat
problem.*

2 Existence Problem

We recall below some basic properties of the most significant parabolic problems: the heat equation and backward heat equation.

2.1 The Heat and Backward Heat Problems

Consider an open domain $\Omega \subset I\!\!R^N$ with regular frontier $\partial\Omega$, and denote by
$Q = (0, T) \times \Omega$, and by $\Sigma = (0, T) \times \partial\Omega$. Then it is well known that the
following heat system:

$$\begin{cases} z' - \Delta z = v & \text{in } Q, \\ \qquad\quad z = 0 & \text{on } \Sigma, \end{cases} \tag{1}$$

and

$$z(0) = 0 \quad \text{in } \Omega, \tag{2}$$

is well-posed. Here, $z = z(t,x)$ is the state solution and $v = v(t,x) \in L^2(Q)$. Moreover, any solution z of (1)–(2) is a.e. continuous from $[0,T]$ to $H^{-1}(\Omega)$ and we have

$$z' \in L^2\left(]0,T[; H^{-2}(\Omega)\right) \quad \text{and} \quad z|_{\Sigma} \in H^{-1}\left(]0,T[; H^{-\frac{1}{2}}(\partial\Omega)\right).$$

But, the above system does not admit a solution for arbitrary data. Indeed, in the case of final data, i.e. replacing (2) by

$$z(T) = 0 \quad \text{in } \Omega, \tag{3}$$

there is no solution for the *backward* heat problem, even for regular control v, as we can see in the following one-dimensional example:

For $\Omega =]0,\pi[$, $T = 1$, consider the backward heat system:

$$\begin{cases} \dfrac{\partial z}{\partial t} - \dfrac{\partial^2 z}{\partial x^2} = v & \text{in }]0,\pi[\times]0,1[, \\ z(0,t) = z(\pi,t) = 0 & \text{in }]0,1[, \\ z(x,1) = 0 & \text{in }]0,\pi[, \end{cases} \tag{4}$$

where $v \in L^2(]0,1[; L^2(]0,\pi[))$ is the uniformly convergent series

$$v(t,x) = \sqrt{\frac{2}{\pi}} \sum_{m \geq 1} \frac{\sin mx}{m^2}.$$

If $z \in L^2(]0,1[; L^2(]0,\pi[))$ is a solution to (4), such that $z(t,x) = \sum_{m \geq 1} z_m(t) w_m(x)$, where $w_m(x) = \sqrt{\dfrac{2}{\pi}} \sin mx$ (w_m is an eigenvector for $-\dfrac{\partial^2}{\partial x^2}$ related to the eigenvalue m^2), then we have

$$\begin{cases} \dfrac{dz_m}{dt}(t) + m^2 z_m(t) = \dfrac{1}{m^2} & \text{in }]0,1[, \\ z_m(1) = 0, \end{cases}$$

so that

$$z_m(t) = \frac{1}{m^2} \int_1^t e^{m^2(s-t)} ds = \frac{1}{m^4}\left(1 - e^{m^2(1-t)}\right).$$

For every $t \in [0,1[$, we then obtain $\|z\|^2_{L^2(]0,\pi[)} = \sum_{m \geq 1} \left| \dfrac{1}{m^4}\left(1 - e^{m^2(1-t)}\right)\right|^2.$

But,

$$\lim_{m \to +\infty} \left| \frac{1}{m^4}\left(1 - e^{m^2(1-t)}\right)\right|^2 = +\infty, \qquad \forall t \in [0,1[.$$

Hence, the series diverges and the solution z of (4) does not exist.

2.2 Existence of a Solution for the Ill-Posed Heat Problem

The prototype of ill-posed problems is the following backward heat problem:

$$\left|\begin{array}{rl} z' - \Delta z &= v \ \ \text{in } Q, \\ z &= 0 \ \ \text{on } \Sigma, \\ z(T) &= 0 \ \ \text{in } \Omega. \end{array}\right. \tag{5}$$

This problem has however a unique solution z in some dense subset of $L^2(Q)$ that we explicit now: Consider the vector space

$$V = \left\{ w = \sum_{i=1}^{N} \lambda_i w_i \ : \ -\Delta w_i = 0, \ w_i = 0 \text{ on } \Gamma, \quad \text{and} \quad w_i \in L^2(\Omega) \right\}. \tag{6}$$

Then, there exist $f \in L^2(]0, T[)$ and $w \in V$ such that

$$v(t, x) = f(t) \left(\sum_{i=1}^{N} w_i(x) \right),$$

for given $v \in L^2(]0, T[) \otimes V$ (which is dense in $L^2(Q)$). It suffices to take z of the form $z(t, x) = \zeta(t)w(x)$, $\zeta = (\zeta_1, ..., \zeta_N)$. So, ζ_i is solution of

$$\begin{cases} \dfrac{\partial \zeta_i}{\partial t} - \lambda \zeta_i = f \ \ \text{in }]0, T[, \\ \zeta_i(0) = 0 \ \ \text{in } \Omega, \end{cases}$$

which defines ζ in a unique manner.

3 Optimal Control of the Backward Heat Problem

3.1 Preliminaries

Consider $v \in \mathcal{U}_{\text{ad}}$, $\mathcal{U}_{\text{ad}}$ a non-empty closed convex subset of the Hilbert space of controls $L^2(Q)$, and the quadratic function

$$J(v, z) = \left\| z - z_d \right\|^2_{L^2(Q)} + N \left\| v \right\|^2_{L^2(Q)}, \tag{7}$$

where $z_d \in L^2(Q)$, $N > 0$, and where $\left\| . \right\|_X$ is the norm on the corresponding Hilbert space X.

If a pair $(v, z) \in \mathcal{U}_{\text{ad}} \times L^2(Q)$ satisfying to (5) exists, then it is called a control-state admissible pair. Denote by $\mathcal{X}_{\text{ad}}$ the set of admissible pairs. We suppose in what follows that $\mathcal{X}_{\text{ad}}$ is non-empty. Then for every $(v, z) \in \mathcal{X}_{\text{ad}}$, we associate the *cost function* defined by (7), and we consider the *optimal control problem*:

$$\inf \ J(v, z), \qquad (v, z) \in \mathcal{X}_{\text{ad}} \tag{8}$$

which admits a unique solution (u, y) that we should characterize.

Lemma 3.1. *The problem (8) admits a unique solution (u, y) called the optimal pair.*

Proof. The functional $J : L^2(Q) \times L^2(Q) \longrightarrow \mathbb{R}$ is a lower semi-continuous function, strictly convex, and coercitive. Hence there is a unique admissible pair (u, y) solution to (8). □

A classical method to control the system (5) and (7) is the well-known penalization method, which consists in approximating (u, y) by a penalized problem. More precisely, for $\varepsilon > 0$ we define the penalized cost function

$$J_\varepsilon(v, z) = J(v, z) + \frac{1}{2\varepsilon}\|z' - \Delta z - v\|_{L^2(Q)}^2.$$

The optimal pair $(u_\varepsilon, y_\varepsilon)$ then converges to (u, y).

The optimality conditions of Euler-Lagrange for $(u_\varepsilon, y_\varepsilon)$ are the following:

$$\frac{d}{dt}J_\varepsilon(u_\varepsilon, y_\varepsilon + t(z - y_\varepsilon))_{|t=0} = 0, \qquad \forall z \in \mathcal{F} \tag{9}$$

and

$$\frac{d}{dt}J_\varepsilon(u_\varepsilon + t(v - u_\varepsilon), y_\varepsilon)_{|t=0} \geq 0 \qquad \forall v \in \mathcal{U}_{ad}, \tag{10}$$

then an optimality system is obtained by the introduction of the *adjoint state*

$$p_\varepsilon = -\frac{1}{\varepsilon}(y_\varepsilon' - \Delta y_\varepsilon - u_\varepsilon).$$

A priori estimates (consisting in bounding p_ε in $L^2(Q)$) have to be obtained, which allows the passage to the limit under some hypothesis: For the problem (5) and (7)–(8), Lions obtained in [4] a singular optimality system, under the supplementary hypothesis of Slater type:

$$\mathcal{U}_{ad} \quad \text{has a non-empty interior.} \tag{11}$$

The following theorem is due to Lions: [3, 4]:

Theorem 3.2. *Under hypothesis (11), there is a unique $(u, y, p) \in \mathcal{U}_{ad} \times L^2(Q) \times L^2(Q)$, solution to the optimal control problem (5) and (7)–(8). Moreover, this solution is characterized by the following singular optimality system (SOS):*

$$\begin{cases} y' - \Delta y = u, \quad -p' - \Delta p = y - z_d \text{ in } Q, \\ \quad y(T) \;\; = 0, \quad p(0) \;\; = 0 \qquad \text{in } \Omega, \\ \quad y \;\;\;\;\; = 0, \quad\;\; p \;\;\;\;\; = 0 \qquad \text{on } \Sigma \end{cases} \tag{12}$$

with the variational inequality:

$$(p + Nu, v - u)_{L^2(Q)} \geq 0 \quad \forall v \in \mathcal{U}_{ad}. \tag{13}$$

Proof. For a proof of this theorem see [3, 4]. □

Remark 3.3. In some applications, the Slater hypothesis (11) is not satisfied. It is the case when $\mathcal{U}_{ad} = \left(L^2(Q)\right)^+$ which has an empty interior. In what follows, we propose another approach which avoids the use of (11).

When $\mathcal{U}_{ad} = L^2(Q)$, the hypothesis (11) is satisfied and the above theorem holds.

4 The Low-Regret Optimal Control

In this section, an elliptic regularization of the ill-posed parabolic problem (5). We obtain a well-posed problem but with a new *unknown* data. We then let the classical control notion away, to consider the one of no-regret and low-regret control as introduced by Lions [6], and recently developed by Nakoulima et al. in [9, 10].

4.1 The Regularization Approach

For any $\varepsilon > 0$, we consider the regularized problem:

$$\left| \begin{array}{rll} z'_\varepsilon - \varepsilon z''_\varepsilon - \Delta z_\varepsilon & = v \text{ in } & Q, \\ z_\varepsilon(0) & = g \text{ in } & \Omega, \\ z_\varepsilon(T) & = 0 \text{ in } & \Omega, \\ z_\varepsilon & = 0 \text{ on } & \Sigma, \end{array} \right. \tag{14}$$

where $g \in L^2(\Omega)$.

It is clear that for any $\varepsilon > 0$, and any given (v, g), there is a unique state solution $z_\varepsilon = z_\varepsilon(v, g)$ of (14) for which we associate a cost function given by

$$J_\varepsilon(v, g) = \left\| z_\varepsilon(v, g) - z_d \right\|^2_{L^2(Q)} + N \|v\|^2_{L^2(Q)}, \qquad g \in L^2(\Omega). \tag{15}$$

We are concerned with the optimal control of the problem (14)–(15). Clearly we want

$$\inf_{v \in \mathcal{U}_{ad}} J_\varepsilon(v, g) \qquad \forall g \in L^2(\Omega).$$

The above minimization problem has no sense since $L^2(\Omega)$ is infinite! One natural idea is to consider the following minimization problem:

$$\inf_{v \in \mathcal{U}_{ad}} \left(\sup_{g \in L^2(\Omega)} J_\varepsilon(v, g) \right),$$

but J_ε is not upper bounded since $\sup_{g \in L^2(\Omega)} J_\varepsilon(v, g) = +\infty$. The idea of Lions is then to look for controls v – if they exist – such that

$$J_\varepsilon(v, g) \le J_\varepsilon(0, g) \qquad \forall g \in L^2(\Omega),$$

and thus

$$J_\varepsilon(v,g) - J_\varepsilon(0,g) \leq 0 \qquad \forall g \in L^2(\Omega).$$

Those controls *doing better than* $v = 0$ for every pollution g are called no-regret controls. We have precisely the

Definition 4.1. *We say that* $u \in \mathcal{U}_{\mathrm{ad}}$ *is a no-regret control for* (14)–(15) *if* u *is a solution to the following problem:*

$$\inf_{v \in \mathcal{U}_{\mathrm{ad}}} \left(\sup_{g \in L^2(\Omega)} (J_\varepsilon(v,g) - J_\varepsilon(0,g)) \right). \tag{16}$$

Lemma 4.2. *For any* $v \in \mathcal{U}_{\mathrm{ad}}$ *we have*

$$J_\varepsilon(v,g) - J_\varepsilon(0,g) = J_\varepsilon(v,0) - J_\varepsilon(0,0) + 2\langle \xi_\varepsilon{}'(0), g \rangle_{L^2(\Omega)} \qquad \forall g \in L^2(\Omega), \tag{17}$$

where ξ_ε *satisfies to:*

$$-\xi_\varepsilon' - \varepsilon\xi_\varepsilon'' - \Delta\xi_\varepsilon = y_\varepsilon(v,0) \ in \ Q, \qquad \xi_\varepsilon(0) = \xi_\varepsilon(T) = 0 \ in \ \Omega, \qquad \xi_\varepsilon = 0 \ on \ \Sigma. \tag{18}$$

Proof. A simple calculus gives

$$J_\varepsilon(v,g) - J_\varepsilon(0,g) = J_\varepsilon(v,0) - J_\varepsilon(0,0) + 2\langle z_\varepsilon(v,0) ; z_\varepsilon(0,g) \rangle_{L^2(Q)}.$$

Using the Green formula we find

$$\langle z_\varepsilon(v,0) ; z_\varepsilon(0,g) \rangle = \varepsilon \langle \xi_\varepsilon{}'(0) ; g \rangle_{L^2(Q)},$$

where ξ_ε is given by (18).   □

Remark 4.3. Of course the problem (16) is defined only for the controls $v \in \mathcal{U}_{\mathrm{ad}}$ such that

$$\sup_{g \in L^2(\Omega)} (J_\varepsilon(v,g) - J_\varepsilon(0,g)) < \infty.$$

From (17) this is realized iff $v \in K$, where $K = \{w \in \mathcal{U}_{\mathrm{ad}}, \langle \xi_\varepsilon(w), g \rangle = 0 \ \forall g \in L^2(\Omega)\}$. This set is difficult to characterize. As in [9], for any $\gamma > 0$, we introduce then the low-regret control.

Definition 4.4. *The low-regret control for* (14)–(15), *is the solution to the following perturbed system:*

$$\inf_{v \in \mathcal{U}_{\mathrm{ad}}} \left(\sup_{g \in L^2(\Omega)} \left(J_\varepsilon(v,g) - J_\varepsilon(0,g) - \gamma\|g\|^2_{L^2(\Omega)} \right) \right). \tag{19}$$

We notice the following :

$$\inf_{v \in \mathcal{U}_{\mathrm{ad}}} \left(\sup_{g \in L^2(\Omega)} \left(J_\varepsilon(v,g) - J_\varepsilon(0,g) - \gamma\|g\|^2_{L^2(\Omega)} \right) \right)$$

$$= \inf_{v \in \mathcal{U}_{\mathrm{ad}}} \left(J_\varepsilon(v,0) - J_\varepsilon(0,0) + \left(\sup_{g \in L^2(\Omega)} 2\langle z_\varepsilon(v,0) ; z_\varepsilon(0,g) \rangle_{L^2(Q)} - \gamma\|g\|^2_{L^2(\Omega)} \right) \right).$$

Thanks to the conjugate, we obtain the classical control problem:

$$\inf_{v \in \mathcal{U}_{\text{ad}}} \mathcal{J}_\varepsilon^\gamma(v), \tag{20}$$

where

$$\mathcal{J}_\varepsilon^\gamma(v) = J_\varepsilon(v,0) - J_\varepsilon(0,0) + \frac{\varepsilon^2}{\gamma} \left\| \xi_\varepsilon'(T,v) \right\|_{L^2(\Omega)}^2, \tag{21}$$

and where ξ_ε satisfies to (18).

Remark 4.5. Then as we can see, the low-regret control method allows us to transform systematically a problem with uncertainty to a standard control problem. Hence, we can use the Euler–Lagrange method.

We can replace now (19) by (20) and (21) for the low-regret control.

Lemma 4.6. *The problem (14) and (20)–(21) has a unique solution u_ε^γ, called the 'approximate' low-regret control.*

Proof. We have $\mathcal{J}_\varepsilon^\gamma(v) \geq -J_\varepsilon(0,0) = -\|z_d\|_{L^2(\Omega)}^2 \ \forall v \in \mathcal{U}_{\text{ad}}$. Then $d = \inf_{v \in \mathcal{U}_{\text{ad}}} \mathcal{J}_\varepsilon^\gamma(v)$ exists. Let v_n be a minimizing sequence such that $d = \lim_{n \to \infty} \mathcal{J}_\varepsilon^\gamma(v_n)$. We have

$$-\|z_d\|_{L^2(\Omega)}^2 \leq \mathcal{J}_\varepsilon^\gamma(v_n) = J_\varepsilon(v_n,0) - J_\varepsilon(0,0) + \frac{1}{\gamma} \left\| \xi_\varepsilon'(0) \right\|_{L^2(\Omega)}^2 \leq d_\gamma + 1.$$

Then we deduce the bounds

$$\left\| v_n \right\|_{L^2(Q)} \leq c, \qquad \frac{1}{\sqrt{\gamma}} \left\| \xi_\varepsilon'(v_n)(0) \right\|_{L^2(\Omega)} \leq c, \qquad \left\| y_\varepsilon(v_n,0) - z_d \right\|_{L^2(Q)} \leq c,$$

where the constant c is independent of n.

There exists $u_\varepsilon^\gamma \in \mathcal{U}_{\text{ad}}$ such that $v_n \rightharpoonup u_\varepsilon^\gamma$ weakly in $\mathcal{U}_{\text{ad}}$ (which is closed). Also, $y_\varepsilon(v_n,0) \to y_\varepsilon(u_\varepsilon^\gamma,0)$ (continuity w.r.t the data). We also deduce from the strict convexity of the cost function $\mathcal{J}_\varepsilon^\gamma$ that u_ε^γ is unique. $\quad\square$

Proposition 4.7. *The 'approximate' low-regret control u_ε^γ is characterized by the unique quadruplet $\{u_\varepsilon^\gamma, y_\varepsilon^\gamma, \rho_\varepsilon^\gamma, p_\varepsilon^\gamma\}$, solution to the system:*

$$\begin{vmatrix} y_\varepsilon^{\gamma\prime} - \varepsilon y_\varepsilon^{\gamma\prime\prime} - \Delta y_\varepsilon^\gamma = u_\varepsilon^\gamma, \quad \rho_\varepsilon^{\gamma\prime} - \varepsilon\rho_\varepsilon^{\gamma\prime\prime} - \Delta\rho_\varepsilon^\gamma = 0, \ \ \text{and} \\[4pt] -p_\varepsilon^{\gamma\prime} - \varepsilon p_\varepsilon^{\gamma\prime\prime} - \Delta p_\varepsilon^\gamma = y_\varepsilon^\gamma - z_d + \rho_\varepsilon^\gamma \quad \text{in } Q, \\[10pt] y_\varepsilon^\gamma(0) = y_\varepsilon^\gamma(T) = 0, \quad \rho_\varepsilon^\gamma(T) = 0, \quad \rho_\varepsilon^\gamma(0) = \frac{\varepsilon}{\gamma}\xi_\varepsilon^{\gamma\prime}(0), \\[4pt] p_\varepsilon^\gamma(0) = p_\varepsilon^\gamma(T) = 0 \quad \text{in } \Omega, \\[10pt] y_\varepsilon^\gamma = 0, \quad \rho_\varepsilon^\gamma = 0, \quad p_\varepsilon^\gamma = 0 \quad \text{on } \Sigma, \end{vmatrix}$$

with (18), and the variational inequality:

$$\langle p_\varepsilon^\gamma + N u_\varepsilon^\gamma, v - u_\varepsilon^\gamma \rangle \geq 0, \qquad \forall v \in \mathcal{U}_{\text{ad}}.$$

Proof. The Euler condition of first order to (20) and (21) gives:

$$\langle y_\varepsilon^\gamma - z_d, y_\varepsilon(w,0)\rangle_{L^2(Q)\times L^2(Q)} + N\langle u_\varepsilon^\gamma, w\rangle_{L^2(Q)\times L^2(Q)}$$
$$+ \langle \frac{\varepsilon^2}{\gamma}\xi_\varepsilon^{\gamma\prime}(0), \xi_\varepsilon{}'(0,w)\rangle_{L^2(\Omega)\times L^2(\Omega)} \geq 0,$$

where $y_\varepsilon^\gamma = y_\varepsilon(u_\varepsilon^\gamma, 0)$, and $\xi_\varepsilon^\gamma = \xi_\varepsilon(u_\varepsilon^\gamma, 0)$. We then introduce $\rho_\varepsilon^\gamma = \rho_\varepsilon(u_\varepsilon^\gamma, 0)$ solution to $\rho_\varepsilon^{\gamma\prime} - \varepsilon\rho_\varepsilon^{\gamma\prime\prime} - \Delta\rho_\varepsilon^\gamma = 0$, $\rho_\varepsilon^\gamma(0) = (\varepsilon/\gamma)\xi_\varepsilon^{\gamma\prime}(T)$, $\rho_\varepsilon^\gamma(T) = 0$, and $\rho_\varepsilon^\gamma = 0$ on Σ, such that:

$$\left\langle \frac{\varepsilon^2}{\gamma}\xi_\varepsilon^{\gamma\prime}(0), \xi_\varepsilon{}'(0,w) \right\rangle_{L^2(\Omega)\times L^2(\Omega)} = \langle \varepsilon\rho_\varepsilon^\gamma(0), \xi_\varepsilon{}'(0,w)\rangle_{L^2(\Omega)\times L^2(\Omega)}$$
$$= \langle \rho_\varepsilon^\gamma, y_\varepsilon(w,0)\rangle_{L^2(Q)\times L^2(Q)}$$

using the Green formula.

Introduce now the adjoint state $p_\varepsilon^\gamma = p_\varepsilon(u_\varepsilon^\gamma, 0)$ as follows : we solve $-p_\varepsilon^{\gamma\prime} - \varepsilon p_\varepsilon^{\gamma\prime\prime} - \Delta p_\varepsilon^\gamma = y_\varepsilon^\gamma - z_d + \rho_\varepsilon^\gamma$, $p_\varepsilon^\gamma(T) = p_\varepsilon^\gamma(0) = 0$, and $p_\varepsilon^\gamma = 0$ on Σ. Hence we have

$$\langle y_\varepsilon^\gamma - z_d + \rho_\varepsilon^\gamma, y_\varepsilon(w,0)\rangle_{L^2(Q)\times L^2(Q)} = \langle p_\varepsilon^\gamma, w\rangle_{L^2(Q)\times L^2(Q)}.$$

Finally,

$$\langle p_\varepsilon^\gamma + Nu_\varepsilon^\gamma, w\rangle_{L^2(Q)\times L^2(Q)} \geq 0. \qquad \square$$

4.2 A Priori Estimates

In this section we give the S.O.S for the low-regret control of the backward heat equation. We first show the following estimates:

Proposition 4.8. *There is a positive constant C, and, for any small $\eta > 0$, there is a constant $C_\eta > 0$ such that:*

$$\|u_\varepsilon^\gamma\|_{L^2(Q)} \leq C, \qquad \|y_\varepsilon^\gamma\|_{L^2(Q)} \leq C, \qquad \frac{\varepsilon}{\sqrt{\gamma}}\|\xi_\varepsilon^{\gamma\prime}(0)\|_{L^2(\Omega)} \leq C, \quad (22)$$

$$\varepsilon\|y_\varepsilon^{\gamma\prime}\|_{L^2(Q)} + \|y_\varepsilon^\gamma\|_{L^2(Q)} \leq C, \qquad \|y_\varepsilon^{\gamma\prime}\|_{L^2(]0,T-\eta[;H^{-1}(\Omega))} \leq C_\eta, \quad (23)$$

and,

$$\varepsilon\|\xi_\varepsilon^{\gamma\prime}\|_{L^2(Q)} + \|\xi_\varepsilon^\gamma\|_{L^2(Q)} \leq C, \qquad \|\xi_\varepsilon^{\gamma\prime}\|_{L^2(]\eta,T[;H^{-1}(\Omega))} \leq C_\eta, \quad (24)$$

$$\varepsilon\|\rho_\varepsilon^{\gamma\prime}\|_{L^2(Q)} + \|\rho_\varepsilon^\gamma\|_{L^2(Q)} \leq C, \qquad \|\rho_\varepsilon^{\gamma\prime}\|_{L^2(]0,T-\eta[;H^{-1}(\Omega))} \leq C_\eta, \quad (25)$$

$$\varepsilon\|p_\varepsilon^{\gamma\prime}\|_{L^2(Q)} + \|p_\varepsilon^\gamma\|_{L^2(Q)} \leq C, \qquad \|p_\varepsilon^{\gamma\prime}\|_{L^2(]\eta,T[;H^{-1}(\Omega))} \leq C_\eta. \quad (26)$$

Proof. We know that

$$J_\varepsilon^\gamma(u_\varepsilon^\gamma) \leq J_\varepsilon^\gamma(v) \qquad \forall v \in \mathcal{U}_{\mathrm{ad}}.$$

We then have for the particular case $v = 0$,

$$J_\varepsilon(u_\varepsilon^\gamma, 0) - J_\varepsilon(0,0) + \frac{\varepsilon^2}{\gamma} \left\| \xi_\varepsilon'(u_\varepsilon^\gamma)(0) \right\|_{L^2(\Omega)}^2 \leq \frac{\varepsilon^2}{\gamma} \left\| \xi_\varepsilon'(u_\varepsilon^\gamma)(0) \right\|_{L^2(\Omega)}^2 .$$

But $y_\varepsilon(0,0)(t,x) = \xi_\varepsilon(0)(t,x) = 0$ in $[0,T] \times \overline{\Omega}$, hence:

$$\left\| y_\varepsilon(u_\varepsilon^\gamma, 0) - z_d \right\|_{L^2(Q)}^2 + N \left\| u_\varepsilon^\gamma \right\|_{L^2(Q)}^2 + \left\| \frac{\varepsilon}{\sqrt{\gamma}} \xi_\varepsilon'(u_\varepsilon^\gamma)(0) \right\|_{L^2(\Omega)}^2$$

$$\leq \left\| z_d \right\|_{L^2(Q)}^2 = \text{constant}, \tag{27}$$

so we have (22).

Now, from the Poincaré formula, there exists a constant $C_1 > 0$ such that:

$$\left(\varepsilon \left\| \frac{\partial y_\varepsilon^\gamma}{\partial t} \right\|_{L^2(Q)} + \| y_\varepsilon^\gamma \|_{L^2(Q)} \right) \| y_\varepsilon^\gamma \|_{L^2(Q)} \leq C_1 \| u_\varepsilon^\gamma \|_{L^2(Q)} \| y_\varepsilon^\gamma \|_{L^2(Q)},$$

thus, there is another constant $C_2 > 0$ such that

$$\varepsilon \left\| \frac{\partial y_\varepsilon^\gamma}{\partial t} \right\|_{L^2(Q)} + \| y_\varepsilon^\gamma \|_{L^2(Q)} \leq C_2,$$

that is the first part of (23). Now, we start from $-\varepsilon \dfrac{\partial^2 y_\varepsilon}{\partial t^2} + \dfrac{\partial y_\varepsilon}{\partial t} = v + \Delta y_\varepsilon$.
Denote by $g_\varepsilon = v + \Delta y_\varepsilon$, then g_ε remains in a bounded subset of $L^2(Q)$.

Introduce a function $\varphi \in C^1([0,T])$ such that $\varphi(0) = 1$ and $\varphi(T) = 0$, then we multiply

$$-\varepsilon \frac{\partial^2 y_\varepsilon}{\partial t^2} + \frac{\partial y_\varepsilon}{\partial t} = g_\varepsilon$$

by $\varphi \dfrac{\partial y_\varepsilon}{\partial t}$ and we integrate over Q. We have

$$-\frac{\varepsilon}{2} \int_0^T \varphi \frac{d}{dt} \left\| \frac{\partial y_\varepsilon}{\partial t} \right\|_{L^2(\Omega)}^2 dt + \int_0^T \varphi \left\| \frac{\partial y_\varepsilon}{\partial t} \right\|_{L^2(\Omega)}^2 dt = \int_0^T \varphi \left(g_\varepsilon, \frac{\partial y_\varepsilon}{\partial t} \right)_{L^2(\Omega)} dt. \tag{28}$$

And,

$$\int_0^T \varphi \frac{d}{dt} \left\| \frac{\partial y_\varepsilon}{\partial t} \right\|_{L^2(\Omega)}^2 dt = - \left\| \frac{\partial y_\varepsilon}{\partial t}(0) \right\|_{L^2(\Omega)}^2 - \int_0^T \frac{d\varphi}{dt} \left\| \frac{\partial y_\varepsilon}{\partial t} \right\|_{L^2(\Omega)}^2 dt,$$

thus

$$\frac{\varepsilon}{2} \left\| \frac{\partial y_\varepsilon}{\partial t}(0) \right\|_{L^2(\Omega)}^2 + \frac{\varepsilon}{2} \int_0^T \frac{d\varphi}{dt} \left\| \frac{\partial y_\varepsilon}{\partial t} \right\|_{L^2(\Omega)}^2 dt + \int_0^T \varphi \left\| \frac{\partial y_\varepsilon}{\partial t} \right\|_{L^2(\Omega)}^2 dt$$

$$= \int_0^T \varphi \left(g_\varepsilon, \frac{\partial y_\varepsilon}{\partial t} \right)_{L^2(\Omega)} dt.$$

The second term is of $O(1)$, so

$$\int_0^T \varphi \left\| \frac{\partial y_\varepsilon}{\partial t} \right\|_{L^2(\Omega)}^2 dt = \int_0^T \varphi \left(g_\varepsilon, \frac{\partial y_\varepsilon}{\partial t} \right)_{L^2(\Omega)} dt + O(1).$$

We finally deduce (23) from the triangular inequality.

The estimates (24), (25) and (26) follow easily.   □

Theorem 4.9. *The low-regret control $u^\gamma = \lim\limits_{\varepsilon \to 0} u_\varepsilon^\gamma$ for the backwards heat equation (5) is characterized by the unique $\{u^\gamma, y^\gamma, \xi^\gamma, \rho^\gamma, p^\gamma\}$, solution to the system:*

$$\left| \begin{array}{l} y^{\gamma\prime} - \Delta y^\gamma = u^\gamma, \quad -\xi^{\gamma\prime} - \Delta \xi^\gamma = y^\gamma, \quad \rho^{\gamma\prime} - \Delta \rho^\gamma = 0, \quad and \\ -p^{\gamma\prime} - \Delta p^\gamma = y^\gamma - z_d + \rho^\gamma \quad in \ \ Q, \\[2mm] y^\gamma(0) = 0, \quad \xi^\gamma(T) = 0, \quad and \\ \rho^\gamma(0) = \lambda^\gamma(0), \quad p^\gamma(T) = 0 \quad in \ \Omega, \\[2mm] y^\gamma = 0, \quad \xi^\gamma = 0, \quad \rho^\gamma = 0, \quad p^\gamma = 0 \quad on \ \Sigma, \end{array} \right.$$

with the following weak limits

$$y^\gamma = \lim_{\varepsilon \to 0} y_\varepsilon^\gamma, \quad \xi^\gamma = \lim_{\varepsilon \to 0} \xi_\varepsilon^\gamma, \quad \rho^\gamma = \lim_{\varepsilon \to 0} \rho_\varepsilon^\gamma, \quad p^\gamma = \lim_{\varepsilon \to 0} p_\varepsilon^\gamma,$$

and the variational inequality:

$$\langle p^\gamma + Nu^\gamma, v - u^\gamma \rangle \geq 0 \qquad \forall v \in \mathcal{U}_{\mathrm{ad}},$$

where

$$u^\gamma, \ y^\gamma, \ p^\gamma, \ \rho^\gamma, \ \xi^\gamma \ \in \ L^2(]0, T[; L^2(\Omega)), \quad \lambda^\gamma(0) \in L^2(\Omega).$$

Proof. We use the estimates of proposition 4.8. From (27), we deduce the following limits:

$$\begin{array}{lll} u_\varepsilon^\gamma & \rightharpoonup u^\gamma & \text{weakly in } \mathcal{U}_{\mathrm{ad}}, \\ y_\varepsilon^\gamma & \rightharpoonup y^\gamma & \text{weakly in } L^2(Q), \\ \dfrac{\varepsilon}{\sqrt{\gamma}} \dfrac{\partial \xi_\varepsilon(u_\varepsilon^\gamma)}{\partial t}(0) & \rightharpoonup \lambda^\gamma(0) & \text{weakly in } L^2(\Omega), \end{array} \qquad (29)$$

(up to extract a subsequences (u_ε^γ), (y_ε^γ) and $\left(\dfrac{\varepsilon}{\sqrt{\gamma}} \dfrac{\partial \xi_\varepsilon(u_\varepsilon^\gamma)}{\partial t}(T) \right)$).

For every fixed $\gamma > 0$, the adjoint state p_ε^γ is also bounded in ε (from (26)).   □

5 The Now-Regret Optimal Control

For the no-regret optimal control to the original problem, we now introduce the notion of corrector of order 0 of Lions [5] for elliptic regularizations, instead of using the Slater hypothesis (11).

5.1 Corrector of Order Zero

It is well known, the passage to the limit gives no information on $y^\gamma(T)$.
Moreover, $y^\gamma(T) \neq 0$ in general.

We make the following hypothesis:

$$y^\gamma(T) \in H_0^1(\Omega), \ y^{\gamma\prime} \in L^2(Q). \tag{30}$$

Denote by

$$V = \{\varphi \in L^2(]0, T[; H_0^1(\Omega)) \text{ such that } \varphi' \in L^2(Q)\},$$

and by

$$V_0 = \{\varphi \in V \text{ such that } \varphi(0) = 0, \ \varphi(T) = 0\}.$$

Definition 5.1. *We say that a function* $\theta_\varepsilon^\gamma \in V$ *is a corrector of order 0 iff*

$$\left| \begin{array}{ll} \varepsilon(\theta_\varepsilon^{\gamma\prime}, \varphi')_{L^2(Q)} + (\theta_\varepsilon^{\gamma\prime}, \varphi)_{L^2(Q)} + (\nabla\theta_\varepsilon^\gamma, \nabla\varphi)_{L^2(Q)} = \sqrt{\varepsilon}\,(f_\varepsilon, \varphi)_{L^2(Q)} \ \forall V_0, \\[2mm] \theta_\varepsilon^\gamma(T) + y^\gamma(T) \qquad\qquad\qquad\qquad\qquad = 0, \end{array} \right.$$

$$\tag{31}$$

where we suppose that

$$\|f_\varepsilon\|_{L^2(]0,T[;H^{-1}(\Omega))} \leq C. \tag{32}$$

5.1.1 Calculus of a Corrector or Order 0

We recall how to calculate a corrector of order 0 :

We define $\varphi_\varepsilon^\gamma$ by writing

$$\left\{ \begin{array}{rl} -\varepsilon\varphi_\varepsilon^{\gamma\prime\prime} + \varphi_\varepsilon^{\gamma\prime} = & 0, \\ \varphi_\varepsilon^\gamma(T) = & -y^\gamma(T), \end{array} \right.$$

$\varphi_\varepsilon^\gamma$ decreasing rapidly when $t \to -\infty$,

then

$$\theta_\varepsilon^\gamma(t) = -y^\gamma(T)e^{-\frac{T-t}{\varepsilon}}. \tag{33}$$

If we suppose that $y^\gamma(T) \in H_0^1(\Omega)$, the function

$$\theta_\varepsilon^\gamma = m\varphi_\varepsilon^\gamma \quad \left| \begin{array}{ll} m = 1 & \text{in the neighbourhood of} \quad t = T, \\ m = 0 & \text{in the neighbourhood of} \quad t = 0 \end{array} \right.$$

is a corrector of order 0.

We then satisfy the variational equation, the main term being

$$m \, \Delta \, y^\gamma(T) \, e^{-\frac{T-t}{\varepsilon}} = \sqrt{\varepsilon}\, h_\varepsilon^\gamma.$$

Hence, under the above hypothesis we have:

$$\int_0^T \|h_\varepsilon^\gamma\|_{H^{-1}(\Omega)}^2 \, dt \leq C\,\varepsilon^{-1} \int_0^T e^{-\frac{2(T-t)}{\varepsilon}} \, dt = 0(1).$$

We then have the theorem:

Theorem 5.2. *Let be $\theta_\varepsilon^\gamma$ a corrector of order 0 defined by (31) and (32). We then have*

$$\|y_\varepsilon^\gamma - (y^\gamma + \theta_\varepsilon^\gamma)\|_{L^2(]0,T[;H_0^1)} \leq C\sqrt{\varepsilon}. \tag{34}$$

Moreover,

$$\frac{d}{dt}[y_\varepsilon^\gamma - (y^\gamma + \theta_\varepsilon^\gamma)] \rightharpoonup 0 \qquad \text{weakly in} \quad L^2(Q), \tag{35}$$

when ε tends to 0.

Proof. If we put $w_\varepsilon^\gamma = y_\varepsilon^\gamma - (y^\gamma + \theta_\varepsilon^\gamma)$, then

$$\varepsilon(w_\varepsilon^{\gamma\prime}, \varphi')_{L^2(Q)} + (w_\varepsilon^{\gamma\prime}, \varphi)_{L^2(Q)} + (\nabla w_\varepsilon^\gamma, \nabla\varphi)_{L^2(Q)}$$
$$= -\varepsilon(y^{\gamma\prime}, \varphi')_{L^2(Q)} - \sqrt{\varepsilon}(f_\varepsilon, \varphi)_{L^2(Q)} \quad \forall \mathcal{V}_0. \tag{36}$$

Particularly, if $\varphi = w_\varepsilon^\gamma$, then

$$\varepsilon\|w_\varepsilon^{\gamma\prime}\|_{L^2(Q)}^2 + \|w_\varepsilon^\gamma\|_{L^2(Q)}^2 \leq C\sqrt{\varepsilon}\left[\sqrt{\varepsilon}\|w_\varepsilon^{\gamma\prime}\|_{L^2(Q)} + \|w_\varepsilon^\gamma\|_{L^2(Q)}\right].$$

Thus the inequalities (34) and (35) hold.   □

5.2 Passage to the Limit

We use the regularity properties of the heat equation, the *well posed* one, as follows:

First, we notice that $\|y_\gamma\|_{L^2(Q)} \leq C$ by the above proposition. It remains to see that ξ_γ (resp. p_γ) formally satisfies to the well-posed system:

$$(*) \begin{vmatrix} -\xi^{\gamma\prime} - \Delta\xi^\gamma = y^\gamma \in L^2(Q), \\ \xi^\gamma = 0, \\ \xi^\gamma(T) = 0, \end{vmatrix} \qquad \left(resp. \ (**) \begin{vmatrix} -p^{\gamma\prime} - \Delta p^\gamma = y^\gamma - z_d, \\ p^\gamma = 0 \quad \text{on } \Sigma, \\ p^\gamma(T) = 0, \end{vmatrix} \right)$$

with the mean of a zero corrector. But $(*)$ implies that

$$\|\xi^\gamma\|_{L^2(0,T;\,H_0^1(\Omega))} + \|\xi^{\gamma\prime}\|_{L^2(0,T;\,H^{-1}(\Omega))} \leq C,$$

(resp. $(**)$ gives $\|p^\gamma\|_{L^2(0,T;\,H_0^1(\Omega))} + \|p^{\gamma\prime}\|_{L^2(0,T;\,H^{-1}(\Omega))} \leq C$.)

Then $\xi^\gamma \rightharpoonup \xi$ (resp. $p^\gamma \rightharpoonup p$) weakly in $L^2(0,T;\,H_0^1(\Omega))$, and by compactness $\xi^\gamma \to \xi$ (resp. $p^\gamma \to p$) stronly in $L^2(0,T;\,L^2(\Omega))$.

Also,

$$\begin{vmatrix} \rho^{\gamma\prime} - \Delta\rho^\gamma = 0, \\ \rho^\gamma = 0, \\ \rho^\gamma(0) = \lambda^\gamma(0), \end{vmatrix}$$

implies that $\rho^\gamma \to \rho$ strong in $L^2(0,T;\,L^2(\Omega))$ by the same arguments, because

$$\|\lambda^\gamma(0)\|_{L^2(\Omega)} \leq \lim_{\varepsilon\to 0} \frac{\varepsilon}{\sqrt{\gamma}}\|\xi_\varepsilon{'}(u_\varepsilon^\gamma)(0)\|_{L^2(\Omega)} \leq C, \quad \text{then} \quad \lambda^\gamma(0) \rightharpoonup \lambda(0) \in L^2(\Omega).$$

We then can announce the theorem:

Theorem 5.3. *The no-regret control u for the backward heat ill-posed problem (5), is characterized by the unique quadruplet $\{u, \xi, \rho, p\}$ solution to the optimality system:*

$$\left| \begin{array}{l} y' - \Delta y = u, \quad -\xi' - \Delta \xi = y, \quad \rho' - \Delta \rho = 0, \\ \text{and} \quad -p' - \Delta p = y - z_d + \rho \quad \text{in} \ Q, \\[2mm] y(0) = 0, \quad \xi(T) = 0, \quad \text{and} \\ \rho(0) = \lambda(0), \quad p(T) = 0 \quad \text{in} \ \Omega, \\[2mm] y = 0, \quad \xi = 0, \quad \rho = 0, \quad p = 0 \quad \text{on} \ \Sigma, \end{array} \right.$$

and the variational inequality:

$$\langle p + Nu, v - u \rangle \geq 0 \quad \forall v \in \mathcal{U}_{\text{ad}},$$

with $u \in \mathcal{U}_{\text{ad}}$, $y \in L^2(0, T; L^2(\Omega))$ and

$$p, \ \rho, \ \xi \ \in \ L^2(0, T; H^2(\Omega) \cap H_0^1(\Omega)), \quad \lambda \in \ L^2(\Omega).$$

Remark 5.4. As we have seen in this work, the hypothesis (11) is replaced by the no-regret notion. This method gives another point of view of solving the control problem of singular distributed systems.

Acknowledgement

The author is thankful for Professor D. Konaté for his many suggestions, especially for his helpful remarks concerning Sect. 5.

References

1. Dorville R., Nakoulima O., Omrane A. (2004) *Low-regret control for singular distributed systems: The backwards heat ill-posed problem.* Appl. Math. Lett., Vol. 17, No. 5, pp. 549–552.
2. Dorville R., Nakoulima O., Omrane A. (2004) *Contrôle optimal pour les problèmes de contrôlabilité de systèmes distribués à données manquantes.* C. R. Acad. Sci. Paris Ser. I Math., Vol. 338, pp. 921–924.
3. Lions J. L. (1969) *Contrôle optimal des systèmes gouvernés par des équations aux dérivées partielles.* Dunod, Paris.
4. Lions J. L. (1983) *Contrôle optimal pour les systèmes distribués singuliers.* Gauthiers-Villard, Paris.
5. Lions J. L. (1973) *Perturbations singulières dans les problèmes aux limites et en contrôle optimal.* Lecture Notes in Mathematics, Springer, Berlin Heidelberg, New York

6. Lions J. L. (1992) *Contrôle à moindres regrets des systèmes distribués.* C. R. Acad. Sci. Paris Ser. I Math., Vol. 315, pp. 1253–1257.

7. Lions J. L. (1994) *No-regret and low-regret control,* Environment, Economics and Their Mathematical Models, Masson, Paris.

8. Lions J. L. (1999) *Duality Arguments for Multi Agents Least-Regret Control.* Collège de France, Paris.

9. Nakoulima O., Omrane A., Velin J. (2000) *Perturbations à moindres regrets dans les systèmes distribués à données manquantes.* C. R. Acad. Sci. Paris Ser. I Math., Vol. 330, pp. 801–806.

10. Nakoulima O., Omrane A., Velin J. (2003) *On the pareto control and no-regret control for distributed systems with incomplete data.* SIAM J. Control Optim. Vol. 42, No. 4, pp. 1167–1184.

11. Savage L. J. (1972) *The Foundations of Statistics.* (2nd edition). Dover, New York.

A Parametric Study of Low Reynolds Number Blood Flow in a Porous, Slowly Varying, Stenotic Artery with Heat Transfer

A. Ogulu

Summary. Blood vessels are modelled as porous media in this study. Asymptotic series expansions about a small parameter, ε, is employed to obtain the axial velocity and temperature distributions from which shear stresses at the tube wall and rates of heat transfer are evaluated. Results obtained, which compare favourably with previously reported studies, show that the shear stress increases as the distance, while the rate of heat transfer increases initially, attains a peak value before dropping to equilibrium value.

1 Introduction

The flow of blood in the cardiovascular system of humans and animals has been studied by quite a few workers, see for instance, Bestman [1], Latinopoulos and Ganoulis [2], Misra and Chakravarty [3], Haldar [4], Misra and Chauhan [5], Haldar and Ghosh [6], Cavaleanti [7], Hung and Tsai [8] and Tay and Ogulu [9]. The flow of a viscous fluid in a tube of slowly varying section is of fundamental importance with obvious applications in physiology and physiological fluid dynamics. Some of the more recent works on blood flow include Vajravelu *et al* [10], Filipovic and Kojic [11], Ogulu and Abbey [12], Mandal [13] and Ogulu [14].

In this study we consider the flow of a viscous fluid, which blood is, in a tube of slowly varying section to simulate stenosis of the blood vessel. Studies of this type form an important basis for the early diagnosis and treatment of heart diseases arising from blockages in the circulatory system. Manton [15] considered the steady flow in axis-symmetric rigid tubes of slowly varying radius as a perturbation to lubrication theory.

Ogulu and Bestman [16, 17] proposed a mathematical model for blood flow during deep heat muscle treatment. This study is an extension of Ogulu and Bestman [16, 17]. Here we are concerned with the stresses on the blood vessel arising from heat transfer to the blood vessel regarded as a porous medium with varying cross-section.

2 The Problem

The problem is formulated thus, since we are interested only in flow in one direction (the axial direction) so we consider swell-free blood flow in cylindrical polar coordinate system (r', z') with velocity component w' such that $r' = 0$ is the axis of symmetry of the tube where we regard the vessel as a porous medium and we take into account heat transfer. We know that the wall of our blood vessel is not rigid but distends and contracts so we define the vessel wall as

$$r' = a_0 s \left(\frac{\varepsilon z'}{a_0} \right) \tag{1}$$

In (1) ε is a small parameter and a_0 is a constant, say the radius of the tube. The modified Navier–Stokes equation governing for the flow which we propose here are:

$$\rho \left(\frac{\partial w'}{\partial t'} \right) + \rho w' \frac{\partial w'}{\partial z'} = -\frac{\partial p'}{\partial z'} + \mu \left(\frac{1}{r'} \frac{\partial}{\partial r'} \left(r' \frac{\partial w'}{\partial r'} \right) + \frac{\partial^2 w'}{\partial z'^2} \right)$$
$$- \rho g \beta \left(T - T_\infty \right) \sin \phi \tag{2}$$

$$\rho \, c_p \left[\left(\frac{\partial T}{\partial t'} \right) + w' \frac{\partial w'}{\partial z'} \right] = k \left((\nabla'^2 T) - \nabla . q_r \right) \tag{3}$$

where $\nabla'^2 = \frac{\partial^2}{\partial r'^2} + \frac{1}{r'} \frac{\partial}{\partial r'}$

k is the thermal conductivity, p' is pressure, μ is the molecular viscosity and c_p is the specific heat at constant pressure. In the undisturbed fluid ρ is the density, T is the dimensional temperature, q_r is the radiative flux vector and gravitation, $\hat{g}$, is assumed to make an angle, ϕ with the radial axis of the tube. (2) and (3) are to be solved subject to the boundary conditions

$$w' = 0; \; T = T_w \; on \; r' = a_0 s \left(\varepsilon z' / a_0 \right)$$
$$w, \, T, \, < \infty \quad on \; r' = 0 \tag{4}$$

Assuming Boussinesq approximation is valid we can write the equation of state for a Boussinesq fluid as

$$\rho_\infty - \rho = \rho_\infty \beta \quad (T - T_\infty) \tag{5}$$

where β is the coefficient of volume expansion.

In this study, we describe the heat flux using the general differential approximation for radiation for a non-grey fluid near equilibrium, Elbarbary and Elgazery [19], so that the last term in equation (3) can be written as

$$\nabla . q_r' = 4 \left(T - T_w \right) \int_0^\infty \alpha^2 \left(\frac{\partial B}{\partial T} \right) \partial \lambda \tag{6}$$

α is the absorption coefficient, B is Planck's constant and λ is frequency. For an optically thin fluid[9] which blood is, $\alpha \ll 1$.

We now introduce the following non-dimensional quantities and parameters.

$$r' = a_0 r, \ z' = \frac{a_0 z}{\varepsilon}, \ (\varepsilon w') = \frac{(\varepsilon w)}{U_\infty}, \ t' = \frac{t}{\omega}, \ \theta = \frac{T - T_\infty}{T_w - T_\infty}, \ p = \frac{(p' - p_\infty) a_0}{\mu U_\infty},$$

$$Pr = \frac{\mu c_p}{k}, \ Pe = \frac{\mu c_p}{k} \frac{a_0 U_\infty}{\nu}, \ \sigma = \frac{\omega_1 a_0^2}{\nu}, \ Gr = \frac{g \beta a_0^2 (T_w - T_\infty)}{\nu U_\infty},$$

$$Ra = \frac{16 \sigma a_0 T}{3 \alpha k}. \tag{7}$$

U_∞ is a typical axial velocity, ω is the frequency of the driving pressure pulse, Pr is Prandtl number, Gr is Grashoff number, Ra is the radiation parameter, Re is Reynolds number, Da is Darcy parameter and ε is a small parameter ($\varepsilon \ll 1$). In virtue of (10) and (11) the governing equations become

$$\sigma \frac{\partial w}{\partial t} + Re \, \varepsilon \left(w \frac{\partial w}{\partial z} \right) = -\varepsilon \frac{\partial p}{\partial z} + \left(\frac{1}{r} \frac{\partial}{\partial r} \left(r \frac{\partial w}{\partial r} \right) + \varepsilon^2 \frac{\partial^2 w}{\partial z^2} \right) - \varepsilon \, Gr \, \theta \sin \phi, \tag{8}$$

$$\sigma \, Pr \frac{\partial \theta}{\partial t} + Pe \, \varepsilon \left(w \frac{\partial T}{\partial z} \right) = \frac{1}{r} \frac{\partial}{\partial r} \left(r \frac{\partial \theta}{\partial r} \right) + \varepsilon \frac{\partial^2 \theta}{\partial z^2} + Ra^2 \theta. \tag{9}$$

$\sigma = \frac{a_0^2 \omega_1}{\nu}$ is the Womersley number, and Pe is the Peclet number. The boundary conditions are now

$$w = 0, \ \theta = \theta_w \text{ on } r = s(z),$$

$$w, \theta < \infty \quad on \ r = 0. \tag{10}$$

The problem therefore depends on the Reynolds number of the flow, Re, the Grashoff number (or the free convection parameter), Gr, the Peclet number, Pe, the Womersley parameter, σ, and the radiation parameter Ra.

The mathematical statement of the problem is now complete; it embodies the solution of (8) and (9) subject to the conditions in (10). We see that the problem depends on oscillation (Womersley parameter), the radiation parameter, the Peclet number, the Prandtl number, Reynolds number and the free convection parameter (Grashoff number).

3 Asymptotic Solutions

Since ε is small, we seek a perturbative solution in the form of a power series in ε as in Ogulu [14]. As shown in Bestman [20] a scheme of the type advanced in (11) converges very rapidly. For the velocity and temperature we put

$$w = w_0(r, z) + \varepsilon w_1(r, z) + \cdots\cdots\cdots$$
$$\theta = \theta_0(r, z) + \varepsilon \theta_1(r, z) + \cdots\cdots\cdots$$

While for the pressure we put

$$p = \frac{1}{\varepsilon} p^{(0)}(z) + p^{(1)}(r, z) + \varepsilon p^{(2)}(r, z) \tag{11}$$

Substituting (11) into our leading equations we obtain the following sequence of approximations for the leading terms

$$\sigma \frac{\partial w}{\partial t} = -\frac{\partial p_0}{\partial z} + \frac{1}{r}\frac{\partial}{\partial r}\left(r\frac{\partial w_0}{\partial r}\right) \tag{12}$$

$$\sigma \Pr \frac{\partial \theta_0}{\partial t} = \frac{1}{r}\frac{\partial}{\partial r}\left(r\frac{\partial \theta_0}{\partial r}\right) + Ra^2 \theta_0 \tag{13}$$

Subject to the boundary conditions

$$-\frac{\partial p_0}{\partial z}(r, 0, t) = A_0 + A\sin(\omega_1 t)$$
$$w_0 = 0, \ \theta_0 = \theta_w \quad \text{on} \quad r = s(z)$$
$$w_0, \ \theta_0 > 0 \quad \text{on} \quad r = 0 \tag{14}$$

A_0 is the steady state part of the pressure gradient, El-Shahed [18], A_1 is the amplitude of the oscillatory part, $\omega_1 = 2\pi f_1$, f_1 is the heart pulse frequency.

Continuing our substitutions we obtain the sequence of equations for the higher approximation as

$$\sigma \frac{\partial w_1}{\partial t} + \text{Re } w_0 \frac{\partial w_0}{\partial z} = -\frac{\partial p_1}{\partial z} + \frac{1}{r}\frac{\partial}{\partial r}\left(r\frac{\partial w_1}{\partial r}\right) - Gr\theta_0 \sin\phi \tag{15}$$

$$\sigma \Pr \frac{\partial \theta_1}{\partial t} + Pe\left(w_0 \frac{\partial \theta_0}{\partial z}\right) = \left(\nabla^2 + Ra^2\right)\theta_0 \tag{16}$$

The boundary conditions are now

$$w_1 = 0 = \theta_1 \quad \text{on} \quad r = s(z)$$
$$\frac{\partial p_1}{\partial z}(r, 0, t) = 0 \tag{17}$$

4 Solutions for the Leading Approximations

Obviously, $p_0 = p_0(z)$, only.

We now put

$$w_0 = w_0^{(0)}(r, z) + w_0^{(1)}(r, z, t) + \cdots\cdots\cdots$$
$$\theta_0 = \theta_0^{(0)}(r, z) + \theta_0^{(1)}(r, z, t) + \cdots\cdots\cdots \tag{18}$$

Substituting (18) into (12) and (13) we obtain the following steady state equations

$$\frac{\partial p_0^{(0)}}{\partial z} = \frac{1}{r}\frac{\partial}{\partial r}\left(r\frac{\partial w_0^{(0)}}{\partial r}\right) \tag{19}$$

$$\left(\nabla^2 + Ra\right)\theta_0^{(0)} = 0. \tag{20}$$

With the conditions,

$$-\frac{\partial p_0^{(0)}}{\partial z} = A_0, \, w_0^{(0)} = 0, \quad \theta_0^{(0)} = \theta_w \quad \text{on} \quad r = s(z) \tag{21}$$

And the following oscillatory state equations

$$\sigma\frac{\partial w_0^{(1)}}{\partial t} = -\frac{\partial p_0^{(1)}}{\partial z} + \frac{1}{r}\frac{\partial}{\partial r}\left(r\frac{\partial w_0^{(1)}}{\partial r}\right), \tag{22}$$

$$\sigma\Pr\frac{\partial \theta_0^{(1)}}{\partial t} = \left(\nabla^2 + Ra^2\right)\theta_0^{(1)}. \tag{23}$$

With the conditions

$$\frac{\partial p_0^{(1)}}{\partial z} = A\sin\left(\omega_1 t\right), \, w_0^{(1)} = 0, \quad \theta_0^{(1)} = 0 \quad \text{on} \quad r = s(z). \tag{24}$$

The solution of (19) and (20) subject to the conditions in (21) can be put in the form

$$w_0^{(0)} = \frac{A_0}{4}\left(1 - \left(\frac{r}{s}\right)^2\right), \tag{25}$$

$$\theta_0^{(0)} = \theta_w\frac{J_0\left(Rar\right)}{J_0\left(Ras\right)}, \tag{26}$$

Where, J_n is the Bessel function of the first kind of order n. (25) gives the classic Poiseuille flow velocity.

For the solution of (22) and (23), we follow the method in Bestman [20] writing

$$w_0^{(1)} = \frac{1}{2}\left(h_0 e^{it} + \widehat{h}_0 e^{it}\right),$$
$$\theta_0^{(1)} = \frac{1}{2}\left(\gamma_0 e^{iRat} + \widehat{\gamma}_0 e^{iRat}\right), \tag{27}$$
$$p_0^{(1)} = \frac{1}{2i}\left(p_0^{(1)} e^{it} + \widehat{p}_0^{(1)} e^{it}\right),$$

Where, we have used a tilde to indicate complex conjugate. When we substitute (27) into (22) and (23) subject to the conditions in (24) gives

$$h_0 = \frac{1}{i\sigma} \frac{\partial p_0^{(1)}}{\partial z} \left\{ 1 - \frac{J_0(\zeta r)}{J_0(\zeta s)} \right\}, \tag{28}$$

$$\gamma_0 = \frac{1}{i \Pr Ra \, \sigma} \left\{ 1 - \frac{J_0\left(\sqrt{\Pr Ra}\, \zeta r\right)}{J_0\left(\sqrt{\Pr Ra}\, \zeta s\right)} \right\}. \tag{29}$$

$\zeta^2 = -i\sigma$, $J_0(x)$ is Bessel function of the first kind of order zero.

5 Higher Approximate Solutions

This order of our approximations is given by (15), (16) and (17). In conformity with flow in porous media, Tay and Ogulu [9], we drop the inertial terms, hence we have

$$\sigma \frac{\partial w_1}{\partial t} = -\frac{\partial p_1}{\partial z} + \frac{1}{r} \frac{\partial}{\partial r} \left(r \frac{\partial w_1}{\partial r} \right) - Gr \, \theta_0 \sin \phi, \tag{30}$$

$$\sigma \Pr \frac{\partial \theta_1}{\partial t} + Pe \left(w_0 \frac{\partial \theta_0}{\partial z} \right) = \left(\nabla^2 + Ra^2 \right) \theta_0. \tag{31}$$

Again we put

$$w_1 = w_1^{(0)}(r, z) + w_1^{(1)}(r, z, t) + \ldots \ldots$$
$$\theta_1 = \theta_1^{(0)}(r, z) + \theta_1^{(1)}(r, z, t) + \ldots \ldots \tag{32}$$

Then (32) separates into

$$\sigma \Pr \frac{\partial \theta_1^{(1)}}{\partial t} + Pe \, w_0 \frac{\partial \theta_0^{(1)}}{\partial z} = \left(\nabla^2 + Ra^2 \right) \theta_1^{(1)}, \tag{33}$$

$$Pe \, w_0 \frac{\partial \theta_0^{(0)}}{\partial z} = \left(\nabla^2 + Ra^2 \right) \theta_1^{(0)} \tag{34}$$

And (32) separates into

$$\sigma \frac{\partial w_1^{(1)}}{\partial t} = -\frac{\partial p_1^{(0)}}{\partial z} + \frac{1}{r} \frac{\partial}{\partial r} \left(r \frac{\partial w_1^{(1)}}{\partial r} \right) - Gr \, \theta_0^{(1)} \sin \phi, \tag{35}$$

$$0 = -\frac{\partial p_1^{(0)}}{\partial z} + \nabla^2 w_1^{(0)} - Gr\theta_0^{(0)} \sin \phi. \tag{36}$$

On solution of (33)–(36) subject to appropriate boundary conditions we obtain

$$w_1^{(0)} = -\frac{p_1^{(0)}}{4}\left[1 - \left(\frac{r}{s}\right)^2\right] + \frac{Gr\,\theta_w}{Ra}\left[1 - \frac{J_0\,(Ra\,r)}{J_0\,(Ra\,s)}\right]\sin\phi, \tag{37}$$

$$\theta_1^{(0)} = Pe\,\theta_0^{(0)'}\left\{\frac{A_0}{4}\left[1 - \frac{r^4}{15s^2}\right] - \frac{p_0^{(1)'}}{\zeta^2}\left[1 - \frac{rJ_1\,(\zeta\,r)}{2J_0\,(\zeta\,s)}\right]\left[\frac{1}{\zeta} - 1\right]\right\}$$
$$-Pe\,\theta_0^{(0)'}\frac{J_0\,(Ra\,r)}{J_0\,(Ra\,s)}\left\{\frac{A_0}{4}\left[1 - \frac{s^2}{15}\right] - \frac{p_0^{(1)'}}{\zeta^2}\left[1 - \frac{sJ_1\,(\zeta\,s)}{2J_0\,(\zeta\,s)}\right]\left[\frac{1}{\zeta} - 1\right]\right\}, \tag{38}$$

$$w_1^{(1)} = \frac{p_1^{(1)'}}{3}\left(r^2 - \frac{s^2 I_0\,(\zeta\,r)}{I_0\,(\zeta\,s)}\right) + \frac{Gr}{i\,Pr\,Ra\,\sigma}\left[\frac{1}{\sqrt{Pr\,Ra}\,\zeta} - 1\right]$$
$$\times\left\{\left(1 - \frac{rJ_1\left(\sqrt{Pr\,Ra}\,\zeta\,r\right)}{2J_0\left(\sqrt{Pr\,Ra}\,\zeta\,s\right)}\right) - \frac{I_0\,(\zeta\,r)}{I_0\,(\zeta\,s)}\left(1 - \frac{sJ_1\left(\sqrt{Pr\,Ra}\,\zeta\,s\right)}{2J_0\left(\sqrt{Pr\,Ra}\,\zeta\,s\right)}\right)\right\}\sin\phi, \tag{39}$$

$$\theta_1^{(1)} = Pe\,\theta_0^{(0)'}\left\{\frac{A_0}{4}\left[1 - \frac{r^4}{15s^2}\right] - \frac{p_0^{(1)'}}{\zeta^2}\left[1 - \frac{rJ_1\,(\zeta\,r)}{2J_0\,(\zeta\,s)}\right]\left[\frac{1}{\zeta} - 1\right]\right\}$$
$$-Pe\,\theta_0^{(0)'}\frac{J_0\,(\Omega\,r)}{J_0\,(\Omega\,s)}\left\{\frac{A_0}{4}\left[1 - \frac{s^2}{15}\right] - \frac{p_0^{(1)'}}{\zeta^2}\left[1 - \frac{sJ_1\,(\zeta\,s)}{2J_0\,(\zeta\,s)}\right]\left[\frac{1}{\zeta} - 1\right]\right\}, \tag{40}$$

$\Omega^2 = \left(Ra^2 + \zeta^2 Pr\right)$ and a prime here is used to denote differentiation with respect to z.

Having obtained the expressions for the axial velocity and the temperature we can terminate the solution of the order $1(\varepsilon)$ problem here without loss of generality.

6 Shear Stress and Heat Transfer

We can define the shear stress at the wall τ_w as

$$\tau_w = \left[-\mu\frac{dw}{dr}\right]_{r=s(z)} \tag{41}$$

and the local rate of heat transfer q_w as

$$q_w = \left[\frac{d\theta}{dr}\right]_{r=s(z)}. \tag{42}$$

7 Results and Discussion

In the previous three sections we have formulated and solved asymptotically
the problem of blood flow as obtained during deep heat muscle treatment
modelling the blood vessel as a porous medium. For the purpose of this numerical discussion we shall only consider a locally dilating tube of the form
$s = e^z$. For the wall temperature of the blood vessel we take $\theta_w = 2$, and
$z = 0$ we shall assume is the entrance to the aorta.

  We shall only focus on the shear stress and the rate of heat transfer at the
wall of the vessel since the other effects are discussed in the literature. (See
for instance Ogulu [14].) Figures 1, 2, and 3 show the effect of shear stress
variation with time at different locations where we observe an increase in the
shear stress as z, the position increases. For any chosen position (constant
z), we observe very little variation in the shear stress as time increases for a
normal heart. Further we observe from Fig. 3 that when f_1 is increased from
0.6 to 1.2 the shear stress begins to oscillate for large values of **z**.

  Figures 4 and 5 show the rate of heat transfer at the wall for different times
and locations. The rate of heat transfer increases slightly as time increases but
rapidly as the radiation parameter and the position are increased.

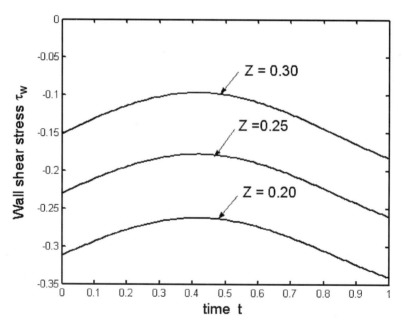

Fig. 1. Variation of wall shear stress with time for different locations, $f_1 = 0.6$

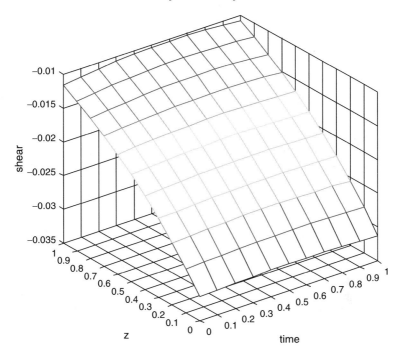

Fig. 2. 3-D Plot of variation of wall shear stress with time for different locations, $f_1 = 0.6\,\text{Hz}$

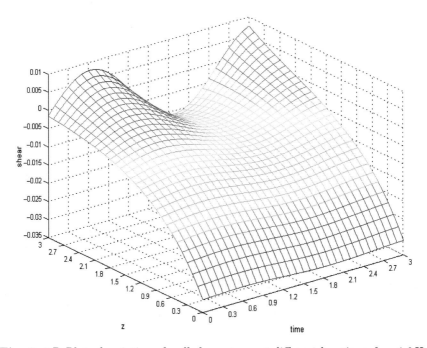

Fig. 3. 3-D Plot of variation of wall shear stress at different locations $f_1 = 1.2\,\text{Hz}$

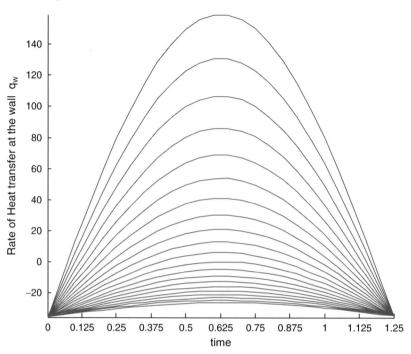

Fig. 4. Variation of wall rate of heat transfer with time for different values of the radiation parameter Ra

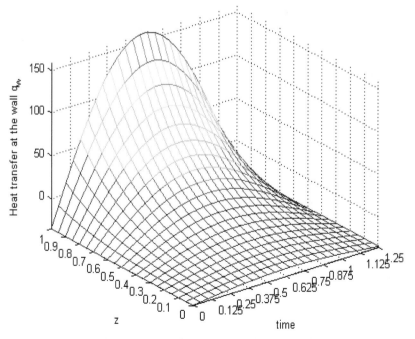

Fig. 5. 3-D plot of wall rate of heat transfer variation with time for different values of the radiation parameter Ra

References

1. A. R. Bestman, Pulsatile flow in heated porous channel, 25(5), (1982), 675–682.
2. P. Latinopoulos and J. Ganoulis, Numerical simulation of pulsatile flow in constricted axi-symmetric tubes, Appl. Math. Modelling, 6 February (1982), 55–60.
3. J. C. Misra and S. Chakravarty, Flow in arteries in the presence of stenosis, J. Biomechamics, 19(11), (1986), 907–918.
4. K. Haldar, Oscillatory flow of blood in a stenosed artery, Bull. Math. Bio., 49(3), (1987), 279–287.
5. J. K. Misra and R. S. Chauhan, A study of pulsatile blood flow in a tube with pulsating walls, Acta Physica Hungarica, 67(1–2), (1990), 123–134.
6. K. Haldar and S. N. Ghosh, Effect of a magnetic field on blood flow through an indented tube in the presence of erythrocytes, Indian J. Pure Appl. Math., 25(3), (1994), 345–352.
7. S. Cavaleanti, Hemodynamics of an artery with mild stenosis, J. Biomech., 28(4), (1995), 387–399.
8. T.-K. Hung and T. M.-C. Tsai, Pulsatile blood flows in stenotic artery, J. Eng. Mech., September (1998), 890–896.
9. G. Tay and A. Ogulu, Low Reynolds number flow in a constricted tube under a time-dependent pressure gradient, J. Fizik Malaysia, 18(3), (1997), 105–112.
10. K. Vajravelu, K. Ramesh, S. Sreenadh and P. V. Arunachalam, Pulsatile flow between permeable beds, Int. J. Nonlinear Mech., 38, (2003), 999–1005.
11. N. Filipovic and M. Kojic, Computer Simulations of blood flow with mass transport through the carotid artery bifurcation, Theoret. App. Mech., 31(1), (2004), 1–33.
12. A. Ogulu and T. M. Abbey, Simulation of heat transfer on an oscillatory blood flow in an indented porous artery, Int. Comm. Heat Mass Transfer, 32, (2005), 983–989.
13. P. K. Mandal, An unsteady analysis of non-Newtonian blood flow through tapered arteries with stenosis, Int. J. Nonlinear Mech., 40, (2005), 151–164.
14. A. Ogulu, Effect of heat generation on low Reynolds number fluid and mass transport in a single lymphatic blood vessel with uniform magnetic field, Int. Comm. Heat Mass Transfer, 33, (2006), 790–799.
15. M. J. Manton, Low Reynolds number flow in a slowly varying axisymmetric tubes, J. fluid Mech., 49, (1971), 451.
16. A. Ogulu and A. R. Bestman, Deep heat muscle treatment: A mathematical model I, Acta Physica Hungarica, 73, (1993), −16.
17. A. Ogulu and A. R. Bestman, Deep heat muscle treatment: A mathematical model II, Acta Physica Hungarica, 73, (1993), 17–27.
18. M. El-Shahed, Pulsatile blood flow through a stenosed porous medium under periodic body acceleration, Appl. Math. Comp., 138, (2003), 479–488.
19. Elbarbary and Elgazery, Chebyshev finite difference method for the effect of variable viscosity on magneto micro-polar fluid flow with radiation, Int. Comm. Heat Mass Transfer, 31(3), (2000), 409.
20. A. R. Bestman, Unsteady low Reynolds number flow in a heated tube of slowly varying section, J. Austral Math Soc. B179, (1988).

Stability of Generalized Convexity and Monotonicity

P.T. An

Summary. It was shown that well-known kinds of generalized convex functions (generalized monotone maps, respectively) are often not stable with respect to the property they have to keep during the generalization. Then the so-called s-quasiconvex functions, s-quasimonotone maps and strictly s-quasiconvex functions were introduced in *Optimization*, vol. **38**, vol. **55** and *Journal of Inequalities in Pure and Applied Mathematics*, vol. **127**, respectively. In this paper, some stability properties of such functions and a use of s-quasimonotonicity in an economics model are presented. Furthermore, an algorithm for finding the stability index for strict s-quasiconvexity of a given continuously twice differentiable function on $I\!R^1$ is presented.

1 Introduction

Convex functions belong to the most important objects investigated in mathematical programming. They have many interesting properties, for instance:

(L) Each lower level set is convex
(M) Each local minimizer is a global minimizer
(S) Each stationary point is a global minimizer
(E) If the considered function attains global maximum on a compact convex set then it attains global one at least at one extreme point of this set.

Definition 1.1. *A function $f : D \subset I\!R^n \to I\!R$ is said to be stable with respect to some property (P) if there exists $\epsilon > 0$ such that $f + \xi$ fulfill (P) for all linear function ξ satisfying $\|\xi\| < \epsilon$.*

It was shown in [14] that well-known kinds of generalized convex functions are often not stable with respect to the property they have to keep during the generalization, for example, quasiconvex functions (pseudoconvex functions, respectively) are not stable with respect to the property (L) ((S), respectively). Then the so-called s-quasiconvex functions were introduced by Phu in [14]. They are stable with respect to the properties (L), (M) and (S).

A subclass of s-quasiconvex functions, namely strictly s-quasiconvex functions which guarantee the uniqueness of the minimizer was introduced in [2].

Various kinds of generalized monotonicity were introduced (see [8, 9]). Among others, pseudomonotonicity was introduced by Karamardian in 1976. This concept is used in economics. We now denote $I\!R_{>0}^n$ the subset of $I\!R^n$ with all positive coordinates. In an n-good competitive economy, $F : D \subset I\!R_{>0}^n \to I\!R^n$ is said to be an excess demand if it is homogeneous in $p \in D$ and satisfies Walras' Law (see [5, 6]). In [11] John showed that a demand which yields a pseudomonotone excess demand if it is combined with an arbitrary supply (in particular, a constant function) is necessarily monotone.

There arises a question: What kind of a demand is it if "arbitrary supply" is replaced by "supply with sufficiently small norm"?

Definition 1.2. *A map* $F : D \subset I\!R^n \to I\!R^n$ *is said to be stable with respect to some property (P) if there exits* $\epsilon > 0$ *such that*

$$F(x) + a \quad satisfies \quad (P) \quad whenever \quad \|a\| < \epsilon$$

$(a \in I\!R^n)$.

The supremum of the set of all ϵ in Definitions 1.1–1.2 is called the *stability index* for the property (P) of f (of F, respectively) and is denoted by s_f (s_F, respectively).

In [1] we showed that quasimonotonicity and pseudomonotonicity are not stable with respect to their first-order characterizations and introduced the notion of s-quasimonotonicity. In [3] we showed that a demand which yields a pseudomonotone excess demand if it is combined with a supply (in particular, a constant function) with sufficiently small norm, is necessarily s-quasimonotone. In this paper some properties of s-quasiconvex functions and s-quasimonotone maps are presented in Sects. 2 and 3. A use of s-quasimonotonicity in an economics model is presented in Sect. 4. The problem to find the stability index for strict s-quasiconvexity of a given continuously twice differentiable function f on $D \subset I\!R^1$ is presented in Sect. 5. Algorithm to find the stability index for s-quasiconvexity of a given f (for s-quasimonotonicity of a given F, respectively) on $D \subset I\!R^1$ can be introduced in the same manner. Some questions and future tasks are given in Sect. 6.

Before we start the analysis, we recall some definitions and properties from [8, 13]. Let $F : D \subset I\!R^n \to I\!R^n$ and D be convex. Denote by ^{T} the matrix transposition. We recall that f is convex (strictly convex, respectively) if for every $x_0, x_1 \in D, \lambda \in]0, 1[$, we have $f(x_\lambda) \le (1 - \lambda)f(x_0) + \lambda f(x_1)$ $(f(x_\lambda) < (1 - \lambda)f(x_0) + \lambda f(x_1)$, respectively). f is quasiconvex (strictly quasiconvex, respectively) if for every $x_0, x_1 \in D, \lambda \in]0, 1[$, we have $f(x_\lambda) \le \max\{f(x_0), f(x_1)\}$ $(f(x_\lambda) < \max\{f(x_0), f(x_1)\}$, respectively), where $x_\lambda : = (1-\lambda)x_0+\lambda x_1$. It is well-known that f is quasiconvex if it has the property (L).

A differentiable function f is pseudoconvex if for every $x_0, x_1 \in D$,

$$f(x_1) - f(x_0) < 0 \quad \text{implies} \quad (x_1 - x_0)^T \bigtriangledown f(x_0) < 0.$$

Obviously, if x^* is a stationary point of f (i.e., $\bigtriangledown f(x^*) = 0$), then x^* is a global minimizer of f. Thus, such a function has the property (S).

F is called quasimonotone if for every $x_0, x_1 \in D$,

$$(x_1 - x_0)^T F(x_0) > 0 \quad \text{implies} \quad (x_1 - x_0)^T F(x_1) \geq 0.$$

F is called pseudomonotone if for every $x_0, x_1 \in D$,

$$(x_1 - x_0)^T F(x_0) \geq 0 \quad \text{implies} \quad (x_1 - x_0)^T F(x_1) \geq 0.$$

In [8], Karamardian and Schaible showed that a differentiable function f is quasiconvex (pseudoconvex, respectively) if gradient $\bigtriangledown f$ is quasimonotone (pseudomonotone, respectively).

2 S-Quasiconvex Functions

We recall the definition of s-quasiconvex functions ("s" stands for "stable").

Definition 2.1. ([14]) $f : D \subset I\!R^n \to I\!R$ is said to be s-quasiconvex if there exists $\sigma > 0$ such that

$$\frac{f(x_0) - f(x_1)}{\|x_0 - x_1\|} \leq \delta \quad \Rightarrow \quad \frac{f(x_\lambda) - f(x_1)}{\|x_\lambda - x_1\|} \leq \delta$$

for $|\delta| < \sigma, x_0, x_1 \in D, x_0 \neq x_1, x_\lambda = (1-\lambda)x_0 + \lambda x_1$ and $\lambda \in [0,1[$.

Clearly, every convex function is s-quasiconvex and a s-quasiconvex function is quasiconvex.

Theorem 2.2. ([14])   Suppose $f : D \subset I\!R^n \to I\!R$.
a) f is s-quasiconvex iff f is stable with respect to quasiconvexity;
b) f is s-quasiconvex iff f is stable with respect to s-quasiconvexity.
Let us recall the concept of strictly s-quasiconvex functions.

Definition 2.3. ([2]) $f : D \subset I\!R^n \to I\!R$ is said to be strictly s-quasiconvex if there exists $\sigma > 0$ such that

$$\frac{f(x_0) - f(x_1)}{\|x_0 - x_1\|} \leq \delta \quad \Rightarrow \quad \frac{f(x_\lambda) - f(x_1)}{\|x_\lambda - x_1\|} < \delta$$

for $|\delta| < \sigma, x_0, x_1 \in D, x_0 \neq x_1, x_\lambda = (1-\lambda)x_0 + \lambda x_1$ and $\lambda \in]0,1[$.

Clearly, a strictly convex function f is strictly s-quasiconvex. Furthermore, every strictly s-quasiconvex function is s-quasiconvex and every strictly s-quasiconvex function is strictly quasiconvex. A complete description of the relations existing between strictly s-quasiconvex function, s-quasiconvex functions and known generalized convex functions was given in [2].

Theorem 2.4. ([2])   *Suppose* $f : D \subset \mathbb{R}^n \to \mathbb{R}$.
a) f is strictly s-quasiconvex if f is stable with respect to strict quasiconvexity;
b) f is strictly s-quasiconvex if f is stable with respect to strict s-quasiconvexity.

It follows from the proof of Theorem 2.4 (see [2]) that the supremum of the set of all σ in Definition 2.3 of the strictly s-quasiconvex function f is s_f. Hence, if f is strictly quasiconvex then $s_f = +\infty$ for strict s-quasiconvexity.

Our next objective is to give a necessary and sufficient condition for a continuously differentiable function to be strictly s-quasiconvex.

Theorem 2.5. ([2])   *A continuously differentiable function f on $D \subset \mathbb{R}^n$ is strictly s-quasiconvex if there exists $\alpha > 0$ such that f is strictly convex on every segment $[x_0, x_1]$ satisfying*

$$\left| \frac{(x_1 - x_0)^T}{\|x_1 - x_0\|} \nabla f(x_\lambda) \right| < \alpha \quad for \; all \quad x_\lambda \in [x_0, x_1].$$

In the proof above, to prove the sufficiency we set $\sigma := \epsilon$ and to prove the necessity we set $\epsilon := \sigma$ (see [2]). Therefore, s_f is the supremum of the set of all α. It follows directly from Theorem 2.5 the following

Corollary 2.6.   *A continuously differentiable function f on $D \subset \mathbb{R}^1$ is strictly s-quasiconvex iff there exists $\alpha > 0$ such that f is strictly convex on the level set*

$$L(|f'|, \alpha) := \{x \in D : |f'(x)| < \alpha\}.$$

Furthermore, $s_f = \sup\{\alpha : f \text{ is strictly convex on } L(|f'|, \alpha)\}.$

3 S-Quasimonotone Maps

Let us recall the concept of s-quasimonotone maps.

Definition 3.1. ([1]) $F : D \subset \mathbb{R}^n \to \mathbb{R}^n$ *is s-quasimonotone if there exists $\sigma_F > 0$ such that*

$$\frac{(x_1 - x_0)^T}{\|x_1 - x_0\|} F(x_0) \geq \delta \quad \Rightarrow \quad \frac{(x_1 - x_0)^T}{\|x_1 - x_0\|} F(x_1) \geq \delta$$

for $|\delta| < \sigma_F, x_0, x_1 \in D$.

Obviously, every monotone map is s-quasimonotone. By setting $\delta = 0$, every s-quasimonotone map is pseudomonotone.

Stability with respect to pseudomonotonicity (or s-quasimonotonicity) is a necessary and sufficient condition for a map to be s-quasimonotone.

Theorem 3.2. ([1]) *1) F is s-quasimonotone iff F is stable with respect to pseudomonotonicity.*
2) F is s-quasimonotone iff F is stable with respect to s-quasimonotonicity.

We now establish a relationship between s-quasimonotonicity and s-quasiconvexity.

Theorem 3.3. ([1]) *Suppose that f is continuously differentiable. Then f is s-quasiconvex iff $\triangledown f$ is s-quasimonotone.*

Note that not all s-quasimonotone maps arise as gradients of s-quasi -convex functions. For example, consider $F(x) := (x_1, x_2 + \phi(x_1))$, $x = (x_1, x_2) \in I\!\!R^2$, where ϕ is a differentiable function of the univariable $x_1 \in I\!\!R^1$ such that $|\phi(x_1) - \phi(x_1')| \leq |x_1 - x_1'|$ for all $x_1, x_1' \in I\!\!R^1$ (see [10], pp. 208–209). Then F is monotone therefore it is s-quasimonote while there is some ϕ such that F does not arise as a gradient of any function.

Proposition 3.4. ([1])

(1) A univariate polynomial F of even degree is s-quasimonotone iff F has no roots.

(2) A univariate polynomial F of odd degree is s-quasimonotone iff the coefficient of highest power in F is greater than 0 and F has a unique root.

It follows directly from Theorem 3.2 and Proposition 3.4 the following

Corollary 3.5.

(1) A univariate polynomial f of odd degree is s-quasiconvex iff $\triangledown f$ has no roots.

(2) A univariate polynomial f of even degree is s-quasiconvex iff the co-efficient of highest power in $\triangledown f$ is greater than 0 and $\triangledown f$ has a unique root.

4 A Use of S-Quasimonotonicity in an Economics Model

In an n-good competitive economy, $Z : P \subset I\!\!R^n_{>0} \to I\!\!R^n$ is said to be an excess demand function if it is homogeneous in $p \in P$, i.e., $Z(p) = Z(\lambda p)$, $\lambda > 0$ and satisfies Walras' Law, $p^T Z(p) = 0$. The following property is called Wald's Axiom:

$$p, q \in P, \ p^T Z(q) \leq 0 \text{ and } q^T Z(p) \leq 0 \ \Rightarrow \ Z(q) = Z(p)$$

(see [5, 6]). Some strong versions of Wald's Axiom were given in [4, 7, 12], . . . A strong version of Wald's Axiom is introduced in [3] as follows:

$\exists \sigma > 0$ such that

$$\left. \begin{array}{l} p, q \in D, \quad |\delta| < \sigma \\ F(q) \neq F(p), q^T F(p) - \delta \leq 0 \end{array} \right\} \Rightarrow p^T F(q) + \delta > 0.$$

By setting $\delta = 0$, the version above implies Wald's Axiom. Moreover, this version is actually stronger than the Wald's Axiom. To see it, in [3] we consider $Z : P \subset \mathbb{R}^2_> \to \mathbb{R}^2$ such that

$$Z(p) = \left(-f(\frac{p_1}{p_2}); \frac{p_1}{p_2} f(\frac{p_1}{p_2}) \right),$$

where $P: =]0, 2] \times]0, 2]$ and $f(x) = x^4 \left(2 + \sin \frac{1}{x} \right), x > 0$. Then Z satisfies Walras' Law and it is homogeneous in $p \in P$. Z satisfies the Wald's Axiom. But Z does not satisfy the strong version of Wald's Axiom (see [3]).

Theorem 4.1. ([3]) *S-quasimonotonicity of $-F$ is equivalent to the strong version of Wald's Axiom.*

By Theorem 4.1, we get directly the following result.

Proposition 4.2. ([3]) *A demand which yields a pseudomonotone excess demand if it is combined with a supply (in particular, a constant function) with sufficiently small norm, is necessarily s-quasimonotone.*

5 Stability Index of Generalized Convex Functions

First of all, it is of interest to know if $s_f = +\infty$ (for strict s-quasiconvexity) implies that f is strictly convex.

Proposition 5.1. $f : D \subset \mathbb{R}^n \to \mathbb{R}$ *is strictly convex iff $f + \xi$ is strictly quasiconvex for all linear functional ξ on $\mathbb{R}^n$.*

Proof. Clearly, if f is strictly convex then $f + \xi$ is strictly convex for all linear functional ξ on $\mathbb{R}^n$ therefore $f + \xi$ is strictly quasiconvex. Conversely, we prove that f is strictly convex if $f + \xi$ is strictly quasiconvex for all linear functional ξ on $\mathbb{R}^n$. Let $x_0, x_1 \in D$, and choose a linear functional ξ on $\mathbb{R}^n$ such that $\xi(x_1 - x_0) = f(x_0) - f(x_1)$. Since $f + \xi$ is strictly quasiconvex, for any $x_\lambda = x_0 + \lambda(x_1 - x_0)$ with $\lambda \in]0, 1[$, it holds

$$(f + \xi)(x_0) > (f + \xi)(x_\lambda) = f(x_\lambda) + \lambda \xi(x_1 - x_0) + \xi(x_0)$$
$$= f(x_\lambda) + \lambda(f(x_0) - f(x_1)) + \xi(x_0).$$

This yields that

$$f(x_\lambda) < (1 - \lambda)f(x_0) + \lambda f(x_1).$$

Hence, f is strictly convex. □

Corollary 5.2. $f : D \subset \mathbb{R}^n \to \mathbb{R}$ *is strictly convex iff* $s_f = +\infty$ *for strict s-quasiconvexity.*

Proof. It follows directly from Theorem 2.2 and Proposition 5.1.   □

It follows from Corollaries 2.6 and 5.2 the following

Corollary 5.3.   *Suppose that continuously differentiable function* $f : D \subset \mathbb{R}^1 \to \mathbb{R}$ *is strictly s-quasiconvex. If it is not strictly convex, then the set of all* $\alpha > 0$ *given in Corollary 2.6 is bounded above.*

Proof. Assume the contrary that the set of all $\alpha > 0$ given in Corollary 2.6 is not bounded above. Then, $s_f = +\infty$ and therefore, by Corollary 5.2, f must be strictly convex, a contradiction.   □

Algorithm
Given continuously twice differentiable function $f : D \subset \mathbb{R}^1 \to \mathbb{R}$, smallest double precision real number z_{min} allowed by a compiler, and step length γ. Find the stability index s_f for strict s-quasiconvexity of f.

1. If f is strictly convex on D (i.e., $f''(x) > 0$ for all $x \in D$) then, by Corollary 5.2, $s_f = +\infty$. STOP.
   Else, choose the initial $\alpha > z_{min}$ and $flag: = 0$.
2. If f is strictly convex on $L(|f'|, \alpha): = \{x \in D : |f'(x)| < \alpha\}$ (i.e., $f''(x) > 0$ for all $x \in L(|f'|, \alpha)$), set $flag: = 1$ and $\alpha: = \alpha + \gamma$, go to 2.
3. If $flag: = 0$, go to 4.
   Else, set $s_f = \alpha - \gamma$. STOP.
4. Set $\alpha = \alpha - \gamma$. If $\alpha \le z_{min}$ then f is not strictly s-quasiconvex, $s_f: = 0$. STOP.
5. If f is strictly convex on $L(|f'|, \alpha)$ (i.e., $f''(x) > 0$ for all $x \in L(|f'|, \alpha)$), set $s_f: = \alpha$. STOP.
   Else, go to 4.

Note that the set $L(|f'|, \alpha)$ above is convex ([14]). By virtue of Corollary 5.3, the algorithm stops after a finite number of steps. Algorithm to find the stability index for s-quasiconvexity of a given f (for s-quasimonotonicity of a given F, respectively) on $D \subset \mathbb{R}^1$ can be introduced in the same manner.

6 Some Questions and Future Tasks

An exhaustive study of stability of generalized convexity and monotonicity is impossible here. Actually, there arises some questions and future tasks:

1. A suitable replacement of inequalities by strict inequalities, as in Definition 3.1 of s-quasimonotonicity, gives rise to new type of generalized monotonicities which is also stable.

2. To find the stability index in case $D \subset I\!R^n$, $n > 1$, is still open.
3. What kinds of generalized convex functions are stable with respect to the property (E)?
4. In Definition 1.1 (Definition 1.2, respectively) the disturbance is linear (constant, respectively). Stability investigation in case of nonlinear disturbances (nonconstant disturbances, respectively) will be considered.
5. Stability investigation for generalized convex vector valued functions will be considered.

References

1. An, P.T. (2006): Stability of generalized monotone maps with respect to their characterizations. Optimization, **55**, pp. 289–299
2. An, P.T. (2006): A new kind of stable generalized convex functions. Journal of Inequalities in Pure and Applied Mathematics, **127(3)**, electronic
3. An, P.T. and Binh, V.T.T.: Stability of excess demand functions with respect to a strong version of Wald's axiom. In Workshop "Small open economies in a globalised world", Rimini, Italy, 29.8–2.9.2006, submitted for publication
4. Brighi, L. (2004): A strong criterion for the Weak Weak Axiom. Journal of Mathematical Economics, **40**, pp. 93–103
5. Debreu, G. (1959): Theory of Value. Yale University Press, New Haven and London
6. Hildenbrand, W. and Jerison, M. (1989): The demand theory of the weak axioms of revealed preference. Economics Letters, **29**, pp. 209–213
7. Houthakker, H.S. (1953): Revealed preference and the utility function. Economica, **17**, pp. 159–174
8. Karamardian, S. and Schaible, S. (1990): Seven kinds of generalized monotone maps. Journal of Optimization Theory and Applications, **66**, pp. 37–46
9. Karamardian, S. Schaible, S. and Crouzeix, J.P. (1993): Characterizations geneenralized monotone maps. Journal of Optimization Theory and Applications, **76**, pp. 399–413
10. Kinderlehrer. D and Stampacchia. G. (1980): An Introduction to Variational Inequalities and Their Applications. Academic, New York
11. John, R. (1998): Variational inequalities and pseudomonotone functions: some characterizations. In: Crouzeix, J.-P. et al. (ed) Generalized Convexity, Generalized Monotonicity. Kluwer, Dordrecht Boston London, 291–301
12. Mossin, A. (1972): A mean demand function and individual demand functions confronted with the weak and the strong axioms of revealed preference: an empirical test. Economica, **40**, pp. 177–192
13. Ponstein, J. (1967): Seven kinds of convexity. SIAM Review, **9**, pp. 115–119
14. Phu, H.X. and An, P.T. (1996): Stable generalization of convex functions. Optimization, **38**, pp. 309–318

Are Viscoelastic Flows Under Control
or Out of Control?

M. Renardy

Summary. We discuss the controllability of viscoelastic shear flows. Both linear and nonlinear results are reviewed. An interesting aspect of the problem is that the nonlinear problem introduces new stress components which are automatically zero in the linear case. It is therefore a challenging problem to determine to what extent these stresses are controllable.

1 Introduction

Many engineering problems are concerned with steering a system in a desirable direction by using inputs we can control. In the context of fluid dynamics, for instance, the process of filling a mold aims to direct the flow inside the mold by controlling the flow at the inlet. Active control of turbulence aims to suppress turbulence by blowing or sucking air at certain locations.

Mathematical control theory is an abstraction of such problems. In control theory we consider a system governed by differential equations which we want to steer from a given initial condition to a desirable outcome, using an input from a given class. There are a number of possible ways to make this notion precise and formulate mathematically well-defined problems. In this paper, we focus on the notion of controllability. A problem of controllability has the form

$$\dot{x} = \phi(x, f), \tag{1}$$

where x lies in some Banach space X and the control f lies in another Banach space Y or a subset thereof. We are given an initial condition $x(0) = x_0$ and a desired final state $x(T) = x_f$. We want to choose $f(t)$ in such a way that the state x_f is reached at time T. If this is always possible with an admissible control, we call the system controllable.

Linear control problems have the form

$$\dot{x} = Ax + f(t). \tag{2}$$

Of course, if $f(t)$ is completely arbitrary, the system is trivially controllable. In applications, however, we are interested in the case where f is restricted to a subspace. In the context of partial differential equations, f is usually restricted to a subset of the spatial domain or to its boundary.

In problems arising from continuum mechanics, a possible choice of control is a body force added to the equation of momentum balance. An extensive literature has developed on controllability of such systems in classical fields of continuum mechanics, such as the Navier–Stokes equations or elasticity. The author's recent work has been concerned with viscoelastic flows, which raise fundamentally new issues.

Some results on controllability of linear viscoelastic media appeared in the 1980s [2–5]. in those papers, the viscoelastic medium is treated as a perturbation of the elastic case and the variables which are controlled are the displacement (if the material is a solid) and velocity. This misses an important issue, however. In contrast to elasticity, displacement and velocity are not sufficient information to determine the future evolution of a viscoelastic medium; in particular, zero displacement and velocity do not guarantee that the stresses are zero either at the present or at any future time. In practice, controlling stresses can actually be more important than controlling the motion; for instance in a manufacturing process stresses can cause subsequent deformation or damage.

In [1], the authors study linear viscoelastic fluids of Maxwell or Jeffreys type. For those equations, the state of the system is characterized by velocities and viscoelastic stresses, and the authors state an affirmative result on controllability. Unfortunately, their results do not hold as stated. Consider, for instance, the linear Maxwell fluid, with a constitutive law given by

$$\mathbf{T}_t + \lambda \mathbf{T} = \mu(\nabla \mathbf{v} + (\nabla \mathbf{v})^T), \tag{3}$$

where $\mathbf{T}$ is the stress tensor and $\mathbf{v}$ is the velocity. It is obvious from this constitutive equation that, if at any time $\mathbf{T}$ is the symmetric part of the gradient of a vector field, then this will also be so at later times. At best, therefore, $\mathbf{T}$ can be controlled within the subspace of those tensors which are symmetric parts of the gradient of a divergence-free vector field, in contrast to the theorems announced in [1]. It is likely that a suitably corrected version of the results in [1] holds.

This raises new questions, however. For nonlinear models of viscoelastic flow, it is in general not the case that $\mathbf{T}$ is the symmetric part of a gradient. Hence we encounter a situation where the linear problem has an invariant subspace which does not persist in the nonlinear case. Controllability in the linear case is restricted to this invariant subspace, and we are left with no intuition from this what to expect in the nonlinear case. Indeed, elementary examples show that there are many possibilities. Consider, for instance the system

$$\dot{x} = -x + f(t), \quad \dot{y} = x^2, \tag{4}$$

with initial and final conditions

$$x(0) = x_0, \qquad y(0) = y_0, \qquad x(T) = x_f, \qquad y(T) = y_f. \tag{5}$$

If we linearize, i.e. neglect the x^2-term, then it is clear that we can control the x variable, but on the other hand $y_f = y_0$ no matter what f is. On the other hand, if we include the x^2-term, we can reach any final state with $y_f > y_0$, with equality possible only if $x_0 = x_f = 0$. We shall encounter situations in viscoelastic flows which behave essentially like this elementary example.

In the rest of this paper, we shall consider viscoelastic shear flows. The flow domain is the interval $0 < x < L$, and the equation of motion is

$$\rho u_t = \tau_x + f(x, t), \tag{6}$$

where u is the velocity, τ is the shear stress, ρ is the density, and $f(x, t)$ is a body force (the control) which can be prescribed on some subinterval $[a, b]$ of $[0, L]$. For simplicity, we shall consider homogeneous Dirichlet boundary conditions for the velocity: $u(0, t) = u(L, t) = 0$.

We shall consider linear and nonlinear problems. In the linear case, we shall assume that the shear stress is related to the velocity by a multimode Maxwell model, i.e.

$$\tau(x, t) = \sum_{i=1}^{N} \tau_i(x, t), \quad \lambda_i(\tau_i)_t + \tau_i = \mu_i u_x, \tag{7}$$

where $\lambda_i \geq 0$, $\mu_i > 0$, and all the λ_i are different. We allow one of the λ_i to be zero, which corresponds to a Newtonian contribution to the stress.

In the nonlinear case, only single-mode models have so far been analyzed. For these models, we have a system of differential equations of the form

$$\mathbf{T}_t = \mathbf{g}(\mathbf{T}, u_x), \tag{8}$$

and $\mathbf{T}$ is a matrix of the form

$$\mathbf{T} = \begin{pmatrix} \sigma & \tau \\ \tau & \psi \end{pmatrix}. \tag{9}$$

The components σ and ψ, known as normal stresses, are simply zero if the system is linearized, but in the nonlinear problem they become nonzero. It turns out to be a nontrivial problem in general to characterize the values which they can take.

2 Linear Controllability

Controllability of linear shear flows was analyzed in [6], and we refer to this paper for further details and proofs. We note that homogeneous boundary conditions for the velocity imply that

$$\int_0^L u_x(x,t)\,dx = 0,\tag{10}$$

and we shall impose the corresponding condition

$$\int_0^L \tau_i(x,t)\,dx = 0\tag{11}$$

on the shear stress for the rest of this section. We denote by $L_0^2(0,L)$ the space of those functions in $L^2(0,L)$ which satisfy this integral constraint. For the case of a single relaxation mode, we have an exact controllability result analogous to those known for the wave equation. Indeed, the problem can be recast in a form equivalent to a lower order perturbation of the wave equation, which forms the basis (see [9]) of the following result from [6].

Theorem 2.1. *Assume $T > 2\sqrt{\frac{\rho\lambda}{\mu}}\max(a, L-b)$. For any choice of initial conditions*

$$u(x,0) = u_0(x) \in L^2(0,L), \quad \tau(x,0) = \tau_0(x) \in L_0^2(0,L),\tag{12}$$

there exists $f \in L^2((a,b)\times(0,T))$ such that the solution of the problem

$$\rho u_t = \tau_x + f, \quad \lambda\tau_t + \tau = \mu u_x,\tag{13}$$

with initial conditions

$$u(x,0) = u_0(x) \in L^2(0,L), \quad \tau(x,0) = \tau_0(x) \in L_0^2(0,L)\tag{14}$$

and boundary conditions

$$u(0,t) = u(L,t) = 0\tag{15}$$

satisfies

$$u(x,T) = \tau(x,T) = 0.\tag{16}$$

We note that if we can control to the final condition $u_f = \tau_f = 0$, we can control to any other final state, because the initial value problem is well-posed in both directions.

There is also an exact controllability result for several relaxation modes, but only in the case where the control is available on the entire interval $[0,L]$. To state such a result, we need to be careful about identifying the right regularity assumptions. For concreteness, we focus on the case of two relaxation modes; the general case is analogous (as long as none of the relaxation times are zero). We thus consider the system

$$\begin{aligned}\rho u_t &= \tau_x + \sigma_x + f(x,t)\\ \lambda_1\tau_t + \tau &= \mu_1 u_x,\\ \lambda_2\sigma_t + \sigma &= \mu_2 u_x.\end{aligned}\tag{17}$$

For the further discussion, it is convenient to introduce the new variables

$$\tau + \sigma = p, \quad \frac{\lambda_1}{\mu_1}\tau - \frac{\lambda_2}{\mu_2}\sigma = q. \tag{18}$$

In these new variables, we have the following regularity result; note that p and q are treated differently in this result.

Theorem 2.2. *Assume that $f \in L^2((0,T); L^2(0,L))$ and that the initial data satisfy $u_0 \in L^2(0,L)$, $p_0 \in L_0^2(0,L)$, $q_0 \in L_0^2(0,L) \cap H^1(0,L)$. Then (17) has a solution with the regularity $u \in C([0,T]; L^2(0,L))$, $p \in C([0,T]; L_0^2(0,L))$, $q \in C([0,T]; L_0^2(0,L) \cap H^1(0,L))$.*

The corresponding result on controllability is the following.

Theorem 2.3. *For any initial data $u(\cdot,0) = u_0 \in L^2(0,L)$, $p(\cdot,0) = p_0 \in L_0^2(0,L)$, and $q(\cdot,0) = q_0 \in L_0^2(0,L) \cap H^1(0,L)$, and any $T > 0$, there exists $f \in L^2((0,T); L^2(0,L))$ such that the corresponding solution of (17) satisfies $u(\cdot,T) = p(\cdot,T) = q(\cdot,T) = 0$.*

For multiple relaxation modes and control on a part of the interval, we can prove a result on approximate controllability.

Theorem 2.4. *Let*

$$G = \sum_{i=1}^{N} \frac{\mu_i}{\lambda_i}, \tag{19}$$

and assume that

$$T > 2\sqrt{\rho/G}\max(a, L - b). \tag{20}$$

For any given initial data $u(\cdot,0) = u_0 \in L^2(0,L)$, $\tau_i(\cdot,0) = \tau_i^0 \in L_0^2(0,L)$, there is a dense set of final states $u_f \in L^2(0,L)$, $(\tau_i)_f \in L_0^2(0,L)$ which can be reached at time T with some control $f \in L^2((a,b) \times (0,T))$.

3 Nonlinear Controllability

For nonlinear shear flows, the stress has more than one non-zero component. Linear results therefore tell us nothing about controllability of the nonlinear problem, even for small data. In general, we can expect it to be quite difficult to characterize the set of states which can be reached from a given initial condition. The simplest case which can be considered is that of homogeneous shear flow. In [7], this problem was consider for a number of popular constitutive models. In homogeneous shear flow, the velocity gradient is given by

$$\nabla \mathbf{u} = \begin{pmatrix} 0 & \dot{\gamma}(t) \\ 0 & 0 \end{pmatrix}, \tag{21}$$

and the stress tensor is

$$\mathbf{T} = \begin{pmatrix} \sigma & \tau \\ \tau & \psi \end{pmatrix}. \tag{22}$$

We regard the shear rate $\dot{\gamma}(t)$ as the control which we are free to pick, and we want to characterize the set of stresses which can be reached for a given constitutive model.

The models considered in [7] include the following:

1. The upper convected Maxwell (UCM) model:

$$\dot{\mathbf{T}} - (\nabla\mathbf{u})\mathbf{T} - \mathbf{T}(\nabla\mathbf{u})^T + \lambda\mathbf{T} = \mu(\nabla\mathbf{u} + (\nabla\mathbf{u})^T). \tag{23}$$

2. The Phan-Thien-Tanner (PTT) model:

$$\dot{\mathbf{T}} - (\nabla\mathbf{u})\mathbf{T} - \mathbf{T}(\nabla\mathbf{u})^T + \lambda\mathbf{T} + \kappa(\operatorname{tr}\mathbf{T})\mathbf{T} = \mu(\nabla\mathbf{u} + (\nabla\mathbf{u})^T). \tag{24}$$

3. The Giesekus model:

$$\dot{\mathbf{T}} - (\nabla\mathbf{u})\mathbf{T} - \mathbf{T}(\nabla\mathbf{u})^T + \lambda\mathbf{T} + \kappa\mathbf{T}^2 = \mu(\nabla\mathbf{u} + (\nabla\mathbf{u})^T). \tag{25}$$

4. The Johnson–Segalman (JS) model:

$$\dot{\mathbf{T}} - \frac{a+1}{2}((\nabla\mathbf{u})\mathbf{T} + \mathbf{T}(\nabla\mathbf{u})^T) - \frac{a-1}{2}((\nabla\mathbf{u})^T\mathbf{T} + \mathbf{T}(\nabla\mathbf{u}))$$
$$+ \lambda\mathbf{T} = \mu(\nabla\mathbf{u} + (\nabla\mathbf{u})^T). \tag{26}$$

For the case of homogeneous shear flow, the UCM model assumes the following form:

$$\dot{\sigma} - 2\tau\dot{\gamma} + \lambda\sigma = 0,$$
$$\dot{\tau} - \psi\dot{\gamma} + \lambda\tau = \mu\dot{\gamma},$$
$$\dot{\psi} + \lambda\psi = 0. \tag{27}$$

Clearly, ψ is unaffected by the shear rate and will be zero for all time if we assume that the flow history is also one of shear flow. Concerning the remaining stress components, we note that

$$\frac{d}{dt}(\mu\sigma - \tau^2) = -\lambda(\mu\sigma - \tau^2) + \lambda\tau^2. \tag{28}$$

It follows that

$$\mu\sigma(t) - \tau(t)^2 = e^{-\lambda t}(\mu\sigma(0) - \tau(0)^2) + \lambda\int_0^t e^{-\lambda(t-s)}\tau(s)^2\,ds, \tag{29}$$

and hence, at the final time T, we have

$$\mu\sigma(T) - \tau(T)^2 \geq e^{-\lambda T}(\mu\sigma(0) - \tau(0)^2), \tag{30}$$

with equality possible only if τ is zero throughout the interval $(0, T)$. On the other hand, we can pick a function $\tau(t)$ such that

$$\tau(0) = A, \qquad \tau(T) = B, \qquad \int_0^T e^{-\lambda(T-s)} \tau(s)^2 \, ds = C \qquad (31)$$

with arbitrary choices of A, B and $C > 0$ (with $C = 0$ also possible if $A = B = 0$). Having chosen such a τ, we can then define $\dot{\gamma}$ by the second equation of (27). Hence the set of attainable states $(\sigma(T), \tau(T))$ is defined precisely by the inequality (30).

Hence the attainable stresses for the UCM model form a subset of a two-dimensional manifold (i.e. $\psi = 0$), which is defined by the inequality (30). For the PTT, Giesekus, and JS models, the result is qualitatively similar, although the analysis is more complicated. That is, the attainable stresses are also a subset of a two-dimensional manifold which is characterized by an inequality. For specifics, I refer to [7].

Inhomogeneous shear flows are far more difficult to analyze. Only the upper convected Maxwell model has been studied at this point [8]. That is, we consider the system

$$\rho u_t = \tau_x + f(x, t),$$
$$\tau_t = -\lambda\tau + \mu u_x,$$
$$\sigma_t = -\lambda\sigma + 2\tau u_x, \qquad (32)$$

with homogeneous boundary conditions for u, initial data at time $t = 0$ and final conditions at time $t = T$. The body force f is the control we can adjust. Clearly, a necessary condition is that the inequality (30) is satisfied pointwise for every x, and it is natural to ask whether this condition is sufficient. In [8], it is shown that this is indeed the case if the control f is applied over the entire spatial interval $[0, L]$. If, on the other hand, the control is applied only on a subinterval, then the attainable stresses satisfy other constraints which are not of a pointwise nature.

References

1. A. Doubova, E. Fernandez-Cara, M. Gonzalez-Burgos, (2000) *Controllability results for linear viscoelastic fluids of the Maxwell and Jeffreys kinds*, C. R. Acad. Sci. Paris I 331, pp. 537–542.
2. I. Lasiecka, (1989) *Controllability of a viscoelastic Kirchhoff plate*, in: Control and Estimation of Distributed Parameter Systems (Vorau, 1988), pp. 237–247, Internat. Ser. Numer. Math. 91, Birkhäuser.
3. G. Leugering, (1987) *Time optimal boundary controllability of a simple linear viscoelastic liquid*, Math. Meth. Appl. Sci. 9, pp. 413–430.
4. G. Leugering, (1987) *Exact boundary controllability of an integro-differential equation*, Appl. Math. Optim. 15, pp. 223–250.

5. G. Leugering, (1984) *Exact controllability in viscoelasticity of fading memory type*, Applic. Anal. 18, pp. 221–243.

6. M. Renardy, (2005) *Are viscoelastic flows under control or out of control?* Syst. Cont. Lett. 54, pp. 1183–1193.

7. M. Renardy, (2005) *Shear flow of viscoelastic fluids as a control problem*, J. Non-Newt. Fluid Mech. 131, pp. 59–63.

8. M. Renardy, (2007) *On control of shear flow of an upper convected Maxwell fluid*, Z. angew. Math. Mech. 87, pp. 213–218.

9. J. Vancostenoble, (2000) *Exact controllability of a damped wave equation with distributed controls*, Acta Math. Hung. 89, pp. 71–92.

On Topological Optimization and Pollution in Porous Media

I. Faye, A. Sy, and D. Seck

Summary. Using the tools of topological optimization, we propose a method for the location of pollution in the porous media. Thus we propose a modeling of the problem and we study the nonlinear partial differential equation arising in the model.

1 Introduction

In this paper we deal about the location of the pollution in a porous medium. This type of problem may arise in agriculture. More generally, it may be meet when we manipulate chemical products. For illustration let us consider the water under the soil. If some chemical product is injected in the soil, the chemical body goes through the soil until reaching the water. In our world, drinkable water is nowadays rare resources. And it would be interesting to try to understand the evolution of the pollution in the water. It is obvious that this topic is studied by scientists. But we are not aware that people study these problems by using topological optimization. It is important to remark that these problems can be interpreted as inverse problems. In fact the problem is to quantify the rate of pollution in the water under the soil after measuring the quantity of the pollution on the accessible boundary.

Typically our aim is to locate the concentration of the pollution. For this, we use tools of topological optimization to get the distribution of the pollution in the porous media.

What is topological optimization? It is a topic belonging to the family of shape optimization. The goal in shape optimization is to optimize a criteria depending on the domain (or the design of the domain) in which state equations are verified. Most of the time these state equations are partial differential equations.

In classical shape optimization, the aim is to get an optimal condition without modifying the topology of the domain.

Unfortunately it is not always possible to get the best shape of the domain conserving however its topology.

There are some problems in classical shape optimization which don't get solution if there are not additional restrictive assumptions on the class of domain.

In relaxation theory and homogenization theory the problems studied could allow to overcome this gap that classical shape optimization lies around in some cases. But it is important to note the large progress that the classical shape optimization provide in industry (for example the improvement of the shape of wings of planes, cars, etc.), in mechanics, etc.

To take account other problems which cannot be treated by classical shape optimization, there is the advent of topological optimization. Here, it is possible to optimize the domain (to get an optimality condition) by changing the geometry and the topology of the domain. And it is important to emphasize that in that topic it is possible to change the topology by putting a hole. And on the boundaries of the holes on can even put Dirichlet or Neumann conditions.

In this paper, we focus our analysis on the pollution problem. But it would be interesting to use topological optimization in thermo-elasticity, evolution of tumor, image processing. We are about to end some works on these topics cf. [6, 12, 15].

This paper is organized as follows: In Sect. 2 we give a model from which we will do mathematical analysis. In Sect. 3, we study nonlinear partial differential equations coming from the model in the stationary case. Section 4 is devoted to topological optimization. This section gives theoretical results in asymptotic analysis. These results allows us to get ideas and information about the topological variation of the domain. In Sect. 5, we will present numerical simulations which give the location of the pollution.

2 A Model

Let $\mathcal{D}$ be a porous medium. Let us introduce, for $x \in \mathcal{D}$ and $t \in (0, T_1)$ $T_1 > 0$ is a fixed time

$\varepsilon(x, t)$ the effective porosity given by

$$\varepsilon(x, t) = \frac{dV_l}{dV_{total}},$$

where dV_l is an element of the volume of the fluid and dV_{total} an element of the total volume

$\sigma(x, t)$ the porosity given by

$$\sigma(x, t) = \frac{dV_v}{dV_{total}},$$

where dV_v is an element of the volume of the vacuum and q the Darcy velocity vector given by

$$q = \varepsilon V \,,$$

where V is the velocity vector of the fluid.

Ω is considered as an elementary domain of a porous domain $\mathcal{D}$.

We have $M(\Omega, t) = \int_{\Omega} dm$; dm is an element of the mass of the fluid.

$dm = \rho(x, t)\varepsilon(x, t)$; $\rho(x, t)$ is the fluid density of the solution.

For our model we will use these notations: $\rho_s (\text{kg m}^{-3})$ the fluid density of the solution given by

$$\rho_s = \frac{dm_{solution}}{dv_{solution}} \,.$$

$W(x,t)$ the fraction of the mass (concentration):

$$W(x, t) = \frac{dm_{solute}}{dm_{solution}}$$

$dm_{solution}$ is an element of the mass of the solution and dm_{solute} is an element of the mass of the pollutant.

2.1 The Conservation of the Mass of the Solution

We have

$$dm_{solution} = \rho_s dv_{solution} = \rho_s \frac{dv_{solution}}{dv_{total}} dv_{toal}$$

$$= \rho_s \varepsilon(x, t) dv_{total}$$

$$M_{solution}(\Omega, t) = \int_{\Omega} dm_{solution} = \int_{\Omega} \rho_s \varepsilon dx.$$

The principle of conservation of the mass stipulate that the variation of the mass in Ω is equal to the flux through the boundary of Ω with velocity V

$$\frac{dM_{solution}(\Omega, t)}{dt} = -\int_{\partial\Omega} \rho_s \varepsilon V \nu d\sigma.$$

Hence,

$$\int_{\Omega} \frac{\partial}{\partial t}(\rho_s \varepsilon) + \int_{\partial\Omega} \rho_s \varepsilon V \nu d\sigma = 0 \,.$$

By the Green formula we obtain

$$\int_{\Omega} (\frac{\partial}{\partial t}(\rho_s \varepsilon) + div(\rho_s \varepsilon V)) dx = 0 \quad \forall \Omega \subset \mathcal{D} \,.$$

Hence

$$\frac{\partial(\rho_s \varepsilon)}{\partial t} + div(\rho_s q) = 0 \quad \text{in } \mathcal{D} \,. \tag{1}$$

2.2 Conservation of the Mass of Pollutant Liquid

Here we consider for example that our pollutant liquid is: water + chemical concentration (it is homogeneous). By the formula given $W(x,t)$ we have

$$dm_{solute} = W(x,t)dm_{solution} = W(x,t)\rho_s(x,t)\varepsilon(x,t)dv_{total}$$

$$M(\Omega,t) = \int_\Omega dm_{solute} = \int_\Omega W\rho_s\varepsilon dx.$$

We use the principle conservation of the mass. This imply that

$$\int_\Omega (\frac{\partial}{\partial t}(W\rho_s\varepsilon) + div(W\rho_s q + J))dx = 0,$$

where J is the flux of dispersion diffusion. Hence

$$\frac{\partial}{\partial t}(W\rho_s\varepsilon) + div(W\rho_s q + J) = 0 \quad \forall \Omega \subset \mathcal{D}$$

$$\frac{\partial}{\partial t}(W\rho_s\varepsilon) + div(W\rho_s q + J) = 0 \quad \text{in } \mathcal{D}. \tag{2}$$

2.3 Conservation of the Momentum

If the porous medium is homogeneous the Darcy law is given by

$$q = -\frac{K}{\mu}(\nabla p + \rho_s g e_3) ,$$

where e_3 is third vector of the canonical basis of $\mathbb{R}^3$; p is the pressure, $ge_3 = \vec{g}$ is the gravity field, K the intrinsic permeability tensor, μ the dynamic viscosity and K/μ hydraulic conductivity. We will assume the following ellipticity condition:
$\frac{K}{\mu}(x)\xi.\xi \geq \alpha_1\|\xi\|^2$; α_1 is a positive constant.

If we have some weak concentration the flux of dispersion diffusion J is determined by the Fick law

$$J = -\rho_s D\nabla W ,$$

where D be the tensor of dispersion diffusion. We assume also the ellipticity condition $D = (d_{ij})_{1\leq i,j\leq n}$; $d_{ij}\xi_i\xi_j \geq \alpha_2\|\xi\|^2$, where α_2 a positive constant.

Remark 2.1. $\rho_s = \rho_s(T,p,W)$. For our study we suppose that ρ_s satisfy the relation

$$\rho_s = \rho_0 \exp(\beta_T(T - T_0) + \beta_p(p - p_0) + \gamma W),$$

here β_T, β_p et γ are constants; p designates the pressure of the fluid, T the temperature and W the concentration. $\rho_0 = \rho(T_0, p_0, 0)$ is a reference density; T_0 and p_0 are respectively the reference temperature and the reference pressure. This expression is used in engineering science see for instance [14].

Finally we have a system of equations

$$
\begin{cases}
\frac{\partial \varepsilon \rho_s}{\partial t} + div(\rho q) = 0 \\
\frac{\partial (\varepsilon \rho_s W)}{\partial t} + div(\rho_s W q + J) = 0 \\
J = -\rho_s D \nabla W \\
\rho_s = \rho_0 \exp[\beta_T(T - T_0) + \beta_p(p - p_0) + \gamma W] \\
q = -\frac{K}{\mu}(\nabla p + \rho_s g e_3)
\end{cases}
\tag{3}
$$

Remark 2.2. The porosity ε of the medium can be given by many laws. We can quote [16]

1. The Garner law(1958) given by

$$
\varepsilon = \frac{\varepsilon_s - \varepsilon_r}{1 + (\alpha h)^\beta} + \varepsilon_r \qquad \text{for} \quad h \leq 0
$$

$$
\varepsilon = \varepsilon_s \qquad \text{for} \quad h > 0
$$

2. The Brooks and Correy law (1964) where

$$
\varepsilon = (\varepsilon_s - \varepsilon_r)(\frac{h}{h_0})^\beta + \varepsilon_r \qquad \text{for} \quad h \leq h_l
$$

$$
\varepsilon = a.h^5 + bh^4 + \varepsilon_s \qquad \text{for} \quad h_l < h \leq 0
$$

$$
\varepsilon = \varepsilon_s \qquad \text{for} \quad h > 0
$$

3. The Van Genuchten Law (1980) where

$$
\varepsilon = (\varepsilon_s - \varepsilon_r)(1 + (\alpha h)^\beta)^\tau + \varepsilon_r \qquad \text{for} \quad h \leq 0
$$

$$
\varepsilon = \varepsilon_s \qquad \text{for} \quad h > 0
$$

with $\tau = 1 - 1/\beta$

h is the pressure measured relatively at the atmospheric pressure and expressed in columns of water.

To fix the idea we will use the Van Genuchten law for our model.

Solving these equations in the porous medium is very difficult. To overcome these difficulties, the following hypothesis:

- **H-1** ρ_s is a constant. Replacing q by its expression in the first equation of (3) we obtain the expansion of the divergence

$$
\frac{\partial}{\partial t}\varepsilon - div(\frac{K}{\mu}\nabla p) - \rho_s g div(\frac{K}{\mu}e_3) = 0 \quad \text{in} \quad \Omega \times (0, T_1),
\tag{4}
$$

where T_1 is a fixed time.

Replacing q and J by their expressions in the second equation (3) and after simplifications we have

$$
\frac{\partial}{\partial t}(\varepsilon W) - div(W\frac{K}{\mu}\nabla p) - \rho_s g div(\frac{K}{\mu}W e_3) - div(D \nabla W) = 0 \quad \text{in} \quad \Omega \times (0, T_1).
\tag{5}
$$

- **H-2** The hydraulic conductivity tensor is a constant positive: ($\frac{K}{\mu} = \beta I d_3, \beta > 0$) and D is a constant positive ($D = a I d_3, a > 0$).

Using the hypothesis (**H-2**), (4) and (5) become respectively

$$\frac{\partial \varepsilon}{\partial t} - \frac{K}{\mu} \Delta p = 0 \quad \text{in} \quad \Omega \times (0, T_1). \tag{6}$$

$$\frac{\partial}{\partial t}(\varepsilon W) - \frac{K}{\mu} div(W \nabla p) - \rho_s g \frac{K}{\mu} \frac{\partial W}{\partial z} - a \Delta W = 0 \quad \text{in } \Omega \times (0, T_1). \tag{7}$$

Using (6), (7) becomes

$$\varepsilon \frac{\partial}{\partial t} W - \frac{K}{\mu} \nabla W \nabla p - \rho_s g \frac{K}{\mu} \frac{\partial W}{\partial z} - a \Delta W = 0 \quad \text{in} \quad \Omega \times (0, T_1). \tag{8}$$

To (6) and (8) we are going to add boundaries conditions adapted to pollution in porous medium. We obtain finally some boundaries and initial value problems given by

$$\begin{cases} \frac{\partial \varepsilon}{\partial t} - \beta \Delta p = 0 & \Omega \times (0, T_1) \\ \varepsilon(x, 0) = \varepsilon_0 & \text{in} \quad \Omega \times \{t = 0\} \\ \varepsilon = \varepsilon_1 & \partial \Omega \setminus \Gamma_1 \times (0, T_1) \\ \varepsilon = \varepsilon_s & \Gamma^1 \times (0, T) \end{cases} \tag{9}$$

and

$$\begin{cases} \varepsilon \frac{\partial W}{\partial t} - \frac{k}{\mu} \nabla W \nabla p - \frac{k}{\mu} \rho_s g \frac{\partial W}{\partial z} - D \Delta W = 0 & \text{in} \quad \Omega \times (0, T_1) \\ \frac{\partial W}{\partial n} = 0 & \partial \Omega \setminus \Gamma_1 \times (0, T_1) \\ W = V & \Gamma_1 \times (0, T_1) \\ W(x, 0) = W_0 & \text{in} \quad \Omega \times \{t = 0\} \end{cases} \tag{10}$$

Other assumptions to get the model which we will study in the next section are:

- **H-3** In the porous medium we have a steady state, this means that $\frac{\partial}{\partial t} = 0$.
- **H-4** The evolution is isotherm.

Let us recall that by hypothesis (**H-1**) ρ_s is constant and is given by the expression

$$\rho_s = \rho_0 \exp[\beta_T(T - T_0) + \beta_p(p - p_0) + \gamma W].$$

Using hypotheses (**H-1**) and (**H-4**) we can find a relation between p the pressure and W the concentration:

$$\log \frac{\rho_s}{\rho_0} = \beta_p(p - p_0) + \gamma W$$

then,

$$p = p_0 + \frac{1}{\beta_p}[\log \frac{\rho_s}{\rho_0} - \gamma W].$$

We deduce

$$\nabla p = -\frac{\gamma}{\beta_p} \nabla W.$$

Applying the hypotheses (**H-1**)–(**H-4**) and replacing ∇p by its value in (10) we obtain

$$\begin{cases} -\Delta p = 0 & \text{in } \Omega_1 \\ p = \frac{[(\frac{\varepsilon_1 - \varepsilon_r}{\varepsilon_s - \varepsilon_r})^{-\frac{1}{m}} - 1]^{\frac{1}{n}}}{\alpha} & \text{on } \partial\Omega \setminus \Gamma_1 \\ p = 0 & \text{on } \Gamma_1 \end{cases} \qquad (11)$$

and

$$\begin{cases} +\beta|\nabla W|^2 - \beta\rho_s g \frac{\partial W}{\partial z} - \frac{D_0}{\rho_0}\Delta W = 0 & \text{in } \Omega \\ \frac{\partial}{\partial n} W = 0 & \text{on } \partial\Omega \setminus \Gamma^1 \\ W = V & \text{on } \Gamma^1 \end{cases} \qquad (12)$$

Remark 2.3. The boundary condition of (11) is obtained by the Van Genuchten law.

In the following section we will study the above boundaries value problems.

3 Study of Partial Differential Equation (PDE)

After our model we fall on a second-order partial differential equation.

3.1 Fixed Point Theorem Approach

In this section we want to study this PDE (12) via a Schauder fixed point theorem which can be found in [2].

3.1.1 Solution of PDE in $H_0^1(\Omega)$

First of all we are interested by the resolution of the boundary value problem

$$\begin{cases} -\Delta u + |\nabla u|^2 - \frac{\partial u}{\partial x_N} = 0 & \text{in } \Omega \\ u \in H_0^1(\Omega) \end{cases}, \qquad (13)$$

where $u : \Omega \subset \mathbb{R}^N \to \mathbb{R}$, with Ω a bounded open set of $\mathbb{R}^N$, $N \geq 2$.

Proposition 3.1. *Let Ω be a bounded open set of $\mathbb{R}^N$. There exist a solution of (13).*

Proof. We are going to use a fixed point theorem to prove the Proposition 3.1. The objective here is "to disappear" the nonlinearity of the term $|\nabla u|^2$. For this we introduce, for $v \in L^2(\Omega)$, the following boundary value problem:

$$\begin{cases} -\Delta u + \nabla u \nabla v - \frac{\partial u}{\partial x_N} = 0 & \text{in } \Omega \\ u \in H_0^1(\Omega) \end{cases}. \qquad (14)$$

We transform the nonlinear problem into a linear boundary value problem for all $v \in L^2(\Omega)$. Multiplying the first equation by a test function $\varphi \in H_0^1(\Omega)$ and by integrating we have

$$\int_\Omega \nabla u \nabla \varphi + \int_\Omega \nabla u \nabla v \varphi - \int_\Omega \frac{\partial u}{\partial x_N} \varphi = 0 \quad , \forall u, \varphi \in H_0^1(\Omega).$$

Let

$$a_v(u, \varphi) = \int_\Omega \nabla u \nabla \varphi + \int_\Omega \nabla u \nabla v \varphi - \int_\Omega \frac{\partial u}{\partial x_N} \varphi$$

be the variational form of the boundary value problem. We look for $u \in H_0^1(\Omega)$ such that $a_v(u, \varphi) = 0$ for all $v \in L^2(\Omega)$ and $\varphi \in H_0^1(\Omega)$. It is easy to show that $a_v(\cdot, \cdot)$ is a bilinear form and is continuous. It remains to show that $a_v(., .)$ is coercive, i.e., $\exists \alpha > 0$ such that $a_v(u, u) \geq \alpha \|u\|_{H_0^1(\Omega)}$. For $\varphi = u$ we have

$$a_v(u, u) = \int_\Omega |\nabla u|^2 + \int_\Omega \nabla u \nabla v \, u - \int_\Omega \frac{\partial u}{\partial x_N} u.$$

$$\int_\Omega \nabla u \nabla v \, u = \int_\Omega \nabla v \nabla u \, u = \int_\Omega \nabla v \nabla (\frac{1}{2} u^2)$$

$$= \sum_{i=1}^N \int_\Omega \frac{\partial v}{\partial x_i} \cdot \frac{\partial}{\partial x_i} (\frac{1}{2} u^2).$$

Using the Green formula we obtain

$$= \sum_{i=1}^N \int_{\partial \Omega} \frac{\partial v}{\partial x_i} \cdot (\frac{1}{2} u^2) - \sum_{i=1}^N \int_\Omega \frac{\partial^2 v}{\partial x_i^2} \cdot \frac{1}{2} u^2.$$

As $u \in H_0^1(\Omega)$ the first term of the right-hand side disappear and

$$\int_\Omega \nabla u \nabla v. u = -\frac{1}{2} \int_\Omega div(\nabla v) u^2.$$

For the last term of $a_v(u, u)$ we have

$$\int_\Omega \frac{\partial u}{\partial x_N} . u = \int_{\partial \Omega} u. u n_N - \int_\Omega u. \frac{\partial u}{x_N} ,$$

where $n = (n_1, \dots, n_N)$ is the unit exterior normal. This imply

$$2 \int_\Omega \frac{\partial u}{\partial x_N} . u = \int_{\partial \Omega} u. u. n_N = 0$$

because $u \in H_0^1(\Omega)$. For $v \in H_0^1(\Omega)$, we have

$$a_v(u, u) = \int_\Omega |\nabla u|^2 - \frac{1}{2} \int_\Omega div(\nabla v) u^2.$$

If we introduce the set

$$K = \{v \in H_0^1(\Omega) \quad \text{such that} \quad m < div(\nabla v) \le 0\},$$

where m is a negative constant, $a_v(u, u) \ge \int_\Omega |\nabla u|^2$ which is a square norm in $H_0^1(\Omega)$, then $a_v(\cdot, \cdot)$ is coercive. By Lax–Milgram theorem for all $v \in L^2(\Omega)$, there exists $u = u(v)$ solution of the variational problem

$$\text{find } u \in K \quad \text{such that} \quad a_v(u, \varphi) = 0 \quad \forall \varphi \in K.$$

Let us introduce the operator

$$T : L^2(\Omega) \to L^2(\Omega)$$

$$v \mapsto T(v)$$

and let $u = T(v) \in K$ be the solution of the linear boundary value problem (14). T is continuous because is a composition of continuous functions. In fact we have

$$T : L^2(\Omega) \to L^2(\Omega) \to K \to L^2(\Omega)$$

$$v \mapsto v \mapsto u_v \mapsto u_v = T(v).$$

For $u = T(v)$ the variational formulation becomes

$$\int_\Omega \nabla T(v) \nabla \varphi + \int_\Omega \nabla v \nabla T(v) \varphi - \int_\Omega \frac{\partial T(v)}{\partial x_N} \varphi = 0.$$

For $\varphi = T(v)$, we have

$$\int_\Omega |\nabla T(v)|^2 + \int_\Omega \nabla v \nabla T(v).T(v) - \int_\Omega \frac{\partial T(v)}{\partial x_N} T(v) = 0.$$

If we consider the trilinear form

$$b(v, u, \varphi) = \int_\Omega \nabla v \nabla u \varphi = \sum_{i=1}^N \int_\Omega \frac{\partial v}{\partial x_i} \frac{\partial u}{\partial x_i} \varphi.$$

Then we have

$$b(v, u, \varphi) + b(v, \varphi, u) = \sum_{i=1}^N \int_\Omega \frac{\partial v}{\partial x_i} \frac{\partial u}{\partial x_i} \varphi + \sum_{i=1}^N \int_\Omega \frac{\partial v}{\partial x_i} \frac{\partial \varphi}{\partial x_i} u$$

$$= \sum_{i=1}^N \int_\Omega \frac{\partial v}{\partial x_i} (\frac{\partial u}{\partial x_i} \varphi + \frac{\partial \varphi}{\partial x_i} u)$$

$$= \sum_{i=1}^N \int_\Omega \frac{\partial v}{\partial x_i} \frac{\partial}{\partial x_i} (u\varphi).$$

Using the Green formula we have

$$b(v, u, \varphi) + b(v, \varphi, u) = -\int_\Omega \Delta v.u\varphi.$$

Taken $\varphi = u$ we obtain

$$b(v, u, u) = -\frac{1}{2}\int_\Omega \Delta v.u^2.$$

For $u = T(v)$

$$\int_\Omega \nabla v \nabla T(v).T(v) = b(v, T(v), T(v)) = -\frac{1}{2}\int_\Omega \Delta T(v).T^2(v).$$

The variational form gives

$$\int_\Omega |\nabla T(v)|^2 = -\frac{1}{2}\int_\Omega \Delta T(v).T^2(v).$$

Hence,

$$\int_\Omega |\nabla T(v)|^2 \leq -\frac{m}{2}\int_\Omega T^2(v) \leq -mc\|T(v)\|_{H_0^1(\Omega)}$$

$$\leq mc'\|\nabla T(v)\|_{L^2(\Omega)}$$

$$\|\nabla T(v)\|_{L^2(\Omega)} \leq mc'.$$

Let us introduce

$$C = \{v \in H_0^1(\Omega), \quad m < \Delta v < 0, \quad \text{et} \quad \|v\|_{L^2(\Omega)} \leq mc'\}.$$

We can say that $T(C) \subset C$. The injection of $H_0^1(\Omega) \hookrightarrow C$ is compact. C is compact and convex by construction. By Schauder fixed theorem there exists a fixed point such that $T(u) = u$.   □

Remark 3.2. For the following problem

$$\begin{cases} \Delta u = 0 & \Omega \\ |\nabla u|^2 - \frac{\partial u}{\partial x_N} = 0 & \Omega \\ u \in H_0^1(\Omega) \end{cases} \tag{15}$$

we consider the space

$$K = \{v \in H_0^1(\Omega), \quad \text{such that} \quad \Delta v = 0\}.$$

By the same approach as above we show that this problem admits a solution. The variational problem gives $\int_\Omega |\nabla u|^2 = 0$. Then $|\nabla u| = 0$ almost everywhere hence u is a constant *a.e.*

3.1.2 The Case of Mixed Boundary Condition

We study here the partial differential equation meted in the proposed model with mixed boundaries value conditions

$$
\begin{cases}
-\alpha \Delta u + \beta |\nabla u|^2 - \gamma \frac{\partial u}{\partial x_N} = 0 & \text{in} \quad \Omega \\
\qquad\qquad\qquad\qquad u = 0 & \text{on} \quad \Gamma^1 \\
\qquad\qquad\qquad\qquad \frac{\partial u}{\partial n} = 0 & \text{on} \quad \Gamma^2
\end{cases}
\tag{16}
$$

where $\partial \Omega = \Gamma^1 \cup \Gamma^2$ and $\overset{.}{\Gamma^1} \cap \overset{.}{\Gamma^2} = \emptyset$ where $\overset{.}{\Gamma}$ denotes the interior of Γ.

Proposition 3.3. *Let Ω be a bounded open set of $\mathbb{R}^N$, $N \geq 2$. Then there exists a solution of problem (16).*

Proof. We use here the Schauder fixed point theorem to prove the proposition. Let us consider the linear mixed boundary values problem derived from (16).

$$
\begin{cases}
-\alpha \Delta u + \beta \nabla u \nabla v - \gamma \frac{\partial u}{\partial x_N} = 0 & \text{in} \quad \Omega \\
\qquad\qquad\qquad\qquad u = 0 & \text{on} \quad \Gamma^1 \\
\qquad\qquad\qquad\qquad \frac{\partial u}{\partial n} = 0 & \text{on} \quad \Gamma^2
\end{cases}
\tag{17}
$$

Multiplying the first equation by a test function and integrating we obtain the variational formula

$$
\alpha \int_\Omega \nabla u \nabla \varphi + \beta \int_\Omega \nabla u \nabla v \varphi - \alpha \int_{\partial \Omega} \frac{\partial u}{\partial n} \varphi - \gamma \int_\Omega \frac{\partial u}{\partial x_N} \varphi = 0,
$$

where α, β et γ are constants strictly positives and n the exterior unit normal. We work with a set of the form

$$
V = \{ v \in H^s(\Omega) \quad \text{such that} \quad v = 0 \quad \text{on} \quad \Gamma^1, \quad \frac{\partial v}{\partial n} = 0 \quad \text{on} \quad \Gamma^2 \}
$$

with $s \geq 2$. The variational problem can be sum up as follows: for all $v, \varphi \in V$ we look for $u \in V$ such that

$$
\alpha \int_\Omega \nabla u \nabla \varphi + \beta \int_\Omega \nabla u \nabla v \varphi - \gamma \int_\Omega \frac{\partial u}{\partial x_N} \varphi = 0 \quad \forall \varphi \in V.
$$

We have already calculated the trilinear form $b(v, u, \varphi)$, in the above section. Using the Green formula in the last integral and taking $u = \varphi$, we have

$$
\int_\Omega u \frac{\partial u}{\partial x_N} = \frac{1}{2} \int_{\partial \Omega} u^2 n_N.
$$

Then

$$
\int_\Omega u \frac{\partial u}{\partial x_N} = \frac{1}{2} \int_{\partial \Omega} u^2 n_N.
$$

220   I. Faye et al.

Then

$$a_v(u, u) = \alpha \int_\Omega |\nabla u|^2 + \frac{\beta}{2} \sum_{i=1}^N \int_{\partial\Omega} u^2 \frac{\partial v}{\partial x_i} n_i - \frac{\beta}{2} \int_\Omega \Delta v . u^2 - \frac{\gamma}{2} \int_{\partial\Omega} u^2 n_N.$$

If we suppose that $\Delta v \in L^\infty(\Omega)$ then the accounts take sense. For that reason we choose $v \in L^4(\Omega)$ and the condition $m \le \Delta v \le \delta < 0$, i.e., Δv negative and bounded. The objective is to obtain a set V_2 in which our bilinear form is continuous and coercive. We can take

$$V_2 = \{v \in W^{1,4}(\Omega) \quad \text{such that} \quad \frac{\partial v}{\partial n} = 0 \quad \text{on} \quad \Gamma^2, \quad v = 0 \quad \text{on} \quad \Gamma^1 \quad \text{and}$$

$$m \le \Delta v \le \delta < 0\}.$$

As Ω is bounded, the set $V_2 \subset L^4(\Omega)$ and $V_2 \subset W^{1,2}(\Omega)$. We use here the norm of $W^{1,2}(\Omega)$ as the norm of V_2; the injection of

$$L^\infty(\Omega) \hookrightarrow W^{1,4}(\Omega)$$

is continuous. To prove this assertion we can see [5]. For all these hypothesis we have $V_2 \subset W^{1,4}(\Omega)$, i.e., all functions of V_2 are bounded. Another hypothesis is to choose n the exterior unit normal so that $n_N(x)$ should be negative for all $x \in \partial\Omega$. This choice is possible because it is a natural condition if Ω is regular. Under these hypothesis, the bilinear form satisfies the inequality

$$a_v(u, u) \ge \alpha \int_\Omega |\nabla u|^2 - \frac{\delta}{2} \int_\Omega u^2$$

$$\ge \min\{\alpha, -\delta/2\} \|u\|^2_{H^1(\Omega)}$$

which proves the coercivity of the bilinear form. By Lax–Milgram theorem there exists a weak solution of the mixed boundary value problem in V_2.

Let $u = \phi(v) \in V_2$ the solution of the mixed boundary problem. ϕ is defined $\phi : L^4(\Omega) \to L^4(\Omega)$. The variational problem gives

$$\alpha \int_\Omega |\nabla\phi(v)|^2 + \beta \int_\Omega \nabla v \nabla\phi(v).\phi(v) - \gamma \int_{\partial\Omega} \frac{\partial\phi(v)}{\partial x_N}\phi(v) = 0.$$

This gives

$$\alpha \int_\Omega |\nabla\phi(v)|^2 - \frac{\beta}{2} \int_\Omega \Delta v.\phi^2(v) = \frac{\gamma}{2} \int_{\Gamma^2} \phi^2(v) n_N$$

$$= \frac{1}{2} \int_{\partial\Omega} \phi(v).\phi(v) n_N$$

$$\le \frac{C}{2} \int_{\partial\Omega} |\phi(v)|$$

$$\le \underbrace{\frac{C}{2}}_{K_0} \|\phi(v)\|_{H^1(\Omega)}$$

hence,

$$\|\phi(v)\|_{H^1(\Omega)} \leq K_0.$$

We choose

$$C = \{v \in V_2, \|v\|_{H^1(\Omega)} \leq K_0\}.$$

Then $\phi(C) \subset C$ and ϕ is continuous. By Schauder fixed theorem the fixed point exists. $\square$

Remark 3.4. For the following problem

$$\begin{cases} \alpha \Delta u = 0 \quad \text{on } \Omega \\ \beta |\nabla u|^2 - \gamma \frac{\partial u}{\partial x_N} = 0 \quad \text{on } \Omega \\ \qquad u = 0 \quad \text{in } \Gamma^1 \\ \qquad \frac{\partial u}{\partial n} = 0 \quad \text{in } \Gamma^2 \end{cases} \tag{18}$$

we consider

$$V_2 = \{v \in W^{1,4}(\Omega) \quad \text{such as} \quad \frac{\partial v}{\partial n} = 0 \quad \text{on} \quad \Gamma^2, \quad v = 0 \quad \text{on} \quad \Gamma^1 \quad \text{and}$$

$$\Delta v = 0\}.$$

We show by the same approach that this problem admits a solution.

3.2 Change of Variables Approach

Let Ω be an open subset of $\mathbb{R}^N$. We consider here an initial-value problem for a quasilinear parabolic equation

$$\begin{cases} u_t - \Delta u + \alpha |\nabla u|^2 + \frac{\partial u}{\partial z} = 0 \text{ in } \Omega \times (0, T_1) \\ \qquad u = f \text{ on } \partial\Omega \times (0, T_1) \\ \quad u(x, 0) = g \text{ in } \Omega \times (t = 0) \end{cases} \tag{19}$$

where $\alpha > 0$. This king of nonlinear partial differential equation arises in pollution as seen in the model.

Proposition 3.5. *The problem (19) has at least one solution.*

Proof. Let us suppose at first that u is the solution of (19). Let us set

$$\omega := \phi(u),$$

where $\phi : \mathbb{R} \to \mathbb{R}$ is a smooth function, not yet specified. We will choose ϕ such that ω solve a linear equation. We have

$$\omega_t = \phi'(u)u_t, \quad \nabla\omega = \phi'(u)\nabla u, \quad \text{and} \quad \Delta\omega = \phi'(u)\Delta u + \phi''(u)|\nabla u|^2$$

and consequently (19) implies

$$
\begin{aligned}
\omega_t = \phi'(u)u_t &= \phi'(u)\left[\Delta u - \alpha|\nabla u|^2 - \frac{\partial u}{\partial z}\right] \\
&= \underbrace{\phi'(u)\Delta u}_{=\Delta\omega - \phi''(u)|\nabla u|^2} \ -\alpha\phi'(u)|\nabla u|^2 \underbrace{-\phi'(u)\frac{\partial u}{\partial z}}_{=\frac{\partial\omega}{\partial z}}
\end{aligned}
$$

$$
\omega_t = \Delta\omega - |\nabla u|^2(\phi''(u) + \alpha\phi'(u)) - \frac{\partial\omega}{\partial z}.
$$

Thus

$$
\omega_t - \Delta\omega + \frac{\partial\omega}{\partial z} = -|\nabla u|^2(\phi''(u) + \alpha\phi'(u)) = 0
$$

provided that we choose ϕ to satisfy $\phi''(u) + \alpha\phi'(u) = 0$. We solve this differential equation and we obtain $\phi(u) = \lambda e^{-\alpha u}$, $\lambda \in \mathbb{R}$. Note that we have just to prove that such ϕ exists. Thus if u satisfies (19), then

$$
\omega = \lambda e^{-\alpha u} \tag{20}
$$

solves this initial-value problem

$$
\begin{cases}
\omega_t - \Delta\omega + \frac{\partial\omega}{\partial z} = & 0 \quad \text{on } \Omega \times (0, T_1) \\
\omega & = \lambda e^{-\alpha f} \text{ in } \partial\Omega \times (0, T_1) \\
\omega & = \lambda e^{-\alpha g} \text{ in } \Omega \times (t = 0)
\end{cases} \tag{21}
$$

Let us remark that using Galerkin's method, one can prove that (21) have a unique weak solution for all $\lambda \in \mathbb{R}$ see for instance [5] for more details. We can choose $\lambda = 1$ for the following.

Conversely, let ω be the solution of (21), and $u = -\frac{1}{\alpha}\log\omega$, then $\omega = e^{-\alpha u}$

$$
\omega_t = -\alpha u_t e^{-\alpha u}, \quad \nabla\omega = -\alpha\nabla u e^{-\alpha u}
$$

$$
\Delta\omega = -\alpha e^{-\alpha u}(\Delta u - \alpha|\nabla u|^2), \quad \frac{\partial\omega}{\partial z} = -\alpha e^{-\alpha u}\frac{\partial u}{\partial z}
$$

thus

$$
\omega_t - \Delta\omega + \frac{\partial\omega}{\partial z} = -\alpha e^{-\alpha u}\left(u_t - \Delta u + \alpha|\nabla u|^2 + \frac{\partial u}{\partial z}\right) = 0
$$

consequently u solve the partial differential equation

$$
\begin{cases}
u_t - \Delta u + \alpha|\nabla u|^2 + \frac{\partial u}{\partial z} = 0 \text{ in } \Omega \times (0, T_1) \\
u = f \text{ on } \partial\Omega \times (0, T_1) \\
u = g \text{ in } \Omega \times \{t = 0\} \quad \square
\end{cases}
$$

Remark 3.6. When $\frac{\partial\omega}{\partial z} = 0$, $\Omega = \mathbb{R}^N$ and $t > 0$ (21) becomes

$$
\begin{cases}
u_t - \Delta u = & 0 \quad \text{in } \mathbb{R}^N \times (0, T_1) \\
u & = e^{-\alpha g} \text{ in } \mathbb{R}^N \times \{t = 0\}
\end{cases} \tag{22}
$$

the unique bounded solution of (22) is

$$\omega(x,t) = \frac{1}{4\pi t} \int_{\mathbb{R}^N} e^{-\frac{|x-y|^2}{4t}} e^{-\alpha g(y)} dy \quad (x \in \mathbb{R}^N, t > 0);$$

and as (20) implies

$$u = -\frac{1}{\alpha} \log \omega,$$

we obtain thereby the explicit formula

$$u(x,t) = -\frac{1}{\alpha} \log \left(\frac{1}{4\pi t} \int_{\mathbb{R}^N} e^{-\frac{|x-y|^2}{4\alpha t}} e^{-\alpha g(y)} dy \right) \quad (x \in \mathbb{R}^N, t > 0); \quad (23)$$

for a solution of (19).

In the stationary case, we have

$$\begin{cases} -\Delta u + \alpha |\nabla u|^2 = 0 \text{ on } \Omega \\ \qquad\qquad u \qquad\quad = g \text{ in } \partial\Omega \end{cases} \cdot \qquad (24)$$

As in above, setting $u = -\frac{1}{\alpha} \log(\omega)$, it follows that ω solve the partial linear differential equation

$$\begin{cases} -\Delta\omega = \quad 0 \quad \text{on } \Omega \\ \quad \omega \;\; = e^{-\alpha g} \text{ in } \partial\Omega \end{cases} \qquad (25)$$

which can be easily solved see for example [5].

4 Topological Optimization

In this section we optimize the functional

$$J(\Omega, W) = \int_\Omega (W - W_1)^2 dx,$$

where W_1 is a given function in $L^2(\Omega)$ which represents a target and W is the solution of the partial differential equation

$$\begin{cases} +\beta|\nabla W|^2 - \beta\rho_s g \frac{\partial W}{\partial z} - a\Delta W = 0 \quad \text{in} \quad \Omega \\ \qquad\qquad \frac{\partial}{\partial n} W = 0 \quad \text{on} \quad \partial\Omega \backslash \Gamma^1 \\ \qquad\qquad W = V \quad \text{on} \quad \Gamma^1 \end{cases} \qquad (26)$$

with α, β and γ are positive constants. In the model we have supposed that

$$\rho_s = \rho_0 \exp(\beta_T(T - T_0) + \beta_p(p - p_0) + \gamma W)$$

is a constant.

The expression of ρ_s gives a relation between the pressure p and the concentration W. It is given by

$$p = p_0 + \frac{1}{\beta_p}[\log \frac{\rho_s}{\rho_0} - \gamma W].$$

The porosity of the medium also is given by the Van Genuchten law

$$\varepsilon = (\varepsilon_s - \varepsilon_r)(1 + (\alpha p)^n)^{-m} + \varepsilon_r.$$

Then we obtain

$$p = \frac{[(\frac{\varepsilon - \varepsilon_r}{\varepsilon_s - \varepsilon_r})^{-\frac{1}{m}} - 1]^{\frac{1}{n}}}{\alpha},$$

the boundary condition of p:

- $\varepsilon = \varepsilon_s$ in Γ_1 then $p = 0$ on Γ_1.
- $\varepsilon = \varepsilon_s$ in $\partial\Omega \setminus \Gamma_1$ then

$$p = \frac{[(\frac{\varepsilon_1 - \varepsilon_r}{\varepsilon_s - \varepsilon_r})^{-\frac{1}{m}} - 1]^{\frac{1}{n}}}{\alpha} \quad \text{on} \quad \partial\Omega \setminus \Gamma_1.$$

We are going to replace this expression by a more general one

$$p = p_1 \quad \text{on} \quad \partial\Omega \setminus \Gamma_1,$$

where p_1 is a given function. Finally p satisfy the boundary value problem

$$\begin{cases} -\Delta p = 0 \text{ in } \Omega \\ p = p_1 \text{ on } \partial\Omega \setminus \Gamma_1 \\ p = 0 \text{ on } \Gamma_1 \end{cases} \qquad (27)$$

W is solution of a nonlinear boundary value problem. As p and W are linked by the law of ρ_s, we transform the problem in W into a problem in p.

In fact using the relation between p and W the topological optimization problem in W is equivalent to look for Ω with

$$\min_{\Omega(\epsilon)} J(\Omega(\epsilon)),$$

where

$$J(\Omega(\epsilon)) = \int_{\Omega(\epsilon)} (1/\gamma[\ln \frac{\rho_s}{\rho_0} - \beta_p(p_\epsilon - p_0)] - W_1)^2 dx, \qquad (28)$$

where $\epsilon \in (0, 1)$ is a small parameter and p_ϵ is solution of

$$\begin{cases} -\Delta p_\epsilon = 0 \text{ in } \Omega(\epsilon) \\ p_\epsilon = p_1 \text{ on } \partial\Omega \setminus \Gamma_1 \\ p_\epsilon = 0 \text{ on } \Gamma_1 \\ p_\epsilon = 0 \text{ on } \partial\omega_\epsilon \end{cases} \qquad (29)$$

under the constraint

$$|\nabla p_\epsilon|^2 - g\rho_s \frac{\partial p_\epsilon}{\partial z} = 0 \quad \text{in} \quad \Omega(\epsilon).$$

Topological optimization appears as a generalization of shape optimization. The topological optimization of a shape functional $J(\Omega)$ is introduced in order to characterize the variation of $J(\Omega)$ with respect to the variation of topology of the domain. It permits us to obtain the new optimality condition

$$J(\Omega_0) = \inf_{\Omega} J(\Omega).$$

In optimization process, we consider a set $\Omega \subset \mathbb{R}^N$ regular in which we dug some small halls ω_ϵ depending on $\epsilon \in (0,1)$ and we introduce the set $\Omega(\epsilon) = \Omega \setminus \bar{\omega}_\epsilon$. We evaluate the difference $J(\Omega(\epsilon)) - J(\Omega)$ to obtain the topological derivative.

It is very difficult to obtain this topological derivative for surface functional. In fact, let us consider a continuous function g and $\omega_\epsilon = B(x_0, \epsilon)$, $x_0 \in \Omega$. Let us consider the functional J

$$J(\Omega) = \int_\Omega g dx + \int_{\partial\Omega} ds,$$

where Ω is a open set of $\mathbb{R}^2$

$$J(\Omega(\epsilon)) - J(\Omega) = -\int_{B(x_0,\epsilon)} g dx + \int_{\partial B(x_0,\epsilon)} ds.$$

Using the mean value theorem, we get

$$\int_{B(x_0,\epsilon)} g dx = g(z_\epsilon) vol(B(x_0, \epsilon)).$$

Then

$$J(\Omega(\epsilon)) - J(\Omega) = -g(z_\epsilon)\pi\epsilon^2 + 2\pi\epsilon.$$

$$\frac{J(\Omega(\epsilon)) - J(\Omega)}{\pi\epsilon^2} = -g(z_\epsilon) + \frac{2}{\epsilon}.$$

As $\epsilon \to 0$, $z_\epsilon \to x_0$ and then we get

$$\frac{J(\Omega(\epsilon)) - J(\Omega)}{\pi\epsilon^2} \to +\infty.$$

Here the topological derivative cannot generate a hole. This prove the principal difficulty for obtaining topological derivative.

4.1 First Approach

We use the topological optimization by using the approach of Nazarov and Sokolowski in [13] in which the operator associated with the boundary value problem is an homogenous linear second order operator for more details cf. [13]. But we are going to give some steps and tools presented in their work.

Let us define the sets w_ϵ and Ω_ϵ. Let Ω and w be two domains in $\mathbb{R}^N$ with compact closures and $\partial\Omega$ and ∂w are regulars boundaries We assume that $0 \in w \subset B_1 \subset B_2 \subset \Omega$ with $B_R = \{x \in \mathbb{R}^N / |x| \le R\}$. We introduce the sets

$$w_\epsilon = \{x \in \mathbb{R}^N ; \xi = \epsilon^{-1}x \in w\}; \quad \Omega(\epsilon) = \Omega \backslash \bar{w}_\epsilon ,$$

where ϵ is a small parameter belonging in $(0,1)$. We obtain the topological derivative of the integral functional if we evaluate

$$T(0) = \lim_{\epsilon \to 0} \frac{J_1(\Omega(\epsilon)) - J_1(\Omega)}{f(\epsilon)} ,$$

where $f(\epsilon) \longrightarrow 0$ if $\epsilon \longrightarrow 0$.

4.1.1 Existence of Solution of Boundary Value Problem

We introduce here two boundary value problems to define the expansion of the solution to problem (29). The first problem is obtained by filing the cavity. We study the problem

$$\begin{cases} -\Delta p = 0 & \text{in} \quad \Omega \\ p = p_1 & \text{on } \partial\Omega \setminus \Gamma_1 \\ p = 0 & \text{on} \quad \Gamma_1 \end{cases} . \tag{30}$$

By the variational formula this problem admits a unique solution in the space

$$V_1 = \{v \in H^1(\Omega) \quad \text{such that} \quad v = 0 \quad in \quad \Gamma_1 \quad \text{and} \quad v = p_1 \quad in \quad \partial\Omega \setminus \Gamma_1\} .$$

The second problem is obtained by replacing the variable x by the fast variable ξ. The boundary of Ω disappears and goes to infinity. We obtain

$$\begin{cases} -\Delta p_2 = 0 \text{ in } \mathbb{R}^N \backslash \bar{w} \\ p_2 = g \quad \partial w \end{cases} . \tag{31}$$

The set $\mathbb{R}^N \backslash w$ is unbounded and there is no limit condition at infinity. This is the exterior problem of Laplace. We look for a solution satisfying the condition $\lim_{|x| \to \infty} u(x) = c$. If g is a continuous function the boundary value problem (31) has a unique solution u verifying the condition at infinity. If w is a ball $B(x_0, r_0)$ the solution of (31) is given by

$$p_2(x) = (1 - (\frac{r_0}{|x - x_0|})^{N-2})c + \frac{1}{r_0 \sigma_N} \int_{\partial B} \frac{|x - x_0|^2 - r_0^2}{|t - x|^N} g(t) d\gamma(t)$$

for all function $g \in C^0(\partial\Omega)$. We see easily that

$$\lim_{|x| \to \infty} u(x) = \frac{1}{2\pi r_0} \int_{\partial B} g(t) d\gamma(t).$$

4.1.2 Power Solutions and Polarization Matrix

In this section we define the power solutions of homogenous system. We recall that in [13] vectors $(u_1, \ldots, u_T)$, $T \geq 1$ are considered. In our case we have a scalar vector, i.e., T is equal to 1. The Dirichlet condition is defined in the boundary of the domain ω. The basis of homogenous polynomials is constituted by a unique vector U verifying the condition $U(zx) = z^{\tau_1} U(x)$, $\tau_1 = 0$ with U constant. Then $U = \delta_{11}$.

The function U^- is defined by $U^- = U(-\nabla_x)\Phi(x) = \Phi(x)$ where Φ is the fundamental solution of the Laplacian operator, i.e., is a solution of $\Delta\Phi = \delta$ in $\mathbb{R}^N$.

Definition 4.1. *The function*

$$\Phi(x) = \begin{cases} -\frac{1}{2\pi}\log|x| & (N = 2) \\ -(N-2)\alpha(N)\frac{1}{|x|^{N-2}} & N \geq 3 \end{cases}$$

defined for all $x \in \mathbb{R}^N$ is the fundamental solution of the Laplace operator. $\alpha(N)$ is the hyper surface of the unit sphere of $\mathbb{R}^N$.

Definition 4.2. *The Green function associated to the domain is given by*

$$G(x, y) = \Phi(x - y) - \Phi^x(y) \quad x, \quad y \in \Omega, \quad x \neq y,$$

Φ^x *is the solution of*

$$\begin{cases} \Delta\Phi^x = 0 & in \quad \Omega \\ \Phi^x = \Phi(x - y) & on \quad \partial\Omega. \end{cases}$$

We recall that here we work with a scalar vector. As the Dirichlet condition is prescribed the polarization m^ω is a 1×1 matrix. And by the Proposition 3.2 in [13] we can claim that the matrix m^ω is negative. m^ω is defined by the following proposition. This proposition gives a description of solution of homogenous problem (31).

Proposition 4.3. *The linear space of solution of boundary value (31) is given by a linear hull of the function*

$$\zeta(\xi) = U(\xi) + z(\xi),$$

where z is the solution of boundary value problem

$$\begin{cases} -\Delta z = 0 & in \quad \mathbb{R}^N \backslash \bar{\omega} \\ z = U & on \quad \partial\omega \end{cases}.$$

In addition we have the representation

$$\zeta(\xi) = U(\xi) + m^\omega U^-(\xi) + \tilde{z}(\xi),$$

where $\tilde{z}$ is the remainder in the development of the function z.

The proof of this proposition is similar to the proof of Proposition 3.1 in [13]. In the proof of this proposition the polarization is given by the expression $m^\omega = -(g, \zeta)_{\partial\Omega}$ with

$$\zeta = U + m^\omega U^- + \tilde{\zeta}$$

et

$$g^\omega = g = U.$$

After neglecting the remainders $\tilde{\zeta}$, m^ω becomes

$$
\begin{aligned}
m^\omega &= -(g, \zeta)_{\partial\omega} = -(-U, -\zeta^1)_{\partial\omega} \\
&= -(U, (U + m^\omega U^-))_{\partial\omega} \\
&= -(U, U)_{\partial\omega} - m^\omega (U, U^-)_{\partial\omega} \\
&= -(U, U)_{\partial\omega} - (U, m^\omega U^-)_{\partial\omega}
\end{aligned}
$$

$(.,.)$ is the inner product. We obtain finally

$$m^\omega = -\frac{(U, U)_{\partial\omega}}{1 + (U, U^{-1})_{\partial\omega}} = -\frac{\int_{\partial\omega} dx}{1 + \int_{\partial\omega} U^-(x)dx}.$$

In the case where $\omega \subset \mathbb{R}^3$ the scalar matrix becomes

$$m^\omega = -\frac{\int_{\partial\omega} dx}{1 + \int_{\partial\omega} E_3(x)dx}.$$

We suppose here that ω is ball $B(0, r_0)$ such that $0 < r_0 < 1$. We use here the spherical coordinate to compute the exact value of the integral. Let

$$x_1 = r \sin\theta \cos\varphi; \quad x_2 = r \sin\theta \sin\varphi; \quad x_3 = r \cos\theta,$$

where $0 \le r \le r_0$; $\theta \in (0, \pi)$ and $\varphi \in (-\pi, \pi)$. The Jacobian of the transformation is $r^2 \sin\theta$. We can calculate the integral

$$\int_{\partial\omega} E_3(x)dx = \int_{\partial B} -\frac{1}{\sigma_3} \frac{1}{|x|} dx = \int_0^{r_0} \int_0^\pi \int_{-\pi}^\pi \frac{-1}{\sigma_3} r \sin\theta \, dr \, d\theta \, d\varphi = 2\pi r_0^2.$$

The polarization matrix depending only on ω is

$$m^\omega = -\frac{4\pi r_0}{1 - \frac{r_0^2}{2}}.$$

4.1.3 Asymptotic Expansion of the Functional

In this section we give the principal theorem stating the asymptotic of the functional $J_\epsilon(p) = J(p_\epsilon)$. The expansion of the functional is deduced by the expansion of solution p_ϵ.

Theorem 4.4. *Let p_ϵ be the solution of (29) and J_ϵ the functional*

$$J_\epsilon(p) = J(p_\epsilon) = \int_{\Omega(\epsilon)} (1/\gamma[\ln \frac{\rho_s}{\rho_0} - \beta_p(p_\epsilon - p_0)] - W_1)^2 dx.$$

Then the following expansion holds

$$\left|J_\epsilon(p) - J_0(p) - \epsilon \int_\Omega 2\left(\frac{-\beta_p}{\gamma}\right) (1/\gamma[\ln \frac{\rho_s}{\rho_0} - \beta_p(p - p_0)] - W_1)\eta(x)m^\omega p(0) dx\right| \le C\epsilon^{1+\delta}$$

$\delta > 0$, where p is solution of (30), η is the Green function solution of

$$\begin{cases} -\Delta\eta = \delta & \Omega \\ \eta = 0 & \partial\Omega \end{cases}.$$

In addition the topological derivative is given by the expression

$$\left|J_\epsilon(p) - J_0(p) - \epsilon V(0)m^\omega p(0)\right| \le C\epsilon^{1+\delta}. \tag{32}$$

V is the solution of the adjoint problem

$$\begin{cases} -\Delta V = 2\left(\frac{-\beta_p}{\gamma}\right) (1/\gamma[\ln \frac{\rho_s}{\rho_0} - \beta_p(p_\Omega - p_0)] - W_1) & \Omega \\ V = 0 & \partial\Omega \end{cases}. \tag{33}$$

Proof. Before beginning the proof of the theorem we determine the first and the second expansion of the solution $p(\epsilon, x) = p_\epsilon(x)$ of problem (29). The first and the second expansion are in the following form

$$p(\epsilon, x) = p(x) + a(\epsilon)\eta(x)$$

$$p(\epsilon, x) = b(\epsilon)\zeta(\epsilon^{-1}x),$$

where p solve the problem (30) and η the Green function of the Laplacian operator satisfying the boundary condition $\eta = 0$ on $\partial\Omega$; ζ is defined in Proposition 4.3. Here we apply the method of matched asymptotic expansions. The matching conditions imposed by the method imply that (see Sect. 4.2 [13]) the coefficients $a(\epsilon)$ and $b(\epsilon)$ are given by

$$a(\epsilon) = \{I - \epsilon^{N-2}m^\omega m^\Omega\}^{-1}\epsilon^{N-2}m^\omega v(0)$$

and

$$b(\epsilon) = \{I - \epsilon^{N-2}m^\omega m^\Omega\}^{-1}v(0),$$

where I denotes the identity matrix and m^ω, m^Ω are polarizations matrix defined in $\partial\omega$ and $\partial\Omega$. In our case a p is scalar function and $N = 3$ these expressions becomes

$$a(\epsilon) = \frac{\epsilon m^\omega v(0)}{1 - \epsilon m^\omega m^\Omega} \quad \text{and} \quad b(\epsilon) = \frac{v(0)}{1 - \epsilon m^\omega m^\omega}.$$

To determine the asymptotic of the solution we will use uniquely the first expansion of the solution

$$J_\epsilon(p) = \int_{\Omega(\epsilon)} (p_\epsilon - p_0)^2 + 2(p - p_0)a(\epsilon)\eta(x) + a^2(\epsilon)\eta^2(x)$$

$$- \int_{\bar\omega_\epsilon} \underbrace{(p_\epsilon - p_0)^2 + 2(p - p_0)a(\epsilon)\eta(x) + a^2(\epsilon)\eta^2(x)}_{bounded} .$$

Replacing $a(\epsilon)$ by its value this expression follow

$$J_\epsilon(p) - J_0(p) - \epsilon \int_\Omega 2(p-p_0)\eta(x)m^\omega p(0)dx = \epsilon^2 \int_\Omega \frac{(m^\omega)^2\eta^2(x)}{(1 - \epsilon m^\omega m^\Omega)^2} + mes(\omega_\epsilon)h(x).$$

Finally if ω_ϵ is a ball of radius ϵ, we have $mes(\omega_\epsilon) = \epsilon^2 2\pi$, then we get the desired inequalities.

To obtain the expression (32) we multiply the first equation of (33) by the Green function η and integrate

$$\int_\Omega \nabla V \nabla \eta = \int_\Omega 2\left(\frac{-\beta_p}{\gamma}\right)(1/\gamma[\ln\frac{\rho_s}{\rho_0} - \beta_p(p - p_0)] - W_1)\eta.$$

Using the Green formula in the first integral we have

$$\int_\Omega -\Delta\eta V = \int_\Omega 2\left(\frac{-\beta_p}{\gamma}\right)(1/\gamma[\ln\frac{\rho_s}{\rho_0} - \beta_p(p - p_0)] - W_1)\eta.$$

As $-\Delta\eta = \delta$ in Ω this expression yields

$$\int_\Omega \delta V = V(0) = \int_\Omega 2\left(\frac{-\beta_p}{\gamma}\right)(1/\gamma[\ln\frac{\rho_s}{\rho_0} - \beta_p(p - p_0)] - W_1)\eta,$$

this proof the theorem. □

We obtain the topological derivative of the functional J at the point considered. The topological derivative at the point x_0 is given by the formula

$$T(x_0) = \int_\Omega 2\left(\frac{-\beta_p}{\gamma}\right)(1/\gamma[\ln\frac{\rho_s}{\rho_0} - \beta_p(p-p_0)] - W_1)\eta(x)m^\omega p(x_0)dx = V(x_0)m^\omega p(x_0),$$

where V is the solution of the adjoint problem (33), p the solution of problem (30) and m^ω the polarization matrix. The principal difficulties to get the topological sensitivity is the calculus of the polarization matrix. In our case we calculate explicitly in the preceding section this matrix which is a scalar.

4.2 Second Approach

In this section, we present a general framework for topological sensitivity. This method was introduced by J. Cea, M. Masmoudi and al. For more details see [3, 10, 11]. Meanwhile we are going to present some fundamental steps of this method.

4.2.1 A Generalized Adjoint Method

Let $\mathcal{V}$ be a fixed Hilbert space and $\mathcal{L}(\mathcal{V})$ (resp $\mathcal{L}_2(\mathcal{V})$) denotes the spaces of linear (resp bilinear) forms on $\mathcal{V}$. We make the following hypothesis:

- **H-5**: There exists a real function f, a bilinear form $\delta_a \in \mathcal{L}_2(\mathcal{V})$ and a linear form $\delta_l \in \mathcal{L}(\mathcal{V})$ such that

$$f(\epsilon) \longrightarrow 0, \quad \epsilon \longrightarrow 0^+ \tag{34}$$

$$\|a_\epsilon - a_0 - f(\epsilon)\delta_a\|_{\mathcal{L}_2(\mathcal{V})} = o(f(\epsilon)) \tag{35}$$

$$\|l_\epsilon - l_0 - f(\epsilon)\delta_l\|_{\mathcal{L}(\mathcal{V})} = o(f(\epsilon)). \tag{36}$$

- **H-6**: The bilinear form a_0 is coercive: There exists a constant $\alpha > 0$ such that

$$a_0(u, u) \geq \alpha\|u\|^2, \quad \forall\, u \in \mathcal{V}.$$

According to (35), the bilinear form a_ϵ depend continuously on ϵ, hence there exists ϵ_0 and $\beta > 0$ such that for $\epsilon \in [0, \epsilon_0]$, the following uniform coercivity condition holds

$$a_\epsilon(u, u) \geq \beta\|u\|^2 \quad \forall\, u \in \mathcal{V}.$$

According to Lax–Milgram's theorem, for $\epsilon \in [0, \epsilon_0]$, the problem find $u_\epsilon \in \mathcal{V}$ such that

$$a_\epsilon(u_\epsilon, v) = l_\epsilon(v) \quad \forall\, v \in \mathcal{V} \tag{37}$$

gets a unique solution.

Lemma 4.5. *If hypothesis **H-5** and **H-6** hold, then*

$$\|u_\epsilon - u_0\| = O(f(\epsilon)).$$

For the proof we can refer to [7].
- **H-7**: We consider now a cost function $j(\epsilon) = J(u_\epsilon)$, where the functional J is differentiable: For $u \in \mathcal{V}$ there exists a linear and continuous form $DJ(u) \in \mathcal{L}(\mathcal{V})$ and δ_J such that

$$J(v) - J(u) = DJ(u)(v - u) + f(\epsilon)\delta_J(u) + o(\|v - u\|_\mathcal{V}). \tag{38}$$

For $\epsilon \geq 0$, we define the Lagrangian $\mathcal{L}_\epsilon$ see for example [10]

$$\mathcal{L}_\epsilon(u, v) = a_\epsilon(u, v) - l_\epsilon + J(u) \quad \forall u\; v \in \mathcal{V}.$$

The next theorem gives the asymptotic expansion of $j(\epsilon)$ and the proof can be founds in [1, 7, 8].

Theorem 4.6. *If hypothesis* **H-5, H-6,** *and* **H-7** *are satisfied, then*

$$j(\epsilon) - j(0) = f(\epsilon)\delta\mathcal{L}(u_0, v_0) + o(f(\epsilon)), \tag{39}$$

where u_0 is the solution of (37) with $\varepsilon = 0$, v_0 is the solution to the adjoint problem: find v_0 such that

$$a_0(w, v_0) = -DJ(u_0)w \quad \forall w \in \mathcal{V} \tag{40}$$

and

$$\delta\mathcal{L}(u, v) = \delta_a(u, v) - \delta_l(v) + \delta_J(u).$$

4.2.2 Position of the Problem

The function p_ϵ solution of (29) is defined on the variable open set Ω_ϵ. Thus it belong to a functional space which depend on ϵ. Hence, if we want to derive the asymptotic expansion of the functional

$$j(\epsilon) = J(p_\epsilon), \tag{41}$$

we cannot apply directly the tool of the above section, which require a fixed functional space. However, a functional space independent on ϵ can be constructed by using a domain truncation technique. This technique has been introduced in topological optimization by Masmoudi in [11]. It allows only to do theoretical analysis, and will never be used for practical computation. During optimization process, the two systems which have to be solved are (27) and the adjoint problem associated to the cost function (28).

4.2.3 The Truncated Problem and the Topological Gradient

We present also some steps of the truncation method. For more details cf. [1, 7, 8]. Let $R > 0$ be a real such that the closed ball $\overline{B(x_0, R)}$ is included in Ω. It is also supposed that ϵ remains small so that $\overline{\omega}_\epsilon \subset B(x_0, R)$. The truncated open subset is defined by $\Omega_R = \Omega \setminus \overline{B(x_0, R)}$ and $B(x_0, R) \setminus \overline{\omega}_\epsilon$ is denoted by D_ϵ. For $\varphi \in H^{1/2}(\Gamma_R)$ and $\epsilon > 0$, let p_ϵ^φ be the solution of the problem: find p_ϵ^φ such that

$$\begin{cases} -\Delta p_\epsilon^\varphi = 0 \text{ in } \Omega_R \\ \quad p_\epsilon^\varphi = 0 \text{ in } \partial\omega_\epsilon \\ \quad p_\epsilon^\varphi = \varphi \text{ in } \Gamma_R \end{cases}, \tag{42}$$

where Γ_R is the boundary of $B(x_0, R)$.
For $\epsilon = 0$, p_0^φ is solution to

$$\begin{cases} -\Delta p_0^\varphi = 0 \text{ on } B(x_0, R) \\ \quad p_0^\varphi = \varphi \text{ in } \quad \Gamma_R \end{cases}. \tag{43}$$

For $\epsilon > 0$ we consider the Dirichlet-to-Neumann operator (see [4] or [9] for details) T_ϵ is defined by

$$T_\epsilon : H^{1/2}(\Gamma_R) \longrightarrow H^{-1/2}(\Gamma_R)$$
$$\varphi \longrightarrow T_\epsilon\varphi = \nabla p_\epsilon^\varphi .n \ ,$$

where the normal n is chosen exterior to D_ϵ on Γ_R and D_ϵ.

For $\epsilon \geq 0$, we define p_ϵ^R as the solution of the truncated problem

$$\begin{cases} -\Delta p_\epsilon^R = 0 \text{ on } \Omega_R \\ p_\epsilon^R = p_1 \text{ in } \partial\Omega \setminus \Gamma_1 \\ p_\epsilon^R = 0 \text{ in } \Gamma_1 \\ p_\epsilon^R - T_\epsilon\varphi = 0 \text{ in } \Gamma_R \end{cases} . \tag{44}$$

The variational formulation associated to (44) is: find $p_\epsilon^R \in V_R$ such that

$$a_\epsilon(p_\epsilon^R, v) = l(v), \quad \forall v \in V_R, \tag{45}$$

where the Hilbert space V, the bilinear form a_ϵ and the linear form l_ϵ are defined by

$$V_R = \{p \in H^1(\Omega_R) \ /p = p_1 \text{ on } \partial\Omega \setminus \Gamma_1 \text{ and } p = 0 \text{ on } \Gamma_1\} \tag{46}$$

$$a_\epsilon(p, v) = \int_{\Omega_R} \nabla p \nabla v dx + \int_{\Gamma_R} (T_\epsilon\varphi)v d\sigma(x); \quad l_\epsilon(v) = \int_{\partial\Omega \setminus \Gamma_1} p_1 v dx. \tag{47}$$

$d\sigma$ is the Lebesque measure on the boundary. It is standard to prove that (45) has a unique solution in V_R which is the restriction to Ω_R of the solution of (39).

We have now a fixed Hilbert space, as required in Theorem 4.6.

The main theorem is the following. The proof is based on the single layer potential, the exterior and interior problems and the Dirichlet-to-Neumann (or capacity) operator. As our coast functional is a particulary case of the more general one studied in [8], we refer there for the proof (Proposition 5.3, for the three-dimensional case and Proposition 5.5 for the two-dimensional case).

Theorem 4.7. *Let*

$$J_\Omega(p) = \int_\Omega \left| \frac{1}{\gamma} \left(\ln\left(\frac{\rho_s}{\rho_0}\right) - \beta_p(p - p_0) \right) - W_1 \right|^2 dx$$

be the cost function. Let $V \in V_R$ be the solution to the adjoint equation

$$a_0(V, w) = -DJ(p, w) \quad \forall w \in V_R. \tag{48}$$

Then the function $j(\epsilon) = J_{\Omega_\epsilon}(p_\epsilon)$ have the following asymptotic expansion

$$j(\epsilon) = j(0) + f(\epsilon)\delta_a(p, V) + \delta_J(p) - \delta_l(V) + o(f(\epsilon)). \tag{49}$$

The function $\delta_j(x_0) = \delta_a(p(x_0), V(x_0)) + \delta_J(p(x_0)) - \delta_l(V(x_0))$ is called topological sensitivity or topological gradient and can be used as descent direction in optimization processus. Moreover, as j is independent of R and δ_j is independent of ϵ, it follows from the uniqueness of the asymptotic expansion that δ_j is also independent of R.

Corollary 4.8. *When $w_\epsilon = B(x_0, \epsilon)$ is a ball, δ_a, δ_J can be computed explicitly and we have*

$$\delta j(x_0) = -4\pi(p(x_0).V(x_0)),$$

where p is the solution of the direct state V is the solution of the adjoint equation.

Remark 4.9. In the case of Neumann condition on the boundary of the hole, the topological sensitivity can be computed. When $w_\epsilon = B(x_0, \epsilon)$, we have

$$\delta j(x_0) = -4\pi \left(\nabla p(x_0).\nabla V(x_0) \right.$$

$$\left. + \left| \frac{1}{\gamma} \left(\ln \left(\frac{\rho_s}{\rho_0} \right) - \beta_p(p(x_0) - p_0(x_0)) \right) - W_1(x_0) \right|^2 \right),$$

where p is the solution of the direct state V is the solution of the adjoint equation.

5 Numerical Simulations

In this section we present a numerical solution which illustrate the topological derivative. We consider the square $\Omega = (-1, 1) \times (-1, 1)$. We denote by p the solution of the boundary value problem

$$\begin{cases} -\Delta p = 0 & \Omega \\ p = r & \partial\Omega \end{cases} \tag{50}$$

with r is a continuous function in $\partial\Omega$ given by and V is the solution of the adjoint state. The integral functional is given by

$$J(\Omega, p) = \int_{\Omega(\epsilon)} (1/\gamma[\ln \frac{\rho_s}{\rho_0} - \beta_p(p_{\Omega(\epsilon)} - p_0)] - W_1)^2 dx.$$

After a choice of our constant we compute the topological derivative of the integral functional. We use a finite element method to represent the solution and its adjoint state. We present also the topological derivative of the functional. For all the examples we use $p_0 = |2x - y|$, $\gamma = log(2)$, $\frac{\rho_s}{\rho_0} = 1$ and $\beta_p = 0, 5$.

Examples

The first example concerns Figs. 1 and 2. In this example we set $r = |x - y|$ and $W_1 = |tan(x^2 + y^2 + 1)|$.

The second example concerns Figs. 3 and 4. In this example, we set $r = 1$ and $W_1 = 1/3|x - 2y|$.

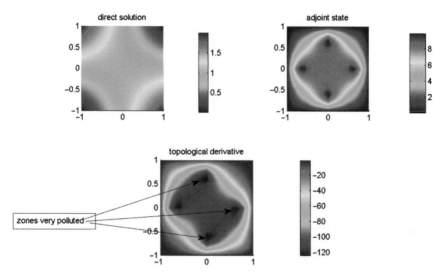

Fig. 1. On the *top*, we have: *left*: the direct solution, *right*: the adjoint solution and in the *bottom*: the topological gradient

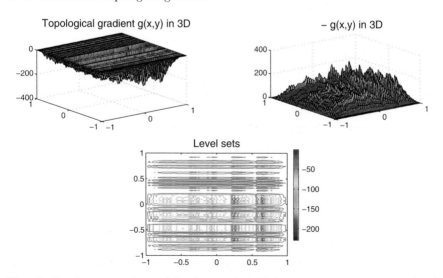

Fig. 2. On the *top*, we have: *left*: the topological derivative in 3D, *right*: $-g(x, y)$ and in the *bottom*: the level sets

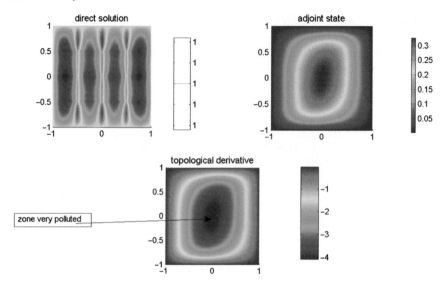

Fig. 3. On the *top*, we have: *left*: the direct solution, *right*: the adjoint solution and in the *bottom*: the topological gradient

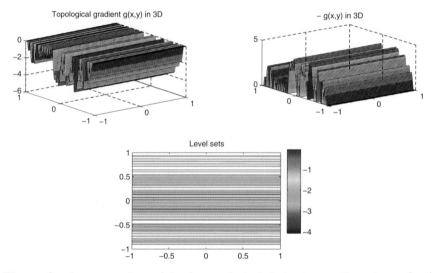

Fig. 4. On the *top*, we have: *left*: the topological derivative in 3D, *right*: $-g(x,y)$ and in the *bottom*: the level sets

In our examples, we see that the topological derivative is negative. It permits us to see the distribution of the pollution in the medium. Then we locate the more polluted zones. The topological gradient in 3D gives the pikes of pollution. This pikes are determined where the gradient is more negative, i.e., where $W - W_1$ is very big.

6 Some Open Problems

(1) Does it possible to compute the topological gradient with the nonlinear equation got in this paper?
   In fact we use the linear equation of the pressure to compute the topological derivative.
(2) What's happen in the case of nonpermanent evolution? Note that we have only studied the problem in the steady case.
(3) Does it possible to give some weak hypothesis and to prove the uniqueness of the solution for the studied nonlinear equations in this paper?
(4) What's happen if ρ_s is not a constant?

References

1. S. AMSTUTZ, *Aspects théoriques et num ériques en optimisation de forme topologique*, Thèse de Doctorat n d'ordre 709, INSA Toulouse 2003.
2. C. BAIOCCHI and A. CAPELO, *Variational and quasivariational inequalities applications to free boundary problems*, Wiley-Interscience, 1984.
3. J. CÉA, *Conception optimale, ou identification de forme, calcul rapide de la dérivée directionnelle de la fonction coût*, M.A.A.N. 20(3), pp. 371–402, 1986.
4. R. DAUTRAY and J.L. LIONS, *Analyse mathématiques et calcul numérique pour les sciences et les techniques*, Volume 2, Masson, 1987.
5. L.C. EVANS, *Partial Differential Equation*, Graduate Studies in Mathematics, Volume 19, AMS, 2002.
6. I. FAYE and D. SECK, *Optimal design in thermo elasticity problems*, preprint in 2007 UCAD.
7. S. GARREAU, Ph GUILLAUME and M. MASMOUDI, *The topological asymptotic for PDE systems: the elastic case*, SIAM J. Control optim. 39(6), pp. 1756–1778, 2001.
8. Ph. GUILLAUME and K. SIDIDRIS, *The topological expansion for the Dirichlet problem*, SIAM J. Control optim. 41(4), pp. 1042–1072, 2002.
9. N. S. LANDKOF, *Foundations of Modern Potential Theory*, Die Grundlehren der mathematischen Wissenschaften in Einzeldarstellungen Band 180, AMS Subject Classification, 1970.
10. J.L. LIONS, *Sur quelques questions d'analyse, de mécanique et de contrôle optimal*, Les presses de l'Université de Montreal, 1976.
11. M. MASMOUDI, *The topological asymptotic, in computational methods for control applications*, H. Kawarada and J. Periaux, eds., GAKUTO Internat. Ser. Math. Sci. Appl. Gakkotōsho, Tokyo, 2002.
12. M. NGOM and D. SECK, *Detection of tumors by shape and topological optimization*, preprint in 2007, UCAD.
13. S.A. NAZAROV and J. SOKOLOWSKI, *Asyptotic analysys of Shape Functionals*, J. Math. Pures Appl. 82, pp 125–196, 2003.
14. R.J. SCHOTTING, *Mathematical Aspects of Sault Transport in Porous Media*, Thesis of Techniische Universiteit Delft, Rotterdam, PROEFSCHRIFT, 1998.
15. A. SY and D. Seck, *Topological optimization with the p-Laplacian operator: an application in image processing*, preprint in 2007, UCAD.
16. J. TOUMA, *Modèle pour tester la représentativité des caractéristiques hydrodynamiquesd'un sol non saturé déterminées in-situ*, décembre 1987.

Dynamical Analysis of Infectious Diseases in Spatially Heterogeneous Environments

J.M. Tchuenche

Summary. Over many decades, mathematical biology has been the subject of many beautiful modeling experiments and a proving ground for a wealth of mathematical theories, both linear and non-linear. We analyze a malaria model first proposed in [10] using functional analytical technique – crude estimates of solutions to the model equations are given, a priori bounds of some parameters obtained as well as stability result. We also study the dynamics of infected individuals in the indirectly transmitted model when the total population equals that of mosquitoes.

1 Introduction

Models of spatially heterogeneous environments are difficult to build and analyze but an understanding of these models can be crucial to a wide variety of problems, ranging from population biology (infectious disease transmission) to chemical dynamics. Recently, Rodriguez and Torres-Sorando [10] have undertaken a detailed model building of such systems and they obtained conditions for the disease to become established.

A major aspect of the spread of infectious diseases is the problem of the geographical distribution of cases. This is in general a very difficult matter, especially in view of the complications that readily arise, even when the spatial aspect is ignored [2]. In general, the solubility of the resulting equations is not always guarantee, whence the necessity to switch to numerical simulation, which we are not dwelling into here.

Homogeneous mixing may be quite reasonable for small groups such as families, and some children's classrooms but there are limitations to its applicability in larger groups. The possibility of allowing non-homogeneous mixing is by recognizing two or more distinct groups (localities) of individuals. A few deterministic results for the case of only two groups were first given many years ago by Wilson and Worceter [17]. Watson [16] studied a more general model for many groups in which the members of any group mix homogeneously amongst themselves, but to a lesser extend with individuals of other groups. However, there is an obvious tendency in many communities for susceptibles who are situated a long way from infected persons, to have a smaller chance of infection than those who are in closer contact with the disease.

In the work at hand, we wish to survey analytic approaches to the problem and to illustrate some new relationships of the parameters of the model equations given in [10], since the authors have formulated the model, but did not attempt any analytical solutions. The notations and model equations are basically theirs. Our purpose differs in the sense that we use functional analytical technique to explain observations by means of biologically plausible assumptions.

We expect that this would give a domain of parameter space that yields outputs within the acceptable ranges.

1.1 Notations and Hypotheses

We assume that:

- N is the total population of humans
- $X(t)$ is the number of infected individuals at time t.
- g is the per capita rate of recovery, so that $\frac{1}{g}$ is the duration of the disease.
- b is the transmission rate per susceptible and per infected individuals.
- $Y(t)$ is the number of infected mosquitoes at time t
- M their total number
- m the per capita death rate of mosquitoes
- α is the number of hosts bitten by a mosquito per unit time.
- β is the proportion of infectious bites on humans that produce infection.
- γ is the proportion of bites of susceptible mosquitoes on infected human that produce an infection.
- $\min \inf(m, g) = \sigma$; $\max \sup(M, N) = \kappa$
- $|X_i| = |X_j|$; $\sum v_{ij} = \sum v_{ji} = v$
- $b + v = 1$

2 The Model

Here, we are mainly interested in the analytic solubility of the model equations proposed by Rodriguez and Torres-Sorando [10].

2.1 Spatially Homogeneous Environment

(a) *Direct transmission.* A simple susceptible-infectious-susceptible (SIS) model equation is the following

$$\frac{dX(t)}{dt} = b(N - X(t))X(t) - gX(t), \tag{1}$$

which also represents the logistic equation of population growth with $X(0) = X_0$. Its solution is simply

$$X(t) = \frac{bN - g}{b + \left(\frac{bN-g}{X_0} - b\right)e^{-(bN-g)t}}. \tag{2}$$

The limiting value $X_\infty = N - \frac{g}{b}$ also represents the value of $X(t)$ at the non-trivial equilibrium X^*, say, and the maximum value $\tilde{X} = \frac{1}{2}X^*$, where $\tilde{X}$ is the point of inflexion, whenever $(bN - g) > 0$.

Three cases present themselves: $X_\infty > 0$, $X_\infty = 0$ and $X_\infty < 0$. The latter case is biologically irrelevant.

(b) *Indirect transmission.* In [10], the authors considered the malaria model of Ross–Macdonald. The classical equations are of the form

$$\frac{dX(t)}{dt} = \left(\frac{\alpha}{N}\beta\right)(N - X(t))Y(t) - gX(t),$$

$$\frac{dY(t)}{dt} = \left(\frac{\alpha}{N}\gamma\right)(M - Y(t))X(t) - mY(t). \tag{3}$$

Equation $(3)_1$ can readily be solved by obtaining approximate values as follows:

$$X(t) = \exp(-gt)\int_0^t \left(\frac{\alpha}{N}\beta\right)\exp(g\tau)(N - X(\tau))Y(\tau)d\tau + X(0) \tag{4}$$

By letting $X(0) = X_0$, an a priori estimate of (4) yields after some algebraic manipulations

$$|X(t)| \leq \frac{\frac{\alpha\beta}{g}(e^{gt} - 1)|Y(t)| + X_0}{1 - \frac{\alpha\beta}{Ng}(e^{gt} - 1)|Y(t)|}. \tag{5}$$

Similarly, an estimate for $Y(t)$ can be obtained.

It is assumed that the denominator of (5) is not identically zero, but positive. Hence, the disease would persist in the population as long as

$$t < \frac{\ln\left(\frac{gN}{\alpha\beta\bar{Y}} + 1\right)}{g}, \tag{6}$$

with $\bar{Y} := |Y(t)|$. Equation (5) can be interpreted as: infected humans receive treatment and apply protection measures to avoid reinfection (since there is no immunity).

By further assuming that $\dfrac{\alpha\beta}{N} = \dfrac{\alpha\gamma}{N} = b$.

Equation (6) takes the form

$$t < \frac{\ln\left(\frac{g}{b\bar{Y}} + 1\right)}{g}, \tag{7}$$

while (3) can be written as

$$\frac{dX}{dt} = bNY - bXY - gX,$$

$$\frac{dY}{dt} = bMX - bXY - mY.$$
(8)

If at a particular instant t the total number of humans equals that of mosquitoes while the per capita death rate of mosquitoes equals the per capita rate of recovery of humans, i.e., $N = M$ and $m = g$, then

$$(X - Y)(t) = (X_0 - Y_0)e^{-(bM+g)t}$$
(9)

If $M < \dfrac{-g}{b}$, then it is obvious that $\alpha\beta < -g$ while $X(t) - Y(t) \to +\infty$ as $t \to \infty$ implying that the number of humans exceeds that of mosquitoes. This last assumption is not realistic in malaria endemic regions where it is often posited that the number of humans relative to the number of mosquitoes is low, this seems to be a more logical assumption [10]. Also, if $bM + g > 0$, then, $X(t) = Y(t)$ for t large and the host population might go extinct if closed, since the disease is fatal. The limiting value of (9) suggests the following:

(i) $X_0 > Y_0$, in this case we have more infected humans than infected mosquitoes at the beginning of the process. This can be ascertain by the fact that a mosquito can infect more than one individual (although in [10], it was hypothesized that a mosquito has a limiting capacity of biting humans.

(ii) $X_0 < Y_0$, a human can infect many mosquitoes if he does not protect himself from bites.

These two cases seem realistic if we take into account the slowness of recovery, the absence of immunity in the model equations [2]. The duration of the disease may be short, but actual infectiousness may persist for a long time. Because the model equations are coupled, closed form solutions are difficult to come by, but without loss of generality, we can estimate each equation separately in order to have a little insight of some key parameters that drives the disease dynamics.

From (4), let $\min \inf(m, g) = \sigma$, and $\max \sup \sum(M, N) = \kappa$. Then,

$$X(t) + Y(t)e^{\sigma t} \le b \int_0^t e^{\sigma\tau}\{\kappa(X + Y) - 2XY\}d\tau + X_0 - Y_0,$$

$$\Rightarrow |(X + Y)|(t) \le (X_0 + Y_0)e^{(b\kappa - \sigma)t}.$$
(10)

A sharper estimate gives

$$|X(t) + Y(t)| \le \left(1 + \frac{g}{\sigma}\right)|X_0 + Y_0|e^{(b\kappa - \sigma)t},$$
(11)

where $|XY| \le |X_0 + Y_0|$.

In a close environment, as times goes on, the number of infected humans as well as infected mosquitoes increases whenever $b\kappa > \sigma$.

Let $\bar{Y} \simeq N$, then

$$|X(t)| \le \frac{\frac{\alpha\beta}{g}(e^{gt}-1)\bar{Y}(t) + X_0}{1 - \frac{\alpha\beta}{g}(e^{gt}-1)}. \tag{12}$$

Substituting the value of $|Y(t)|$ into (11) above we obtain

$$|X(t)| \le \frac{\left(1 + \frac{\alpha\beta}{m}\right)X_0 + \frac{\alpha\beta}{g}Y_0 e^{gt} - \alpha\left(\frac{\alpha}{m}e^{mt}X_0 + \frac{\beta}{g}Y_0\right)}{1 - \alpha\left\{\frac{\beta}{g}(e^{gt}-1) + \frac{\gamma}{m}(e^{mt}-1)\right\}}. \tag{13}$$

The following heuristic assumption $\dfrac{\gamma}{m} = \dfrac{\beta}{g} = 1$ leads to

$$|X(t)| \le \frac{(1 + \alpha - \alpha e^{mt})X_0 + Y_0(\kappa e^{gt} - 1)}{1 - \alpha(e^{gt} + e^{mt} - 2)}, \tag{14}$$

which is feasible if

$$D := 1 - \alpha(e^{gt} + e^{mt} - 2) > 0.$$

That is,

$$\cosh(gt) < 1 + \frac{1}{2\alpha}; \quad m = -g,$$

or

$$t < \frac{2\alpha + 1}{\alpha(g+m)}. \tag{15}$$

In deriving (14) we have made use of the fact that $|e^x + e^y| > |x + y|$. It is more interesting to look at (12) by substituting it into the following equation

$$|Y(t)| \le \frac{\alpha\gamma}{N}\int_0^t e^{m\tau}(M - Y(\tau))|X(\tau)|d\tau + Y_0. \tag{16}$$

Choosing t such that $e^{gt} - 1 < \infty$, for instance let $g = \dfrac{\ln 2}{t}$ then

$$|Y(t)| \le \frac{\alpha\gamma}{N}\int_0^t e^{m\tau}(M - Y(\tau))\left(\frac{\alpha Y(\tau) + X_0}{1 - \alpha}\right)d\tau + Y_0$$

$$\le \frac{\alpha^2\gamma}{|1-\alpha|N}\int_0^t e^{m\tau}\left\{-Y^2(\tau) - \left(\frac{X_0}{\alpha} - M\right)Y(\tau) + \frac{MX_0}{\alpha}\right\}d\tau + Y_0$$

$$\leq \frac{\alpha^2 \gamma}{|1 - \alpha|N} \int_0^t e^{m\tau} \left[- \left\{ Y(\tau) + \frac{1}{2} \left(\frac{X_0}{\alpha} - M \right) \right\}^2 + \frac{1}{4} \left(\frac{X_0}{\alpha} - M \right)^2 \right.$$

$$\left. + \frac{MX_0}{\alpha} \right] d\tau + Y_0$$

$$\leq \frac{\alpha^2 \gamma}{|1 - \alpha|N} \int_0^t e^{m\tau} \left\{ \frac{1}{4} \left(\frac{X_0}{\alpha} - M \right)^2 + \frac{MX_0}{\alpha} \right\} d\tau + Y_0$$

$$\leq \frac{\alpha^2 \gamma}{|1 - \alpha|N} \left(\frac{X_0}{\alpha} + M \right)^2 \int_0^t e^{m\tau} d\tau + Y_0$$

$$\leq \frac{\alpha b}{|1 - \alpha|Nm} \left(\frac{X_0}{\alpha} + M \right)^2 + Y_0, \tag{17}$$

where m is scaled such that $|e^{mt} - 1| \approx 1$. Equation (17) can be rewritten as follows.

$$|Y(t)| \leq \frac{1}{|1 - \alpha|Nm} (X_0 + \alpha M)^2 + Y_0. \tag{18}$$

If the number of hosts bitten by a mosquito per unit time is negligible, then

$$|Y(t)| \leq \frac{X_0^2}{Nm} + Y_0. \tag{19}$$

This suggests that the total population of humans must always be very large so as to outweigh that of mosquitoes. Also, if the initial number of infected humans is small, then the estimated number of infected mosquitoes can be kept to its barest minimum by protection from bites. A similar reasoning can be given for $|X(t)|$, but we are not going to delve into it as the argument seems trivial from (19). The ultimate boundedness of $Y(t)$ is related to the persistence of mosquitoes in endemic areas. In finding the points of inflexion of (8), we obtained

$$b^2(N - X) = (bX + m)(bY + g), \tag{20}$$

which cannot be moved further.

Without loss of reality, we could use this to find an estimate for $|N - X|$. By following the earlier arguments and assumptions, we have

$$|N - X(t)| \leq \left(1 + \frac{\sigma}{b} \right) (X_0 + Y_0), \tag{21}$$

assuming mg negligible.

Also, on applying the method of coupled differential equations developed in [1, 7], we have from (8) that

$$Y(t) = Y_0 + bN \int_0^t e^{-m\tau} X(\tau) d\tau$$

$$= Y_0 + bN \int_0^t e^{-m\tau} \left(X_0 + bN \int_0^\tau e^{-g\alpha} Y(\alpha) d\alpha \right) d\tau$$

$$= Y_0 + \frac{bN}{m} X_0 (1 - e^{-mt}) + (bN)^2 \int_0^t e^{-m\tau} \left(\int_0^\tau e^{-g\alpha} Y(\alpha) d\alpha \right) d\tau$$

$$= Y_0 + \frac{\alpha\beta}{m} X_0 (1 - e^{-mt}) + (bN)^2 \int_0^t e^{-m\tau} \left(\int_0^\tau e^{-g\alpha} Y(\alpha) d\alpha \right) d\tau.$$

Let

$$\int_0^\tau e^{-g\alpha} Y(\alpha) d\alpha < M,$$

then

$$Y(t) \le Y_0 + \frac{\alpha\beta}{m} X_0 (1 - e^{-mt}) + (bN)^2 \frac{M}{m} (1 - e^{-gt})$$

$$= Y_0 + \frac{\alpha\beta}{m} \left[X_0 (1 - e^{-mt}) + \alpha\beta M (1 - e^{-gt}) \right]. \tag{22}$$

For the population of infected mosquitoes to be kept below its initial value at any time t, the total number of mosquitoes must equal

$$M = \frac{X_0 (e^{-mt} - 1)}{\alpha\beta (1 - e^{-gt})}. \tag{23}$$

If α and β are small enough, so that $(\alpha\beta)^2$ is negligible, then, from (23)

$$Y_\infty \le Y_0 + \alpha\beta X_0. \tag{24}$$

(c) *Indirect transmission with time delay.* Consider again the malaria model above. The previous formulation assumes that there is no time lag between a mosquito bite and the development of malaria. This is somewhat restrictive. It has long been noted that a population often takes a finite time τ to react to changes in the environment [15] or to respond to changes in population size [6]. Therefore, we assume that the product term XY is of convolution type. This assumption enables us to carry over the operation of Laplace transform.
Equation (8) becomes

$$\frac{dX}{dt} = bNY(t) - bX(t - \tau)Y(\tau) - gX(t)$$

$$\frac{dY}{dt} = bMX(t) - bY(\tau)X(t - \tau) - mY(t) \tag{25}$$

Since the Laplace transform technique is suitable for unknown functions and their derivatives whose coefficients are independent of time [11], let

p be the transform variable and $\mathcal{L}$ the Laplace transform operator with $\hat{X}$ the transform of $X(t) \in L^1_{[0,\infty)}$ (which is sufficiently smooth or well-behaved), then applying this operator on $(25)_1$, we obtain

$$\mathcal{L}[\dot{X};p] = p\hat{X}(p) - X_0 = bN\hat{Y}(p) - b\hat{X}(p)\hat{Y}(p) - g\hat{X}(p), \qquad (26)$$

i.e.,

$$\hat{X}(p) = \frac{bN\hat{Y}(p) + X_0}{p + g + b\hat{Y}(p)}. \qquad (27)$$

As assumed earlier, if $\hat{X}(p) = \hat{Y}(p)$, then

$$\hat{X}(p) = \frac{bN - (p+g) \pm \sqrt{(p+g-bN)^2 + 4bX_0}}{2b}. \qquad (28)$$

It is worth noting here that in general p is a complex number, but for many applications such as above, it is enough to regard it as a real parameter [5]. Isn't it fascinating and perhaps exhilarating finding the inverse Laplace transform of (28)? We shall keep this for future study, together with approach of finding the inverse Laplace transform of a constant, if it exists.

2.2 Spatially Heterogeneous Environment

In the regulation of population growth, boundedness and stability are two concepts most likely to be given prominence [14]. We now consider that the habitat is partitioned into k localities. The subscript i denotes the ith locality $(i = 1, 2, \ldots, k)$. Rodriguez and Torres-Sorando [10] considered three patterns of contact between localities. We shall concentrate only on the patterns of contact between localities with visitation between those localities. In the case of no contact between localities, the model of the directly transmitted disease and malaria model equation are the same, except for the fact that N, M, X and Y are replaced by N_i, M_i, X_i and Y_i, $i = 1, 2, \ldots, k$. If we assume that a fraction v_{ij} of the time devoted by humans to reside in locality i per unit time is devoted to visit locality j $(i \neq j; i, j = 1, \ldots, k)$, and that after the visit these humans return to their locality of origin, the malaria model (8) becomes (see [10])

$$\frac{dX_i(t)}{dt} = b(N_i - X_i(t))Y_i(t) - gX_i(t) + b(N_i - X_i(t))\sum_{j\neq i} v_{ij}Y_j(t),$$

$$\frac{dY_i(t)}{dt} = b(M_i - Y_i(t))X_i(t) - mY_i(t) + b(M_i - Y_i(t))\sum_{j\neq i} v_{ji}X_j(t). \quad (1)$$

It is explicitly assumed that hosts have a constant residence time, during which they stay in their locality of origin. During the rest of the time unit, there are visitations to other localities. This is similar to the migratory effects on the spread of AIDS [8]. In the above malaria model, it is further assumed that human can travel but mosquitoes cannot.

An analytic solution of (1) is difficult to obtain, but an estimate can be given.

Equation $(1)_1$ can readily be transformed to give

$$X_i(t) = \exp(-gt) \int_0^t \exp(g\tau)(bY_i(t) + \sum_{j \neq i} v_{ij}Y_j(\tau))(N_i - X_i(\tau))d\tau + X_{i0}. \quad (2)$$

By denoting $|Y_i(t)| = K = |Y_j(t)|$ and $|X_i(t)| = |X_j(t)| = Q$, and taking the estimate of (2), where it is further assumed that $\sum v_{ij} = \sum v_{ji} = v$, we obtain the following:

$$|X_i(t)| \leq \int_0^t e^{gt}(b|Y_i(\tau)| + \sum_{j \neq i} v_{ij}|Y_j(\tau)|)(N_i + |X_i(\tau)|)d\tau + X_{i0}$$

$$\leq (b+v)(N_i + Q)K\frac{(e^{gt} - 1)}{g} + X_{i0}. \quad (3)$$

A simple and straightforward algebra and a little rearrangement yields

$$|X_i(t)| \leq \frac{(b+v)|Y_i(t)|N_i + gX_{i0}e^{-gt}}{ge^{-gt} - (b+v)|Y_i(t)|}. \quad (4)$$

Equation (4) is positive provided

$$t > -\frac{1}{g}\ln\left\{\frac{(b+v)K}{g}.\right\} \quad (5)$$

The above estimate in (4) is not sharp enough because of the assumptions made earlier, that is, $|X_i| = |X_j|$ and $|Y_i| = |Y_j|$. This may not be true in general. In a future study, we hope to work out a sharper estimate. The estimate of equation $(1)_2$ can be worked out in a similar manner, thus, substituting its value into (4) yields

$$gQ - gX_{i0} - \frac{[(b+v)QN_ie^{mt} + mY_{i0}][(b+v)N_i + (b+v)Qe^{gt}]}{m - (b+v)Qe^{mt}} \leq 0. \quad (6)$$

A trivial algebraic manipulation gives

$$Q^2[-g(b+v)e^{mt} - (b+v)^2N_ie^{(m+g)t}] + Q[gm - (b+v)^2N_i^2e^{(m+g)t} - m(b+v)Y_{i0}e^{gt}$$

$$+ X_{i0}g(b+v)e^{mt}] - gmX_{i0} - mY_{i0}(b+v)N_ie^{gt} \leq 0, \quad (7)$$

which takes the form

$$AQ^2 + BQ + C \leq 0, \quad (8)$$

with

$$A = -g(b+v)e^{mt} - (b+v)^2 N_i e^{(m+g)t},$$
$$B = gm - (b+v)^2 N_i^2 e^{(m+g)t} - m(b+v)Y_{i0}e^{gt} + gX_{i0}(b+v)e^{mt},$$
$$C = -m(gX_{i0} + Y_{i0}(b+v)N_i e^{gt}).$$

Lemma 1. By taking equality in (8) and letting $t = 0$, then $Q > 0$ whenever $AC > 0$.

Proof. Without loss of reality, let us assume that b and v are scaled such that their sum $b+v = 1$, then AC reduces to

$$m[g^2 X_{i0} + gY_{i0}N_i + gX_{i0}N_i + N_i^2 Y_{i0}], \tag{9}$$

while B^2 takes the form

$$gX_{i0}^2 + 2mgX_{i0} + g^2 m^2 + 2mgX_{i0}Y_{i0} - 2gX_{i0}N_i^2$$
$$+ 2gm^2 Y_{i0} + m^2 Y_{i0}^2 - 2mY_{i0}N_i + N_i^2. \tag{10}$$

Despite all these assumptions on the modified equation (8), it has no equal roots, and this ascertain the fact that $|X_i|$ is not necessarily equal to $|X_j|$. The over assumptions made above lead to loss of generality and we hope to circumvent this difficulty in a nearby future by weakening these assumptions. □

3 Stability

In [10], the authors posited that for the disease to become established, the trivial equilibrium must be unstable. One of the standard approaches for analyzing solutions to (8) consists of constructing a Lyapunov function that is non-increasing along trajectories. The arguments necessary for this result have appeared in several recent literature (and from the time of Lyapunov) and will not be given here. The global asymptotic stability result would be base on our succeeding to construct a suitable Lyapunov function or with a certain abuse of terminology, a Lyapunov structure [4]. The construction is long and tedious, but rigorous enough under stringent conditions.

Now, we define a Lyapunov function $V(X,Y) : C[0,\infty] \to \mathbb{R}^+$ by

$$2V(X,Y) = K_1 X^2 + K_2 Y^2 + 2K_3 XY, \tag{11}$$

where K_1, K_2 and K_3 are to be determined. A more suitable Lyapunov function would probably involve integrals [9].

Theorem 3.1. *A necessary and sufficient condition for the equilibrium (X^*, Y^*) to be asymptotically stable is that $\dot{V}(X,Y)$ be directly proportional to $-X^2$ or $-Y^2$.*

Proof. This proof is basically based on finding a Lyapunov function V such that $\dot{V} < 0$. Calculating the derivative of V along the solution path, we have

$$\dot{V}(X,Y) = -X^2(K_1bY + K_1g - K_3bM + K_3bY) - Y^2(K_2bX - K_2m + K_3b(N-X))$$

$$+XY(K_1bN + K_2bM - K_3g - K_3m). \tag{12}$$

If we assume that $\dot{V}$ is proportional to $-X^2$, then X must satisfies

$$X = \frac{b^2N(K_1N + K_2M) - K_2m(m-g)}{b^2(K_1N + K_2M) - K_2b(m+g)}. \tag{13}$$

This results from the fact that the second and the last expressions in (12) both equal zero, while

$$\frac{K_1}{K_2} \neq \frac{M}{N} - \frac{g+m}{bN}; \quad N \neq X \tag{14}$$

for (13) to be valid.

Since $-X^2$ is a decreasing function of X, then we require that $\exists\ \epsilon > 0$ such that

$$\dot{V}(\cdot,\cdot) \leq -\epsilon X^2. \tag{15}$$

K_2 does not appear explicitly in the first expression on the right-hand side of (12). If for brevity and without any ambiguity we let $K_2 = 0$, then (13) reduces to $X = N$ and the disease invades. If $K_2 = 1$, from (14), we have the following inequality

$$K_1(bY + g) - K_3b(M - Y) = \epsilon > 0, \tag{16}$$

i.e.,

$$\frac{K_1}{K_3} > \frac{b(M - Y)}{bY + g}. \tag{17}$$

All the parameters on the right hand side of (17) are assumed known at time $t = 0$. With this choice of constants K_1 and K_3,

$$\dot{V}(X,Y) \leq -\epsilon.X^2$$

Also from (16) and the fact that

$$K_3 = \frac{K_2(m - bX)}{b(N - X)}, \tag{18}$$

we have,

$$K_1 = \frac{K_2[(g+m)(m - bX) - b^2(N - X)M]}{b^2N(N - X)}. \tag{19}$$

Thus K_3 also satisfies

$$K_3 = \frac{b(K_1N + M)}{g + m}. \tag{20}$$

Substituting K_1 in (19) and solving for X yields

$$X = N - \frac{1}{bN}. \tag{21}$$

Equation (20) shows the possibility of the disease to invade which might be due to the instability of the trivial equilibrium [10]. We note that if $Y > -\frac{g}{b}$, then $X - Y < \frac{m+g}{b}$,

$$\Rightarrow X < \frac{m}{b} \tag{22}$$

This procedure is based on an idea that is closely related to the guessing principle, and it includes (15) as a special case. Nevertheless, we have made use of (16) and (18), with $K_1 = K_2 = 1$, thus obtaining

$$\frac{N - X}{M - Y} > \frac{X - \frac{m}{b}}{Y + \frac{g}{b}}. \tag{23}$$

It should probably be of interest in applications to have a clear picture of what such a lower bound (23) means.

A disease free equilibrium would exist if no endemic equilibrium point exists $X = 0$ [3]; If (15) is satisfies, then every solution $(X, Y) \to 0$, giving rise to a mosquito-free environment.

Stability analysis of the limiting values of (2) reveals that $X_\infty > 0$ is stable, while $X_\infty = 0$ is unstable. Hence, if the disease is introduce into the community and $\frac{g}{b} < N$, it will always remain in the society, reaching a stable situation when $N - \frac{g}{b}$ of its members are infected and the other $\frac{g}{b}$ are healthy. The disease is therefore endemic and persistent within the population for this equilibrium state.

If $\frac{g}{b} \geq N$, then the disease dies out independent of the number who are initially infected. For large values of N, this case corresponds to the situation when recovery rate g is much greater than the contact rate b. □

Sowunmi [12] gave conditions under which there is possibility of asymptotically vanishing sets of infected humans and mosquitoes at a predetermined rate, while in [13], he obtained a threshold theorem. To attempt any quantification of model cost, more realistic models would be required, but the mathematical tractability must be at the forefront in order for such models to be usable and useful. Their realism should be judged within their self-imposed limits. Applications should be cost effective so as to reduce the drawbacks at the time of assessing the implications for control.

We hope the results obtained here together with those in [10] shed more light on more subtlety in model building with heterogeneous mixing.

4 Future Trends

Future research will involve the concept of saturation, with the rate of new infections satisfying the Generalized Law of the Minimum in some form.

Acknowledgement

I wish to acknowledge all those who do research without financial support. This work has been a proving ground of how difficult it is to carry out research without assistance.

References

1. Akinwande, N.I. (1997). A Deterministic Mathematical Model of Yellow Fever Epidemics. Ph.D Thesis, Univ. of Ibadan.
2. Bailey, N.T.J. (1975). The Mathematical Theory of Infectious Diseases and its Applications. Charles Griffin and Co. Ltd. London and High Wycombe.
3. Beretha, E. and Takeuchi, Y. (1995). Global Stability of an SIR Epidemic Model with Time Delays. J. Math. Biol. 33, 250–260.
4. Fitzgibbon, W.E., Morgan, J.J. and Waggoner, S.J. (1995). Generalized Lyapunov Method for Iterative Systems in Biology *in*: Mathematical Population Dynamics. Lecture Notes in Pure and Applied Mathematics 13, O Arino, DE Axelrold and M Kimmel (Eds).
5. Hoskins, R.F. Delta Functions: Introduction to Generalized Functions. Horwood Series in Mathematics and Applications. Horwood Publishing, Chichester, 1999.
6. Lalli, B.S. and Zhang, B.G. (1994). On a Periodic Delay Population Model. Quat. Appl. Math. LL(I), 35–42.
7. Liadi, M.A. (1999). A Qualitative Study of Andrew Dobson's Problem in Helminth Infection. Ph.D. Thesis, Univ. of Ibadan.
8. Luboobi, L.S. (1990). Mathematical Models for the Dynamics of the AIDS Epidemic, in Proc. of Biometric Society Meeting, Nairobi, Kenya, 2–6 April, 76–83.
9. Mackey, M.C. and Rudnick, R. (1994) Global Stability in a Delayed Partial Differential Equation Describing Cellular Replication. J. Math. Biol. 33, 89–109.
10. Rodriguez, D.J. and Torres-Sorando, L. (2001). Models of Infectious Diseases in Spatially Heterogeneous Environments. Bull. Math. Biol. 63, 547–571.
11. Sneddon, I.N. The Use of Integral Transforms (New York: McGraw-Hill) 1972.
12. Sowunmi, C.O.A. (1977). On the Asymptotic Behaviour of G. Adler's Model of Vector-Borne Epidemic in Proc. Int. Symposium on Funct. Anal. and its applications, 171–179. A. Olubummo, S.O. Iyahen and G.O.S. Ekhaguere (Eds).
13. Sowunmi, C.O.A. (1977). G. Adler's Deterministic Model of Vector-Borne Epidemic – A Threshold Theorem Math. Biosci. 35, 47–54.
14. Sowunmi, C.O.A. (2002). Stability of Steady State and Boundedness of a 2-Sex Population Model. Nonlinear Anal. 51, 903–920.
15. Wangersky, P.J. and Cunningham, W.J. (1956). On Time Lags in Equations of Growth. Proc. N.A.S. (42), 699–702.
16. Watson, R.K., (1972) On an Epidemic in a Stratified Population. J. Appl. Prob. 9, 659–666.
17. Wilson, E.B., and Worceter, J., (1945) The Spread of an Epidemic. Proc. Nat. Acad. Sci. Wash. 31, 322–330.

Approximate Scale-Invariant Random Fields: Review and Current Developments

O.I. Yordanov

Summary. During the last several decades, a great variety of irregular time-dependent phenomena and spatial morphologies have been shown to possess stochastic scale-invariance. This led to the development of models based on random fractal processes and, in general, (multi-dimensional) random fractal fields. In contrast to the ideal fractals, commonly assumed to be "scale-free" (reflected for example in the assumption of a simple power-law type correlation functions), the real scale-invariant hierarchies have a finite extend, limited by both a smallest and a largest scales.

In this paper I review a class of random fields which incorporates both the scale invariance and the finite size effects. The fields are constructed in the Fourier space and involve domains where their spectra are represented by power-law functions. The fields' two-point correlation functions are defined over the entire real line and are shown to be analytic. The scaling arises as an asymptotic behavior and therefore is only approximate. The effect of the finite sizes is imbalanced and inflicts systematic biases in the evaluation of the fractal dimensions and other scaling exponents. I also present applications and discuss certain technical and statistical subtleties involved in the construction and validation of models based on approximately scale-invariant fields.

1 Introduction

Scale invariant are structures and processes, which does not change or appears qualitatively the same when examined under different magnifications or respectively time scales. Mathematically, such property satisfy functions, solutions of the homogeneous functional equation, $f(\lambda \mathbf{x}) = \lambda^\chi f(\mathbf{x})$, where $\lambda \in \mathbb{R}$, $\lambda > 0$, $\chi \in \mathbb{R}$ and $\mathbf{x} \in \mathbb{R}^d$. (Sometimes f is called positively homogeneous function; generalizations involving complex functions will not be considered here.) Indeed the equation says that if we magnify/reduce the argument by a factor λ, this is equivalent of re-scaling the function by a factor of λ^χ. In a common jargon we say that $f(\mathbf{x})$ "scales in x with exponent χ". Another frequently used names are self-similar or self-affine functions. An example

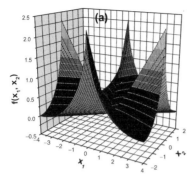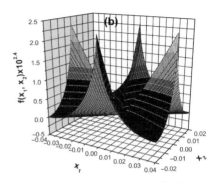

Fig. 1. Illustration of the exactly scale-invariant function $f(x_1, x_2) = (x_1^2/a^2)^\mu +$ $(x_2^2/b^2)^\mu - c\left((x_1^2/a^2)(x_2^2/b^2)\right)^{\mu/2}$. The scaling exponent for this plot is $\chi = 2\mu = 1.2$. Observe that the domain of the graph in panel (**b**) is reduced by a factor $\lambda = 10^{-2}$, consequently $f(x_1, x_2)$ is "magnified" by a factor of $\lambda^{-\chi}$ to obtain the same values as in panel (**a**)

of a solution of the functional equation in $\mathbb{R}^2$, with $\chi = 2\mu$ is the function $f(x_1, x_2) = (x_1^2/a^2)^\mu + (x_2^2/b^2)^\mu - c\left((x_1^2/a^2)(x_2^2/b^2)\right)^{\mu/2}$, where μ, a, b and c are real parameters. This function for $\mu = 0.6$, $a = 2.0$, $b = 1.0$ and $c = 2.0$, is illustrated in both panels (a) and (b) of Fig. 1. The graphs appear to be identical, notice however that the domain in the right panel is reduced by a factor of $\lambda = 10^{-2}$, consequently the values of $f(x_1, x_2)$ are "magnified" by a factor of $\lambda^{-\chi} = 10^{2.4} \approx 251$.

The classical "fundamental" laws of physics – the Newton's law of gravitation, the Coulomb's law of electrostatics – are examples of scale-invariant functions, both as functions of distance and the functions of masses/charges. Scale invariance is frequently "inherited" in relationships derived on basis of scale-invariant laws. One of the earliest discovered such relationship is the third Kepler's law, which could be stated as: "The orbital period of the planets P scales in the average distance to the sun a with exponent $3/2$; $P = ka^{3/2}$, where k originally was thought to be a constant. Another scaling relationship involving planetary motion and hence steaming from the gravitation law is called "rotational curve": The average velocity of a body orbiting around a center of gravity scales as $a^{1/2}$, where a is the average radius of the orbit. In electrostatics or magnetostatic, to take simple examples, the field at large distance from a dipole scales with exponent -3, from a quadruple with -4 and so on.

Scale-invariance arises also in complex phenomena, where the dynamics is affected by many and possibly different factors. The complexity of such phenomena implies a stochastic approach for their description. How then the scale-invariance manifests in a stochastic function $x(t)$? (Traditionally, a stochastic function of a single argument is called random process, i.e., time-dependent, while a stochastic function with domain in $\mathbb{R}^d$, $d > 1$, is called a

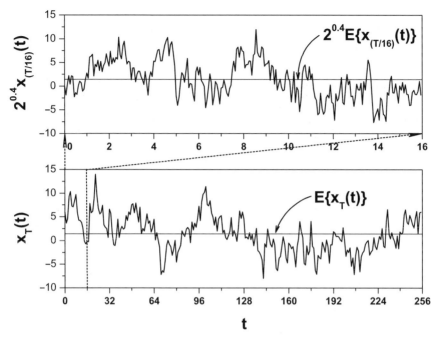

Fig. 2. Illustration of stochastic scale-invariant curve, (1), with power-law, power spectrum with parameters $A = 1$ and $\alpha = 1.2$, see the text. The domain of the realization of the upper panel is reduced by a factor of $\lambda = 1/2^4$, consequently the random curve is rescaled by $\lambda^{-(\alpha-1)/2}$. Although the two realizations are not exactly identical, their behavior is closely similar and in particular the mean of the bottom realization is the same as the rescaled mean of the upper (marked by the horizontal lines)

random field.) It is natural to assume that the scaling applies to each of the realizations of $x(t)$: $x^{(r)}(\lambda t) = \lambda^\chi x^{(r)}(t)$, where the superscript (r) is introduced to distinguish a realization from the random function itself. Since the domain of $x^{(r)}(\lambda t)$ is different from that of $x^{(r)}(t)$, these functions are two different realizations and hence the values of $x^{(r)}(\lambda t)$ are not equal to $\lambda^\chi x^{(r)}(t)$. However, the graphs of the realizations should have similar behavior and the scale-invariance shows up in the statistical moments/functions of $x(t)$. This, referred to as a stochastic scale-invariance, is illustrated in Fig. 2, where two computer realizations of the random process

$$x_T(t) = \sum_{k=0}^{N/2} c_k \cos\left(\omega_k t + \phi_k\right) \qquad (1)$$

are plotted. In (1), referred to as harmonic process, the randomness is brought in solely by the phases ϕ_k, assumed to be independent, uniformly distributed in the interval $[-\pi, \pi]$ random variables. The time domain of $x_T(t)$ is taken to be $2T$, which defines the (circular) frequencies, $\omega_k = k\pi/T$, $k = 0, 1, \ldots, N/2$. The amplitudes c_k are determined from the requirement that $x_T(t)$ has a preelected power spectral density function $S(\omega)$; for this purpose $c_k \propto \sqrt{\Delta\omega S(\omega)}$; details of the construction and the numerical evaluation of (1) are presented in the appendix. In particular, for the graphs in Fig. 2, I have chosen $S(\omega) = A\omega^{-\alpha}$ for $\omega \geq \omega_1$ and $S(0) = S(\omega_1)$, where the parameters A and α are called spectral constant and exponent, respectively. This choice makes $x_T(t)$ stochastic scale-invariant: an easy check shows $x_{\lambda T}(\lambda t) = \lambda^{(\alpha-1)/2} x_T(t)$. In particular, since the mean of the curve is given by $E\{x_T(t)\} = c_0 = \sqrt{\Delta\omega S(0)}$, it follows that $\lambda^{-(\alpha-1)/2} E\{x_{\lambda T}(t)\} = E\{x_T(t)\}$, which is indicated by the position of the horizontal lines in Fig. 2.

The scaling relationships in Nature are not valid universally, that is, not valid for all time or distance scales. Even the Coulomb's law tested in experiments of electron–proton scattering seems to fail at distances less than 10^{-16} m. Due to a much serious experimental hurdles, the law of gravity was tested down to the sub-centimeter range distances only [1, 2] but is expected to break down at distances where the quantum effects become important. The third Kepler's law is meaningless at orbital distances below the radius of the Sun. In the cases where complex structures or interactions are involved, the range of validity of the scaling relationships is limited from both below and above and typically are only approximate. An emblematic example is provided by the rotational curve of a galaxy, where the deviation of the expected scaling served as the first evidence for the presence of an unseen mass. Therefore it is of interest to study structures that inherently have limited range of scale-invariant hierarchy. It turns out that the scale-invariance for such system is approximate.

In this paper I review current studies of approximately scale-invariant random fields. The review begins with a generic example of exact scale-invariant random fields, presented in the next section. This field is representative of an ideal fractal object with non-integer Hausdorff dimension. I consider a more realistic scale-invariant fields in Sect. 3. They are constructed with a limited self-affine hierarchy, which shows up in the leading asymptotic behavior of their second-order statistical functions and therefore are approximately scale-invariant only. In Sect. 4, I focus on some of the applications these field find and discuss directions for future studies. The list of references is by far not complete. A special care is taken to list, whenever possible, entries that are freely available.

2 Exactly Scale-Invariant Stochastic Fields

There is a generic example of a Gaussian stochastic field which is exactly scale-invariant [3]. The example is constructed from the spectral representation, which for an arbitrary homogeneous random field, $f : \mathbb{R}^d \to \mathbb{R}$, is given in terms of the Fourier–Stielties integral [4, 5]

$$f(\mathbf{x}) = \int \exp(i\mathbf{k} \cdot \mathbf{x}) \, dz(\mathbf{k}), \tag{2}$$

where $\mathbf{k} \in \mathbb{R}^d$. The is a complex random field $z(\mathbf{k})$ is assumed with zero mean, $E\{z(\mathbf{k})\} = 0$, and independent,

$$E\{dz^*(\mathbf{k}) \, dz(\mathbf{q})\} = \delta(\mathbf{k} - \mathbf{q})S(\mathbf{k})d\mathbf{k}d\mathbf{q}, \tag{3}$$

for $\mathbf{k} \neq \mathbf{q}$. The function $S(\mathbf{k})$ in (3) is referred to as (power) spectral density function (shortly, spectrum), c.f. with the discrete example discussed in the appendix. The spectrum defines the field (2) up to moments of second order. If $z(\mathbf{k})$ is a Gaussian field, $S(\mathbf{k})$ determines the field throughout arbitrary order. The ideal scale-invariant random fields are generated by choosing in (3):

$$S(k) = Ak^{-\alpha}, \tag{4}$$

for every $k = |\mathbf{k}|$. The (spectral) constant A must be positive and the values of the spectral exponent are restricted within $d < \alpha < d + 2$, where d is the dimension of the domain of the field. Since $S(\mathbf{k})$ depends on $\mathbf{k}$ through its Euclidean length k only, it follows that the field is isotropic, $f = f(x)$, $x = |\mathbf{x}|$.

This field has a number of peculiar properties. For example, its variance and the variances of the higher derivatives of the field are not finite. Its autocovariance function (AcF) also does not exist. However, mean-square-increment function (MSIF), also called structure function, and defined by

$$\mathcal{B}_{(f)}(\mathbf{x}) = E\left\{[f(\mathbf{x}_0 + \mathbf{x}) - f(\mathbf{x}_0)]^2\right\}, \tag{5}$$

is finite. We remark that $\mathcal{B}_{(f)}(\mathbf{x})$ is a second-order, two-point characteristics of the field. MSIF is related to the variance of the f, $\sigma^2_{(f)}$, and its AcF, $\mathcal{A}_{(f)}(\mathbf{x})$, when they exist, through $\mathcal{B}_{(f)}(\mathbf{x}) = 2\sigma^2_{(f)} - 2\mathcal{A}_{(f)}(\mathbf{x})$.

Using (2) and (3), it is straightforward to obtain an integral representation for the function $\mathcal{B}_{(f)}(\mathbf{x})$. In particular, for isotropic fields

$$\mathcal{B}_{(f)}(x) = \frac{4\pi^{d/2}}{\Gamma(d/2)} \int_0^{\infty} \left[1 - {}_0F_1\left(\frac{d}{2}; -\frac{k^2 x^2}{4}\right)\right] S(k)k^{d-1} \, dk, \tag{6}$$

where Γ and ${}_0F_1$ denote the Euler's Gamma function and type $(0,1)$ hypergeometric function, respectively. (${}_0F_1$ from these parameter and argument can be expressed in terms of order $(d/2 - 1)$ Bessel function.) Inserting (4) into (6) and taking the integral we obtain [6]:

$$\mathcal{B}_{(f)}(x) = \tau^{d+2-\alpha} x^{\alpha-d}, \tag{7}$$

where

$$\tau^{d+2-\alpha} = \frac{A\pi^{d/2}2^{d+2-\alpha}\Gamma\left((d+2-\alpha)/2\right)}{(\alpha-d)\Gamma(\alpha/2)}. \tag{8}$$

Equation (7) shows that up to the second order, $f(x)$ is scale-invariant with scaling exponent $H = (\alpha - d)/2$. In dimension $d = 1$, H is called the Hurst exponent. However, since f is Gaussian, (7) determines all statistical properties of the field. Also, Eq. (7) shows that the scale-invariance holds for every x, i.e., the field does not have a characteristic length; such structures are termed "scale-free". Therefore, f is exactly scale-invariant with exponent H. Further, it has been proven by Orey [7], see also [8], that if the MSIF of the random process $f(x)$ behaves at zero as $\mathcal{B}_{(f)}(x) \sim x^\mu$, $x \to 0^+$, with $0 < \mu < 2$, then the realizations of the process have a non-trivial Hausdorff dimension given by $D_H = (4 - \mu)/2$. In dimension d, Orey's result can be generalized to yield $D_H = d + 1 - \mu/2$ and therefore using (7) we find that the Hausdorff dimension of the field is related to the spectral exponent α by,

$$D_H = d + 1 - (\alpha - d)/2. \tag{9}$$

The Orey's theorem shows that the realizations of the random functions defined by (4) are fractals (according to Mandelbrot's tentative definition [9]). We shall refer to this field as the ideal fractal case. Equation (7) furnishes a simple (scaling) method for practical evaluation of D_H: draw a log–log plot of $\mathcal{B}_{(f)}(x)$, which in virtue of the power-law form will be a straight line; the slope of this line is $(\alpha - d)$, and using the obtained value of α, calculate D_H from (9).

When f is used as a model of rough surfaces, in dimension $d = 2$, τ is called topothesy [10]. If, in general, $f(x)$ has a physical dimension identical to x, then τ has the same dimension and (7) shows that over a "distance" $x = \tau$, the mean increment of f is precisely τ. The expression (8) diverges in both $\alpha \to d$ and $\alpha \to d + 2$ limits. Graphs of the behavior of τ as function of the scaling exponent H are depicted in [6].

Equation (4), respectively (7) are considered to be adequate model for variety of natural phenomena and structures. It is easily seen, however, that the ideal fractal models have serious drawbacks. Indeed, the experimentally observed structures always have finite variances. The sample AcFs calculated from experimental data are also finite. Real structures and processes always have finite hierarchy of scales. In terms of the spectrum, the latter translates to a spectrum with limited wave-number/frequency content. Hence, the way toward a more realistic self-affine models is to modify the ideal fractal process by introducing an appropriate form of spectral fall-off regions.

3 Approximate Scale-Invariant Fields

The simplest path towards physically more realistic stochastic fields is to take spectra defined by a power-law with two sharp cut-offs [11]:

$$S(k) = \begin{cases} Ak^{-\alpha}, & \text{if } k_0 \le k \le k_1 \\ 0, & \text{otherwise.} \end{cases} \tag{10}$$

I will show below that such fields are scale-invariant, however only approximate. The latter makes them appropriate basic blocks for building elaborate models possessing approximate scale-invariance.

First, note that the variance of the field, which is represented by half of the first term in the representation (6), is finite. Using (10) one immediately obtains: $\sigma^2_{(f)} = \frac{1}{2}\mathcal{B}_{(f)}(\infty) = \sigma_0^2 - \sigma_1^2$, where

$$\sigma_p^2 = \frac{2A\pi^{d/2}}{\Gamma(d/2)}\frac{k_p^{d-\alpha}}{(\alpha-d)}, \qquad p = 0, 1, \tag{11}$$

see also [6]. Next, the AcF, which is represented by half of the second term in (6), is also finite and can be calculated exactly:

$$\mathcal{A}_{(f)}(x) = \sigma_0^2 As(d, \alpha; k_0 x) - \sigma_1^2 As(d, \alpha; k_1 x), \tag{12}$$

where As is a function, which is expressed in terms of type $(1, 2)$ hypergeometric function $As(\mu, \alpha; z) = {}_1F_2\left(\frac{\mu-\alpha}{2}; \frac{\mu+2-\alpha}{2}, \frac{\mu}{2}; -\frac{z^2}{4}\right)$. It is shown in [12] that $As(\mu, \alpha; z)$ satisfies a third-order linear ordinary differential equation with coefficients involving the parameters μ and α. The point $z = 0$ is a regular singular point of the equation with $As(\mu, \alpha; z)$ representing one of the three linearly independent solutions about zero. The other two being: $w(z) = z^{\alpha-\mu}$, which is scale-invariant, and $w(z) = z^{2-\mu}As(4 - \mu, 2 + \alpha - \mu; z)$, which is scale-invariant for small z only. The function $As(\mu, \alpha; z)$, however, is an entire function with a series representation involving the even powers of z only. In particular, the leading term dominating how $\mathcal{A}_{(f)}(x)$ decreases from $\sigma^2_{(f)}$ for small x, $x < k_1^{-1}$, is quadratic,

$$\mathcal{A}_{(f)}(x) \sim \sigma^2_{(f)} - A\frac{\pi^{d/2}k_1^{d+2-\alpha}\left(1 - \delta^{d+2-\alpha}\right)}{\Gamma(d/2)d(d+2-\alpha)}x^2 + O\left[(k_1 x)^4\right], \tag{13}$$

where the ratio of the spectral cut-offs, $\delta = k_0/k_1 \ll 1$, is introduced. How then the scale-invariance of (12) arises? To understand this, I turn to the asymptotic expansion for large z. The asymptotic expansion of As reads:

$$As(\mu, \alpha; z) \sim \Gamma\left(\frac{\mu}{2}\right)\frac{\Gamma\left[(\mu - \alpha)/2 + 1\right]}{\Gamma(\alpha/2)}\left(\frac{z}{2}\right)^{\alpha-\mu}$$

$$- \left(\frac{\alpha - \mu}{2\sqrt{\pi}}\right)\left(\frac{2}{z}\right)^{(\mu+1)/2}\Gamma\left(\frac{\mu}{2}\right)\cos\left(z - \frac{\mu + 1}{4}\pi\right), \tag{14}$$

where for simplicity I wrote down the leading and the next order terms only. In the applications, we almost always deal with spectra for which $k_1 \gg k_0$. Then for intermediate values of the argument $k_1^{-1} \ll x \ll k_0^{-1}$, we use the series expansion in the first term of (12) and its asymptotic form of As in the second. Up to the leading orders

$$\mathcal{A}_{(f)}(x) \sim \sigma^2_{(f)} - \frac{1}{2}\tau^{d+2-\alpha}x^{\alpha-d} - \sigma_0^2\frac{2(\alpha-d)}{d(d+2-\alpha)}\left(\frac{k_0 x}{2}\right)^2$$

$$+\sigma_1^2\left[1 + \left(\frac{\alpha-d}{2\sqrt{\pi}}\right)\Gamma\left(\frac{d}{2}\right)\left(\frac{2}{k_1 x}\right)^{\frac{d+1}{2}}\cos\left(k_1 x - \frac{\mu+1}{4}\pi\right)\right]$$

$$+O\left[(k_0 x)^4, (k_1 x)^{-\frac{d+3}{2}}\right]. \tag{15}$$

Equation (15) represents a decreasing from $\sigma^2_{(f)}$ function. The ratio of the third (the quadratic in x) over the second (the power-law) term is of order $\sim(k_0 x)^{d+2-\alpha}$. Since $(k_0 x) < 1$, it follows that the third term is significant compared to the second only for α values close to its upper limit of $d+2$. Likewise, the ratio between the forth (involving σ_1^2) and the second term is of order $\sim 1/(k_1 x)^{\alpha-d}$ and hence affects the values of $\mathcal{A}_{(f)}(x)$ for α close to its lower limit d. Thus, the dominant term is the second, power-law one (note it does not depend on the smallest or largest scales of the random field) which constitutes the approximate scale-invariance of the random field. Comparing with (7), we see that the scale-invariance of $\mathcal{A}_{(f)}(x)$ is characterized by exponent identical to the exponent of the ideal fractal case. On this ground one loosely speaks of a fractal dimension D of the band-limited random field with value given again by (9). It should be stressed that this quantity has no meaning of Hausdorff dimension. Indeed, as (13) demonstrates, $\mathcal{B}_{(f)}(x) \sim x^2$ as $x \to 0^+$, and therefore Orey's theorem yields $D_H = d$.

On the other hand, notice that the third and the fourth terms involve only the small and the large cutoffs k_0 and k_1, respectively. The imbalance of the contribution of these terms which depends on the value of α, reflects the imbalance of the two finite size effects for $\alpha \neq d+1$. It systematically alters the slope of the MSIF (or equivalently the AcF) in log–log scale and leads to the errors of fractal dimension estimates [11]. These errors were first found in [13] using numerical simulations.

The asymptotic expansions of $\mathcal{A}_{(f)}(x)$ for large $x > k_0^{-1}$ shows that it approaches zero in an oscillatory manner; the full asymptotic expansion in this regime for $d = 1$ is developed in [11]. Here I provide the leading term in the case of arbitrary dimension d;

$$\mathcal{A}_{(f)}(x) \sim -\frac{A}{\pi k_0^{\alpha-d}}\left(\frac{2\pi}{k_0 x}\right)^{\frac{d+1}{2}}$$

$$\times\left[\cos\left(k_0 x - \frac{\mu+1}{4}\pi\right) - \delta^{\frac{2\alpha+1-d}{2}}\cos\left(k_1 x - \frac{\mu+1}{4}\pi\right)\right]. \tag{16}$$

Such a behavior is often observed in the experimental AcFs, which is an evidence that despite its simplicity, the random processes/fields considered here might be expected to be relatively realistic models in variety of circumstances. It is important to note that the absolutely sharp cutoffs assumption is justified when the largest and the smallest scales are not physical but implicitly imposed within the power branch of the spectrum by not enough length or resolution of the data.

4 Applications: Future Work

The detailed analysis of the approximate scale-invariant fields, which I reviewed in the preceding section, naturally suggests a method for building stochastic models of complex processes and structures that possess scale-invariant symmetry. The method involves: compute the sample AcF (alternatively the sample MSIF) from the experimental data and fit these values with expression (12) using a nonlinear, least-square algorithm. The method have several advantages:

(i)   It is based on exact expressions, which can be evaluated with a controlled precision over wide range of arguments.

(ii) The fit renders the values of all spectral parameters: spectral exponent, spectral constant (strength), and the extent of the wavenumber/frequency band. In contrast, for example the classical spectral analysis provides reliable estimates of the spectral exponent only.

(iii) The methodology explicitly accounts for the finite-size effects and removes the systematic errors in the estimation of the spectral scaling exponents.

(iv) The models can easily be improved by augmenting the spectrum (10) with other power-law spectral segments.

The method does have certain drawbacks stemming chiefly from the fact that the "experimental" AcFs are estimated with certain biases and typically have large sampling variances in their "tails". Throughout the end of this section I present subtleties of the method's implementation.

4.1 Initial Values of the Model's Parameters

Obtaining approximate values for the model parameters makes up an useful preliminary step in several ways. First, it provides initial parameters to start the fitting procedure, already close to the best solution. Second, it gives a rough idea of the overall behavior of the selected model. Third, it suggests how good the final fit we shall hope to be. Fourth, it hints about the experimental points one should put more weight on at the final fit. Finally, since, as we shall see in a moment, the approximation of the parameters rests essentially on the scaling analysis, the displacement of the initial AcF from the best fitted one is a measure of the finite size effects.

The calculation of the approximate values of the model is simple: First, it is almost always good enough to approximate the value of the upper cutoff k_1 by the Nyquist wavenumber $k_{(N)} = \pi/2$, where Δx is the sampling interval of the data. Next, estimate the spectral exponent α. For this purpose calculate the sample MSIF from the sample ACF using the relation $\mathcal{B}_{(f)}(x) = 2\sigma^2_{(f)} - 2\mathcal{A}_{(f)}(x)$ and plot its values in a log–log scale; if the data show up scale-invariance, slope of this graph should be $\alpha - d$. (This is equivalent to some of scaling plots used for determining the fractal dimension.) The spectral constant A and the lower cutoff k_0 are found by solving numerically a system of two equations. The latter are obtained by equating the model's and the sample variances, and the first zero crossings of (12) and the sample ACF. An approximation of the second of these equations, accurate enough provided α is not close to d and $d+2$ is given in [14]. An alternative and simpler approach to estimate k_0 is available when the experimental AcF has a well defined first minimum. Using the leading asymptotic order dictating the AcF oscillatory behavior for $x > k_0^{-1}$, see (16), the position of the minimum is approximately:

$$x_{\min} = \frac{(d+1)\pi}{4k_0}. \tag{17}$$

Taking $x_{\min}$ from the graph, we determine the value of k_0.

4.2 Numerical Evaluation of the Function $As(d, \alpha; k_p x)$

This issue arises due to the need to compute As within an extremely wide interval of arguments and for all admissible values of its parameters, d, α, and k_p, $p = 0, 1$. For relatively small intermediate values of its argument, As is easy to calculate from its series representation. In fact, As is an entire function which converges faster compared to the exponential function from the same argument. For large arguments, however, the "numerical convergence" of the series is limited by the computer precision. It can be shown that the convergent series summation fails above

$$x_p = M \ln 10/k_p, \tag{18}$$

where M is the maximum number of digits not rounded off by the computer. For $x > x_p$, I use the (divergent) asymptotic summation. A helpful scheme is so-called optimal truncation rule [15], which I found always reliable in the evaluation of the As from its asymptotic series. Another thing to be noted is that in the neighborhood of x_p, the series and the asymptotic summations agree to within seven digits. This is especially important since $\mathcal{A}_{(f)}(x)$ is computed as a difference between numbers which are 5–7 orders of magnitude greater that its own value.

4.3 Multi-Segmented Power-Law Spectra

Experimental data often show multiple scaling regimes. That is, scaling with different exponent within different ranges of resolution of the instruments. I sketch next how the results presented in Sect. 3 can easily be generalized to model such complex systems. Let $k_0 = 0, k_1, \ldots, k_{n+1} = k_{(N)}$ be a partition of the spectral domain, where $k_{(N)}$ is the Nyquist wavenumber. I shall refer to k_p, $p = 0, 1, \ldots, n + 1$ as cross-over wavenumbers. Define $S(k)$ as piecewise continuous function, which in each of the intervals $k_p \leq k \leq k_{p+1}$, $p = 1, 2, \ldots, n + 1$, is represented by a power-law function, $S^{(p)} = A_p k^{-\alpha_p}$, where all $A_p > 0$. The leftmost segment, i.e., the segment defined over $0 \leq k \leq k_1$ is given by $S^{(0)}(k) = B_0 + c_0 k^{\alpha_0}$. This form of $S^{(0)}$ with the condition $\alpha_0 > 0$ ensures the finiteness of the spectra at zero and at the same time makes this segment amenable to the same analytical and numerical methods as the other segments, see below. Imposing continuity at k_p, $p = 2, 3, \ldots, n$, I express all spectral constants by A_1:

$$A_p = A_1 k_2^{-\alpha_1} \left(k_2/k_3 \right)^{\alpha_2} \cdots \left(k_{p-1}/k_p \right)^{\alpha_{p-1}} k_p^{\alpha_p}, \quad p > 1 \tag{19}$$

Introducing for convenience $A_0 = B_0 k^{\alpha_1}/A_1$ and using the continuity condition at k_1, we have

$$S^{(0)}(k) = A_1 k_1^{-\alpha_1} \left[A_0 + (1 - A_0) \left(k/k_1 \right)^{\alpha_0} \right]. \tag{20}$$

Note that $A_0 = S(0)/S(k_1)$ and that $\alpha_0 = 0$ produces a flat leftmost branch of the spectrum. The number of the free parameters of the spectrum is $2n + 3$: $n + 1$ spectral exponents α_p, $p = 0, 1, \ldots, n$; n crossover wavenumbers k_p, $p = 1, \ldots, n$ and two spectral constants A_0 and A_1. Not dwelling into further details of the model, I remark that according to the MSIF integral representation, (6), the contribution of the individual segments to $\mathcal{B}_{(f)}$, and thus to the AcF is additive. In this way, the implementation and the verification of a multi-segmented model requires no additional developments than those presented so far.

4.4 Examples and Problems

Single-segment spectra were used as models of morphologies of gold deposits on a quartz substrates [14] and of Cr deposits on a highly polished $BK7$ glass. The latter were further smoothed via ion beam sputtering under glancing incidence [16]. Single-segment spectra, however, with incorporated white noise component, were used for cultivated soil roughness [17]. A model with two-segments, a principle segment in the form of (10) and a high-frequency fall-off region was shown to be a very accurate model of the statistical behavior of the series generated by the X-component of the Lorenz chaotic system [18]. Three-segment spectra were needed to characterize the statistics of two other low-dimensional chaotic systems, N and B systems in the nomenclature of Sprott, see [19]. The last applications, I would like to note here, involve approximate scale-invariant stochastic models of computer generated morphologies mimicking ballistic deposition and molecular beam epitaxy [20, 21].

As already noted, the principle hurdle that needs to be overcomed is associated with the estimates of the experimental AcFs. Two sample estimates that exist [5] can employed each with a tradeoff. The first estimate is unbiased when the mean of the field is known. This estimate typically has large sample variances and in particular for large distance/time lags. The variances can be reduced by averaging the AcFs of several measurements if available. The second estimate is biased but has a lower sample variances. In addition, this estimate has the important advantage of being finite Fourier transform of the sample estimate of the spectrum. An important development would be to construct an estimate, which combines the virtues of both the biased and the unbiased estimates.

I conclude this review by discussing possible directions of future developments. First, the fitting procedure described in this section needs a reliable algorithm for assessment of the confidence intervals of the retrieved parameters. Such an algorithm might be based on the method of synthetic data simulation [22]. For this purpose, the random curve generation presented in the appendix should be extended for wider class of spectra and dimensions $d > 1$. Such a result would not only improve the validation of the models but also will facilitate predictions and large scale simulations of real phenomena. It would be very interesting to construct examples of scale-invariant fields that are not: (i) isotropic, (ii) homogeneous/stationary, and (iii) linear. Such models are certain to find a rich scope of applications.

5 Appendix

In this appendix I present developments leading to (1) and provide details of its numerical implementation. The goal is to obtain an approximation of random curve with a given spectral power density function. Since the numerical realization would necessarily be over a finite time domain, I consider a curve $x_T(t)$ defined for $|t| \leq T$ at the outset of the construction. Accordingly, $x_T(t)$ is represented by a Fourier series rather than a Fourier integral,

$$x_T(t) = \sum_{k=-\infty}^{+\infty} A_k e^{-i\omega_k t}, \qquad (21)$$

where $\omega_k = 2\pi f_k$, $f_k = k/2T$, are discrete circular frequencies. It is easy to check that $\int_{-T}^{T} e^{i(\omega_n - \omega_k)t} dt = 2T\delta_{nk}$, where δ_{nk} denotes the Kronecker delta symbol, and hence the inverse of (21) is

$$A_k = \frac{1}{2T} \int_{-T}^{T} x_T(t) e^{i\omega_k t} dt. \qquad (22)$$

Since $x_T(t)$ is a random function, A_k, $k = 0, \pm 1, \pm 2, \ldots$, is a (complex) random sequence of amplitudes. The "spectral energy" of $x_T(t)$, carried by the component with frequency ω_k, is $\Phi(\omega_k) = (2T^2/\pi)E\{|A_k|^2\}$, where E denotes the expectation operator; c.f. with the general definition energy spectral density, see e.g., [23]. The power spectral density, that is the spectral energy per unit of time, is then defined by

$$S(\omega_k) = \frac{1}{2T}\Phi(\omega_k) = \frac{T}{\pi}E\left\{|A_k|^2\right\}. \tag{23}$$

Note that $\Delta\omega_k = \omega_{k+1} - \omega_k = \pi/T$ and hence $E\left\{|A_k|^2\right\} = \Delta\omega S(\omega_k)$.

To proceed further we have to specify a probability model for A_k. Taking what can be perceived as the simplest choice, I consider $A_k = |A_k|\exp\left(-i\phi_k\right)$, where

$$|A_k| = \sqrt{S(\omega_k)\Delta\omega} \tag{24}$$

and the phases ϕ_k are independent random variables uniformly distributed in the interval $[-\pi, \pi]$. It follows that $E\left\{A_k\right\} = 0$ and (23) holds for any appropriate choice of $S(\omega_k)$. Hence, specifying $S(\omega_k)$ one defines A_k and therefore $x_T(t)$. Observe, however, that for $x_T(t)$ real, (22) requires $A_k^* = A_{-k}$; hence, $S(\omega_k)$ must be an even function, and $\phi_{-k} = -\phi_k$. In particular, $\phi_0 \equiv 0$. Changing $k \mapsto -k$ into the negative part of sum (21) we obtain

$$x_T(t) = \sum_{k=0}^{+\infty} c_k \cos\left(\omega_k t + \phi_k\right), \tag{25}$$

where $c_0 = \sqrt{S(0)\Delta\omega}$ and $c_k = 2\sqrt{S(\omega_k)\Delta\omega}$, $k = 1, 2, \ldots$ Note also that $E\left\{x_T(t)\right\} = c_0$. Another important remark concerning $x_T(t)$: although we begun with a function not defined outside the interval $-T \leq t \leq T$, we ended up with a function periodic over the entire real line with a period of $2T$, $x_T(t + 2nT) = x_T(t)$, for $n = \pm 1, \pm 2, \ldots$. This of course is a consequence of the fact that discrete Fourier transform was used in the construction.

In order to employ (21) or (25) for numerical simulation of random curves, we need to truncate the series and discretize the time. The truncation number, say $N/2$, should ultimately depends on the high-frequency behavior of $S(\omega)$ and is related to the time interval T. Let $\Delta t = 2T/N$ be the discretisation interval, $t_n = n\Delta t$, $n = 0, \pm 1, \pm 2, \ldots$ Then the truncated (21) is

$$x_T(t_n) = \sum_{k=-N/2}^{+N/2} \sqrt{S(\omega_k)\Delta\omega}\, e^{-i\phi_k} e^{-2\pi ikn/N}; \tag{26}$$

note that the discretized process have period N, $x_T(t_{n+N}) = x_T(t_n)$. The exponents in (26) have the same periodicity with respect to k and thus by shifting the negative part of the sum, $k \mapsto k + N$, we can cast $x_T(t_n)$ into the following form

$$x_T(t_n) = \frac{1}{N}\sum_{k=0}^{N-1} X_k e^{-2\pi ikn/N}. \tag{27}$$

In the above expression, $X_0 = N\sqrt{S(0)\Delta\omega}$, $X_k = N\sqrt{S(\omega_k)\Delta\omega}\,e^{-i\phi_k}$ for $k = 1, \ldots, N/2 - 1$, $X_{N/2} = 2N\sqrt{S(\omega_{N/2})\Delta\omega}\cos(\phi_{N/2})$, and

$X_k = N\sqrt{S(\omega_{k-N})\Delta\omega}e^{-i\phi_k}$ for $k = N/2 + 1, \ldots, N - 1$. (Since ϕ_k are independent the indexing ϕ_{N-k} is equivalent to ϕ_k.) Equation (27) is directly amenable to Fast Fourier Transform (FFT), see [22].

References

1. Chen, Y.T., Alan H. Cook, Metherell A.J.F.: An experimental test of the inverse square law of gravitation at range of 0.1 m. Proc. R. Soc. Lond. Ser. A, Math. Phys. Sci., **394**, 47–68 (1984).
2. Long, J.C., Chan, H.W. and Price, J.C.: Experimental status of gravitational-strength forces in the sub-centimeter regime. Nucl. Phys. B, **539**, 23–34 (1999); available at arXiv:hep-ph/9805217.
3. Berry, M.V.: Diffractals. J. Phys. A: Math. Gen. **12**, 781–797 (1979).
4. Panchev, S.: Random Functions and Turbulence. Pergamon Press, Oxford (1971); Monin, A.S. and Yaglom, A.M.: Statistical Fluid Mechanics. MIT Press, Boston (1971).
5. Priestley, M.A.: Spectral Analysis and Time Series. Academic Press, London (1981).
6. Yordanov, O.I. and Atanasov, I.: Self-affine random surfaces. European Phys. J. B, **29**, 211–215 (2002).
7. Orey, S.: Gaussian sample functions and the Hausdorff dimension of level crossings. Z. Wahrsch'theorie verw. Geb. **15**, 249–256 (1970).
8. Falconer, K.J.: The Geometry of Fractal Sets. Cambridge University Press, Cambridge (1985).
9. Mandelbrot, B.B.: Fractals, W.H. Freeman and Company, San Francisco (1979); Mandelbrot, B.B.: The Fractal Geometry of Nature, W.H. Freeman and Company, San Francisco (1982).
10. Sayles, R.S. and Thomas, T.R.: Surface topography as a nonstationary random process, Nature (London) **271**, 431–434 (1978); Berry M.V. and Hannay, J.R.: Nature (London) **273**, 573 (1978).
11. Yordanov, O.I. and Nickolaev, N.I.: Self-affinity of time series with a power-law power spectrum. Phys. Rev. E, Rapid Commun., **49**, R2517–R2520 (1994).
12. Yordanov, O.I.: Properties of a special function pertaining to approximately scale-invariant random fields. In: Dobrev V.K. (ed) Proc. 4th Int. Symp. Quantum Theory and Symmetries IV, Heron Press Int., Sofia, Bulgaria, p. 809. (2005).
13. Greis, N.P. and Greenside, H.S.: Implication of a power-law power-spectrum for self-affinity Phys. Rev. A, **44**, 2324–2334 (1991).
14. Yordanov, O.I. and Ivanova, K.: Description of surface roughness as an approximate self-affine random structure, Surf. Sci., **331–333**, 1043–1049 (1995).
15. Bender, C. M. and Orszag, S. A. Advanced Mathematical Methods for Scientists and Engineers, McGraw-Hill, New York (1978).
16. Holzwarth, M., Wißing M., Simeonova D.S., Tzanev, S., Snowdon, K.J. and Yordanov, O.I.: Preparation of atomically amooth surfaces via sputtering under glancing incidence conditions, Surf. Sci., **331–333**, 1093–1098 (1995).
17. Yordanov, O.I. and Guissard, A.: Approximate self-affine model for cultivated soil roughness, Physica A, **238**/1–4, 49–65 (1997).

18. Yordanov, O.I. and Nickolaev, N.I.: Approximate, saturated and blurred self-affinity of random processes with finite domain power-law power spectrum, Physica D, **101**, 116–130 (1997).

19. Dimitrova, E.S. and Yordanov, O.I.: Statistics of some low-dimensional chaotic flows, Int. J. Chaos Bifur., **11**, 2675–2682 (2001).

20. Vulkova, L.A., Atanasov, I.S. and Yordanov, O.I.: Approximate Self-Affine Characterization of Two-Point Rough Surface Statistics Simulated by Eden Algorithm, Vacuum, **58**, 158–165 (2000).

21. Vulkova, L.A., Atanasov, I.S. and Yordanov, O.I.: Comparison of Numerical Algorithms for Simulation of Molecular Beam epitaxy, Vacuum, **69**, 69–73 (2003).

22. Press W.H., Teukolsky, S.A., Vetterling W.T and Flannery, B.P.: Numerical Recipes in C, Second edition. Cambridge University Press, Cambridge (1992). Book and codes available on-line on http://www.nr.com.

23. See Wikipeadia: http://en.wikipedia.org/wiki/Spectral_density.

Part III

Simulation and Visualization

Non-Stationary Vibrations of Viscoelastic Circular Cylindrical Thick Shell Under the Influence of Temperature

F.A. Amirkulova

Summary. In the represented work, free from hypotheses and preconditions used in known classical and refined theories, the new refined theory of non-stationary vibrations of the circular cylindrical viscoelastic thin and thin walled shells and columns concerning the temperature is developed. The developed method of a deduction of the partial differential vibration equations is based on the 3D problems' exact mathematical formulation of the theory of elasticity and their general solutions in transformations. As the basic unknown functions the displacements of intermediate surface of the shell which can change into median, external or internal surface are accepted. The calculation algorithm of temperature, stress and displacement field by values of the unknown functions is offered, allowing to formulate applied problems on its vibrations. On the basis of the developed approach a number of applied problems are solved and numerical results are received. The thermo-stressed state of the semi-infinite viscoelastic cylindrical thick walled shell excited by kinematic influence at the end face and temperature influence on the shell's surfaces is explored.

1 Introduction

Engineering constructions that can maintain the action of intensive external load have been widely applied in aircraft construction, rocket engineering, mechanical engineering, shipbuilding, civil engineering and other areas of economy. Modern requirements to decrease weight clearance parameters of flying devices and industrial civil constructions under condition of maintenance of necessary durability and reliability have made calculating their stress-strain state a problem of interest in mechanics. Among these problems, the problems of studying of non-stationary interaction of construction element with the environment are rather complex. At present in a number of areas of engineering and construction there is an interest in the problems adjoining to the mentioned above actual problems and connected to dynamic behavior of constructional elements.

During the investigation of stressed-strained state of circular cylindrical layers and shells subject to connected fields use vibration's equations of cylindrical shell of Kirchhoff–Love theory and Timoshenko type theories [9, 10, 12, 14]. For interacting connected field (e.g. temperature) corresponding characteristic equation is used. Then with boundary, physical and other conditions are interconnected parameters, which enter in basic equations and are subject for definition. Above-mentioned theories are based on using of different hypothesis, which simplify the form of basic resolving vibration equation and at the same time lead to essential disadvantages and errors. To avoid such errors, in this work the refined equations of longitudinal-radial vibrations of termoviscoelastic cylindrical shell are developed [1–4, 6–8] and used which are suitable for thick-walled cylindrical layers and more general than Timoshenko type equations and Herrmann–Mirsky [7, 11] equations of longitudinal-radial vibrations of termoviscoelastic cylindrical shell and take into account the effect of transversal shear deformation and the rotary inertia.

2 Derivation the Refined Equations of Longitudinal-Radial Vibrations of Thermoviscoelastic Cylindrical Shell

Let's consider circular cylindrical shell with inner r_1 and outer r_2 radius in cylindrical coordinate system (r, θ, z). Material of shell is viscoelastic, isotropic and homogeneous. Furthermore the change in temperature appearing under deformation of shell is neglected, i.e. formulation of problem is done on the base of disconnected theory. Assume for viscoelastic and thermal isotropic material when temperature increases on $\Delta T = T - T_0$, the changes in length in all directions are equal, i.e. only prolongation appears, there are not shears. So, relationship between tensor's component of stress and strain and also temperature assume the form

$$\sigma_{ii} = L(\varepsilon) + 2M(\varepsilon_{ii}) - \alpha_0 N(T); \quad \sigma_{ij} = M(\varepsilon_{ij}), \quad (i \neq j), \qquad (1)$$

where ε - cubic strain, L, M - viscoelastic operators

$$(L, M)\varsigma = (\lambda, \mu) \left[\varsigma(t) - \int\limits_0^t [K_1(t - \tau)], \; K_2(t - \tau) \right] \varsigma(\tau) d\tau; \quad N = L + \frac{2}{3}M;$$

λ, μ - Lame coefficients; α_0 - coefficient of thermal expansion. Assume that operators L, M are reversible, cores $K_1(t)$ and $K_2(t)$ are arbitrary.

The motion of shell points as $3D$ thermoviscoelastic body under small strain is described by equation

$$(L + 2M) \, grad \, (div \, \vec{u}) - M[rot \, (rot \, \vec{u})] = \rho \ddot{\vec{u}} + \alpha_0 N(grad \, T), \qquad (2)$$

where $\vec{u}$ is displacement vector.

The temperature T on the base of disconnected theory of thermoviscoelasticity and finitude of heat propagation velocity is satisfied to equation:

$$\Delta T - \frac{1}{c_0^2} T - \frac{1}{c^2} \ddot{T} = 0; \quad c_0^2 = \frac{k_0}{\rho c_p}; \quad c^2 = \frac{c_0^2}{\tau_0}, \tag{3}$$

where c - temperature propagation velocity, c_0 - thermal diffusivity, c_p - heat capacity under stationary pressure, τ_0 - relaxation time of heat flows, k_0 - thermal conductivity, ρ - density of shell's material.

Let's assume that longitudinal-radial vibrations of cylindrical shell are excited by external force on its $r = r_i$ surfaces:

$$\sigma_{rr} = f_r^{(i)}(z, t); \quad \sigma_{rz} = f_{rz}^{(i)}(z, t) \tag{4}$$

and temperature condition on shell surfaces has form

$$T = G^{(i)}(z, t) \quad for \quad r = r_i \quad (i = 1, 2). \tag{5}$$

Initial conditions are taken as zero.

Having presented vector of displacement $\vec{u}$ by potentials of longitudinal Φ and transversal Ψ_1 wave and solving jointly (2) and (3) we can easily get

$$T = \frac{1}{\alpha_0} N^{-1} \left[L_1(\Delta\Phi) - \rho\ddot{\Phi} \right]; \quad M(\Delta\Psi_1) = \rho\ddot{\Psi}_1;$$

$$L_1(\Delta^2\Phi) - \left[L_1 \left(\frac{1}{c_0^2} \frac{\partial}{\partial t} + \frac{1}{c^2} \frac{\partial^2}{\partial t^2} \right) + \rho \frac{\partial^2}{\partial t^2} \right] \Delta\Phi + \rho \left(\frac{1}{c_0^2} \frac{\partial^3}{\partial t^3} + \frac{1}{c^2} \frac{\partial^4}{\partial t^4} \right) \Phi = 0, \tag{6}$$

where $L_1 = L + 2M$.

For solving problem potentials Φ, Ψ_1 and temperature T are presented in form

$$[\Phi, \, T] = \int_0^\infty \left. \begin{matrix} \sin kz \\ -\cos kz \end{matrix} \right\} dk \int_{(l)} (\varphi, \, T_0) \, e^{pt} dp; \, \Psi_1 = \int_0^\infty \left. \begin{matrix} \cos kz \\ \sin kz \end{matrix} \right\} dk \int_{(l)} \Psi_{10} e^{pt} dp \tag{7}$$

Substitution of (7) into (6) gives simple differential equations, the general solutions of which subject to T. Boggio theorem [13] and their limitedness for $r \to \infty$ and $r = 0$ have form

$$\Psi_{10}(r) = B_1 I_0(\beta r) + B_2 K_0(\beta r);$$

$$\varphi(r) = A_1 I_0(\alpha_1 r) + A_2 K_0(\alpha_1 r) + A_3 I_0(\alpha_2 r) + A_4 K_0(\alpha_2 r); \tag{8}$$

$$T = \omega_1 [A_1 I_0(\alpha_1 r) + A_2 K_0(\alpha_1 r)] + \omega_2 [A_3 I_0(\alpha_2 r) + A_4 K_0(\alpha_2 r)],$$

where

$$\beta^2 = k^2 + \rho p^2 M_0^{-1}; \quad M_0 = \mu \left[1 - K_{20}(p)\right] \quad \alpha_i^2 = k^2 + \bar{\alpha}_i^2; \quad (i = 1, 2),$$

$$K_{20}(p) = \int_0^\infty K_2(t) e^{-pt} dp; \quad \omega_i = L_{10} \left(\alpha_0 N_0\right)^{-1} \left(\alpha_i^2 - \alpha^2\right); \quad \alpha^2 = k^2 + \rho p^2 L_{10}^{-1};$$

$$\bar{\alpha}_1^2 + \bar{\alpha}_2^2 = p \left[\frac{1}{c_0^2} + \left(\frac{1}{c^2} + \rho L_{10}^{-1}\right) p\right]; \quad \bar{\alpha}_1^2 \cdot \bar{\alpha}_2^2 = \rho p^2 L_{10}^{-1} \left(\frac{p}{c_0^2} + \frac{1}{c^2} p^2\right), (i = 1, 2).$$

$$(9)$$

Constants of integration A_i ($i = 1, 4$), B_1 and B_2 will be expressed in terms of main parts of displacements and temperature. For this purpose displacements U_z, U_r and temperature T are presented as (7) and instead of φ and ψ_{10} their expressions (8) are substituted. Hereinafter in the derived expressions Bessel's functions are transformed into a power series of radial coordinate r and their first term when $r = \xi$ are considered which will be main part of transformed displacement and temperature (ξ - radius of the intermediate surface).

In classical vibration equation of cylindrical shell as basic unknown the displacement of points of median surface are taken. In experimental investigations we get information of displacement of points of external and internal shell surfaces. So considering cylindrical shell it is reasonable as unknown to choose the displacement of points such shell surface which goes for bar into axis line, for thin shells it passes into the meridian surface.

Considering the up given as unknown quantities we choose displacement, stress and temperature in surface points of cylindrical shell, radius of which is defined as

$$\xi = \frac{r_1}{2} \left(\chi - \frac{r_1}{r_2}\right); \quad 2 + \frac{r_1}{r_2} \le \chi \le 2\frac{r_2}{r_1} + \frac{r_1}{r_2},$$

where ξ - radius of the intermediate surface.

Assume

$$U_{r,0}^{(0)} = \alpha_1^2 A_{10} + \alpha_2^2 A_{30} - k\beta^2 B_{20}; \quad U_{r,1}^{(0)} = \frac{1}{\xi} \left(A_2 + A_4 - kB_2\right);$$

$$U_{z,0}^{(0)} = k \left(A_{10} + A_{30}\right) - \beta^2 B_{20}; \quad U_{z,1}^{(0)} = \frac{1}{\xi} \left(k(A_2 + A_4) - \beta^2 B_2\right); \quad (10)$$

$$T_{0,0} = \omega_1 A_{10} + \omega_2 A_{30}; \quad T_{0,1} = \frac{1}{\xi}(\omega_1 A_2 + \omega_2 A_4),$$

where

$$[A_{i0}, B_{20}] = [A_i, B_1] - [A_{i+1}, B_2] \left[\ln \frac{(\alpha_1, \alpha_2, \beta)\xi}{2} - \psi(1) - \frac{1}{2}\right]; \quad (i = \overline{1,3}).$$

Having solved the system (10) relatively to A_{10}, A_{30}, B_{20}, A_2, A_4, B_{22} the following is obtained

$$A_{10} = \frac{\omega_2 \left[U_{r,o}^{(0)} - k\,U_{z,o}^{(0)} \right] - \left(\alpha_2^2 - k^2 \right) T_{0,0}}{\omega_2 \left(\alpha_1^2 - k^2 \right) - \omega_1 \left(\alpha_2^2 - k^2 \right)},$$

$$A_{30} = \frac{\left(\alpha_1^2 - k^2 \right) T_{0,0} - \omega_1 \left[U_{r,o}^{(0)} - k\,U_{z,o}^{(0)} \right]}{\omega_2 \left(\alpha_1^2 - k^2 \right) - \omega_1 \left(\alpha_2^2 - k^2 \right)},$$

$$A_2 = \xi \frac{\omega_2 \left[\beta^2 U_{r,1}^{(0)} - k\,U_{z,1}^{(0)} \right] - \left(\beta^2 - k^2 \right) T_{0,1}}{\left(\beta^2 - k^2 \right) \left(\omega_2 - \omega_1 \right)},$$

$$A_4 = \xi \frac{\left(\beta^2 - k^2 \right) T_{0,1} - \omega_1 \left[\beta^2 U_{r,1}^{(0)} - k\,U_{z,1}^{(0)} \right]}{\left(\beta^2 - k^2 \right) \left(\omega_2 - \omega_1 \right)},$$

$$B_{20} = \frac{\left(\omega_1 \alpha_2^2 - \omega_2 \alpha_1^2 \right) U_{z,o}^{(0)} + k \left(\omega_2 - \omega_1 \right) U_{r,o}^{(0)} + k \left(\alpha_1^2 - \alpha_2^2 \right) T_{0,0}}{\beta^2 \left[\omega_2 \left(\alpha_1^2 - k^2 \right) - \omega_1 \left(\alpha_2^2 - k^2 \right) \right]}, \tag{11}$$

$$B_{22} = \xi \frac{k\,U_{r,1}^{(0)} - U_{z,1}^{(0)}}{\beta^2 - k^2}.$$

By converting boundary conditions (4), (5) and substituting (8) into converted expressions the system six equations relatively to constant A_i and B_j is derived

$$\left[L_{10} M_0^{-1} \left(\alpha_1^2 - k^2 \right) + 2k^2 \right] \left[A_1 I_0 \left(\alpha_1 r_i \right) + A_2 K_0 \left(\alpha_1 r_i \right) \right] +$$

$$+ \left[L_{10} M_0^{-1} \left(\alpha_2^2 - k^2 \right) + 2k^2 \right] \left[A_3 I_0 \left(\alpha_2 r_i \right) + A_4 K_0 \left(\alpha_2 r_i \right) \right] -$$

$$- \frac{2\alpha_1}{r_i} \left[A_1 I_1 \left(\alpha_1 r_i \right) - K_1 \left(\alpha_1 r_i \right) A_2 \right] - \frac{2\alpha_2}{r_i} \left[A_3 I_1 \left(\alpha_2 r_i \right) - A_4 K_1 \left(\alpha_2 r_i \right) \right] -$$

$$- 2k\beta^2 \left\{ \left[I_0 \left(\beta r_i \right) - \frac{1}{\beta r} I_1 \left(\beta r_i \right) \right] B_{21} + \left[K_0 \left(\beta r_i \right) + \frac{1}{\beta r} K_1 \left(\beta r_i \right) \right] B_{22} \right\} =$$

$$= M_0^{-1} \left[f_r^{(i0)} \right], \quad (i = 1, 2),$$

$$2k\alpha_1 \left[A_1 I_1 \left(\alpha_1 r_i \right) - A_2 K_1 \left(\alpha_1 r_i \right) \right] + 2k\alpha_2 \left[A_3 I_1 \left(\alpha_2 r_i \right) - A_4 K_1 \left(\alpha_2 r_i \right) \right] -$$

$$- \beta (\beta^2 + k^2) \left[B_{21} I_1 \left(\beta r_i \right) - B_{22} K_1 \left(\beta r_i \right) \right] = M_0^{-1} \left[f_{rz}^{(i0)} \right], \quad (i = 1, 2),$$

$$\omega_1 \left[A_1 I_0 \left(\alpha_1 r_i \right) + A_2 K_0 \left(\alpha_1 r_i \right) \right] + \omega_2 \left[A_3 I_0 \left(\alpha_2 r_i \right) + A_4 K_0 \left(\alpha_2 r_i \right) \right] =$$

$$= G_0^{(i0)}(k, p), \quad (i = 1, 2). \tag{12}$$

Using standard expansion of Bessel's functions into a power series of radiuses r_1 and r_2 in the derived equations and consecutively substituting into expansion the value of constant by formulas (10) it is got the system of six algebraic equations relatively to the entered functions $U_{r,0}^{(0)}$, $U_{r,1}^{(0)}$, $U_{z,0}^{(0)}$, $U_{z,1}^{(0)}$, $T_{0,0}$, $T_{1,0}$. Step of going from Laplace and Fur'e transforms to the original time and space function is realized and the following system of equations is obtained

$$\chi_{11}U_{r,0} + \chi_{12}U_{z,0} + \chi_{13}T_0 + \chi_{14}U_{r,1} + \chi_{15}U_{z,1} + \chi_{16}T_1 = F_r^{(i)},$$
$$\chi_{21}U_{r,0} + \chi_{22}U_{z,0} + \chi_{23}T_0 + \chi_{24}U_{r,1} + \chi_{25}U_{z,1} + \chi_{26}T_1 = F_{rz}^{(i)},$$
$$\chi_{31}U_{r,0} + \chi_{32}U_{z,0} + \chi_{33}T_0 + \chi_{34}U_{r,1} + \chi_{35}U_{z,1} + \chi_{36}T_1 = F_0^{(i)}, (i = 1, 2).$$
$$(13)$$

Herewith operators $\chi_{lj}(l = 1,\overline{3};\ j = 1,\overline{6})$, $\lambda_q(q = 1,\overline{4})$ and $\overline{Q}_n (n = 1, 2, 3, ...)$ as well as functions $F_r^{(k)}, F_{rz}^{(k)}, F_0^{(k)}$ have form

$$\chi_{11}(r_i) = \sum_{n=0}^{\infty} \left\{ L_1 M^{-1} \left[\lambda_1 \bar{Q}_{n+1} - \left(\lambda_4^2 - \frac{\partial^2}{\partial z^2} \lambda_1 \right) \bar{Q}_n - \frac{\partial^2}{\partial z^2} \lambda_4^2 \bar{Q}_{n-1} \right] - \right.$$

$$-2\frac{\partial^2}{\partial z^2} \left[\lambda_1 \bar{Q}_n - \lambda_4^2 \bar{Q}_{n-1} \right] - \frac{1}{n+1} \times$$

$$\left. \times \left[\lambda_1 \bar{Q}_{n+1} - \lambda_4^2 \bar{Q}_n \right] + \left(2 - \frac{1}{n+1} \right) \frac{\partial^2}{\partial z^2} \lambda_2^n \right\} \frac{(r_i/2)^{2n}}{(n!)^2},$$

$$\chi_{12}(r_i) = \frac{\partial}{\partial z} \sum_{n=0}^{\infty} \left\{ L_1 M^{-1} \left[\lambda_1 \bar{Q}_{n+1} + \left(\lambda_4^2 - \frac{\partial^2}{\partial z^2} \lambda_1 \right) \bar{Q}_n + \frac{\partial^2}{\partial z^2} \lambda_4^2 \bar{Q}_{n-1} \right] - \right.$$

$$\left. -2\frac{\partial^2}{\partial z^2} \left[\lambda_1 \bar{Q}_n - \lambda_4^2 \bar{Q}_{n-1} \right] - 2\lambda_1 \lambda_2^n - \frac{1}{n+1} \left[\lambda_1 \bar{Q}_{n+1} - \lambda_1 \lambda_2^n - \lambda_4^2 \bar{Q}_n \right] \right\} \frac{(r_i/2)^{2n}}{(n!)^2},$$

$$\chi_{13}(r_i) = \sum_{n=0}^{\infty} \left\{ \alpha_0 N \left[M^{-1} \left(\lambda_4^2 + \frac{\partial^2}{\partial z^2} \lambda_3 + \frac{\partial^4}{\partial z^4} \right) \bar{Q}_n + 2\frac{\partial^2}{\partial z^2} L_1^{-1} \times \right. \right.$$

$$\left. \left. \times \left(-\lambda_2^n - \lambda_4^2 \bar{Q}_{n-1} - \frac{\partial^2}{\partial z^2} \bar{Q}_n \right) \right] - \frac{1}{n+1} \left[\frac{\partial^2}{\partial z^2} \bar{Q}_{n+1} + \lambda_4^2 \bar{Q}_n - \frac{\partial^2}{\partial z^2} \lambda_2^n \right] \right\} \frac{(r_i/2)^{2n}}{(n!)^2},$$

$$\chi_{14}(r_i) = -\xi \sum_{n=0}^{\infty} \left\{ \eta_{3,n}(r_i) L_1^{-1} M \left[L_1 M^{-1} \left\{ \lambda_1 \bar{Q}_{n+1} - \left(\lambda_4^2 - \frac{\partial^2}{\partial z^2} \lambda_1 \right) \bar{Q}_n - \right. \right. \right.$$

$$\left. \left. - \frac{\partial^2}{\partial z^2} \lambda_4^2 \bar{Q}_{n-1} \right\} - 2\frac{\partial^2}{\partial z^2} \times \right.$$

$$\left. \times \left\{ \lambda_1 \bar{Q}_n - \lambda_4^2 \bar{Q}_{n-1} - \lambda_2^n \right\} \right] \cdot \lambda_2 - \frac{\eta_{1,n}(r_i)}{n+1} L_1^{-1} M \lambda_2 \left(\lambda_1 \bar{Q}_{n+1} + \lambda_2^n \frac{2}{z^2} - \lambda_4^2 \bar{Q}_n \right) \right\} \times$$

$$\times \frac{(r_i/2)^{2n}}{(n!)^2} + \frac{2\xi}{r_i^2} \rho L_1^{-1} \frac{\partial^2}{\partial t^2},$$

$$\chi_{15}(r_i) = -\xi \frac{\partial}{\partial z} \sum_{n=0}^{\infty} \left\{ \eta_{3,n}(r_i) L_1^{-1} M \left[L_1 M^{-1} \left(\lambda_1 \bar{Q}_{n+1} - \left(\lambda_4^2 - \lambda_1 \frac{\partial^2}{\partial z^2} \right) \right. \right. \right.$$

$$\left. \left. \bar{Q}_n - \frac{\partial^2}{\partial z^2} \lambda_4^2 \bar{Q}_{n-1} \right) - 2\frac{\partial^2}{\partial z^2} \times \right.$$

$$\left. \times \left(\lambda_1 \bar{Q}_n - \lambda_4^2 \bar{Q}_{n-1} \right) - 2\lambda_2^{n+1} \right] - \frac{\eta_{1,n}(r_i)}{n+1} L_1^{-1} M \left(\lambda_1 \bar{Q}_{n+1} - \lambda_4^2 \bar{Q}_{n+1} - 2\lambda_2^{n+1} \right) \right\} \times$$

$$\times \frac{(r_i/2)^{2n}}{(n!)^2} + \frac{2\xi}{r^2} \rho L_1^{-1} \frac{\partial^3}{\partial t^2 \partial z},$$

$$\chi_{16}(r_i) = -\xi \sum_{n=0}^{\infty} \left\{ \eta_{3,n}(r_i)\alpha_0 \rho N L_1^{-2} \frac{\partial^2}{\partial t^2} \left[L_1 M^{-1} \left[\bar{Q}_{n+1} + \frac{\partial^2}{\partial z^2}\bar{Q}_n \right] - 2\frac{\partial^2}{\partial z^2}\bar{Q}_n \right] - \right.$$

$$\left. - \frac{\eta_{1,n}(r_i)}{n+1}\alpha_0^{-1}\rho N_0^{-1}L_1^{-2}\frac{\partial^2}{\partial t^2}\bar{Q}_{n+1} \right\} \frac{(r_i/2)^{2n}}{(n!)^2},$$

$$\chi_{21}(r_i) = -\sum_{n=0}^{\infty} \frac{\partial}{\partial z}\left[2\left(\lambda_1 \bar{Q}_{n+1} - \lambda_4^2 \bar{Q}_n\right) - \left(\lambda_2 - \frac{\partial^2}{\partial z^2}\right)\lambda_2^n \right] \frac{(r_i/2)^{2n+1}}{n!(n+1)!},$$

$$\chi_{22}(r_i) = \sum_{n=0}^{\infty}\left[2\frac{\partial^2}{\partial z^2}\left(\lambda_1 \bar{Q}_{n+1} - \lambda_4^2 \bar{Q}_n\right) - \lambda_1\left(\lambda_2 - \frac{\partial^2}{\partial z^2}\right)\lambda_2^n \right] \frac{(r_i/2)^{2n+1}}{n!(n+1)!},$$

$$\chi_{23}(r_i) = -\sum_{n=0}^{\infty}\alpha_0 N L_1^{-1}\frac{1}{z}\left[2\left(\frac{\partial^2}{\partial z^2}\bar{Q}_{n+1} + \lambda_4^2 \bar{Q}_n\right) - \left(\lambda_2 - \frac{\partial^2}{\partial z^2}\right)\lambda_2^n \right] \frac{(r_i/2)^{2n+1}}{n!(n+1)!},$$

$$\chi_{24}(r_i) = \xi \sum_{n=0}^{\infty}\eta_{1,n}(r_i)\left[2\left(\lambda_1 \bar{Q}_{n+1} - \lambda_4^2 \bar{Q}_n\right) - \left(\lambda_2 - \frac{\partial^2}{\partial z^2}\right)\lambda_2^n \right] \times$$

$$\times L_1^{-1}M \cdot \lambda_2 \frac{\partial}{\partial z} \cdot \frac{(r_i/2)^{2n+1}}{n!(n+1)!} + \frac{\xi}{r_i}\rho L_1^{-1}\frac{\partial^3}{\partial z\partial t^2},$$

$$\chi_{25}(r_i) = \xi \sum_{n=0}^{\infty} L_1^{-1}M \left[-2\frac{\partial^2}{\partial z^2}\left(\lambda_1 \bar{Q}_{n+1} - \lambda_4^2 \bar{Q}_n\right) - \left(\lambda_2 - \frac{\partial^2}{\partial z^2}\right)\lambda_2^{n+1} \right] \times$$

$$\times \eta_{1,n}(r_i)\frac{(r_i/2)^{2n+1}}{n!(n+1)!} - \frac{\xi}{r_i}\rho L_1^{-1}\frac{\partial^2}{\partial t^2},$$

$$\chi_{26}(r_i) = 2\xi \sum_{n=0}^{\infty}\eta_{1,n}(r_i)\alpha_0 \rho N L_1^{-2}\frac{\partial^3}{\partial t^2 \partial z}\bar{Q}_{n+1}\frac{(r_i)^{2n+1}}{n!(n+1)!},$$

$$\chi_{31}(r_i) = -\sum_{n=0}^{\infty}\alpha_0^{-1}L_1 N^{-1}\left(\lambda_4^2 - \lambda_1\lambda_3 + \lambda_1^2\right)\bar{Q}_n\frac{(r_i/2)^{2n}}{(n!)^2},$$

$$\chi_{32}(r_i) = -\sum_{n=0}^{\infty}L_1\left(\alpha_0 N\right)^{-1}\left(\lambda_4^2 - \lambda_1\lambda_3 + \lambda_1^2\right)\frac{\partial}{\partial z}\bar{Q}_n\frac{(r_i/2)^{2n}}{(n!)^2},$$

$$\chi_{33}(r_i) = \sum_{n=0}^{\infty}\left[\frac{\partial^2}{\partial z^2}\bar{Q}_{n+1} + \left(\lambda_4^2 - \frac{\partial^2}{\partial z^2}\lambda_1\right)\bar{Q}_n - \lambda_1 \cdot \lambda_4^2 \bar{Q}_{n-1} \right] \frac{(r_i/2)^{2n}}{(n!)^2},$$

$$\chi_{34}(r_i) = \xi \sum_{n=0}^{\infty}\eta_{3,n}(r_i)M\left(\alpha_0 N\right)^{-1}\left(\lambda_4^2 - \lambda_1\lambda_3 + \lambda_1^2\right)\lambda_2 \bar{Q}_n\frac{(r_i/2)^{2n}}{(n!)^2},$$

$$\chi_{35}(r_i) = \xi \sum_{n=0}^{\infty}\eta_{3,n}(r_i)M\left(\alpha_0 N\right)^{-1}\left(\lambda_4^2 - \lambda_1\lambda_3 + \lambda_1^2\right)\frac{\partial}{\partial z}\bar{Q}_n\frac{(r_i/2)^{2n}}{(n!)^2},$$

$$\chi_{36}(r_i) = -\xi \sum_{n=0}^{\infty}\eta_{3,n}(r_i)\rho L_1^{-1}\frac{\partial^2}{\partial t^2}\left(\bar{Q}_{n+1} - \lambda_1 \bar{Q}_n\right)\frac{(r_i/2)^{2n}}{(n!)^2},$$

$$\lambda_1^n = \left[\rho L_1^{-1}\left(\frac{\partial^2}{\partial t^2}\right) - \frac{\partial^2}{\partial z^2}\right]^n, \quad \lambda_2^n = \left[\rho M_1^{-1}\left(\frac{\partial^2}{\partial t^2}\right) - \frac{\partial^2}{\partial z^2}\right]^n,$$

$$\lambda_3^n = \left\{\frac{\partial}{\partial t}\left[a_0^2 + \left(a_1^2 + \rho L_1^{-1}\right)\frac{\partial}{\partial t}\right] - 2\frac{\partial^2}{\partial z^2}\right\}^n,$$

$$\bar{Q}_n^{(0)} = \sum_{i=0}^{n-1} \alpha_2^{2(n-i-1)}\alpha_1^{2i}, \quad \lambda_4^2 \cdot \bar{Q}_{-1} = -1,$$

$$F_r^{(i)} = \rho L_1^{-1}M^{-1}\left(\frac{\partial^2}{\partial t^2}f_r^{(i)}\right), \quad F_{rz}^{(i)} = \rho L_1^{-1}M^{-1}\left(\frac{\partial^2}{\partial t^2}f_{rz}^{(i)}\right),$$

$$F_0^{(i)} = \rho L_1^{-1}\left(\frac{\partial^2}{\partial t^2}G_0^{(i)}\right).$$

Equation (13) is the system of general equations of longitudinal radial vibrations of cylindrical shell with initial stresses. According to form of operators λ_i^n, a_{ij}, b_{ij} this system of equations contains the n-th time and directional derivatives of functions $U_{r,0}$, $U_{r,1}$, $U_{z,0}$, $U_{z,1}$. Therefore it is reasonable to restrict a number of terms in equations for using them in applied problems. Here we can get different approximate and refined vibration equation and consider various limited cases. Thus bounding the first approximation in (13) the following system of integral-differential equations is derived

$$\frac{4\nu-1}{1-2\nu}\tilde{M}^{-1}\left[\frac{\partial^2 U_{r,0}}{\partial t^2}\right] + \frac{2(1-4\nu)}{1-2\nu}\tilde{M}^{-1}\left[\frac{\partial^3 U_{z,0}}{\partial t^2 \partial z}\right] -$$

$$-\frac{1}{2}\left(\tilde{M}^{-2}\left[\frac{\partial^4 U_{r,1}}{\partial t^4}\right] - \tilde{M}^{-1}\left[\frac{\partial^4 U_{r,1}}{\partial t^2 \partial z^2}\right] + \frac{2}{r_i^2}\cdot\tilde{M}^{-1}\left[\frac{\partial^2 U_{r,1}}{\partial t^2}\right]\right) + \frac{3}{2}\tilde{M}^{-1}\left[\frac{\partial^3 U_{z,1}}{\partial t^2 \partial z}\right] -$$

$$-\frac{\partial^3 U_{z,1}}{\partial z^3} - \frac{1+\nu}{3(1-2\nu)}\tilde{M}^{-1}\left[\frac{\partial^2 T_1}{\partial t^2}\right] = \frac{\rho}{\lambda\cdot\mu}\tilde{M}^{-2}\left(\frac{\partial^2 f_r^{(i)}}{\partial t^2}\right); \qquad (i=1,\,2)$$

$$\frac{1-4\nu}{1-2\nu}r\tilde{M}^{-1}\left[\frac{\partial^3 U_{r,0}}{\partial t^2 \partial z}\right] + \frac{r}{2}\cdot\left(\frac{4-6\nu}{1-2\nu}\cdot\tilde{M}^{-1}\left[\frac{\partial^4 U_{z,0}}{\partial t^2 \partial z^2}\right] - \frac{8\nu}{1-2\nu}\cdot\frac{\partial^4 U_{z,0}}{\partial z^4} -$$

$$-\tilde{M}^{-2}\left[\frac{\partial^4 U_{z,0}}{\partial t^4}\right] - \frac{1}{r}\tilde{M}^{-1}\left[\frac{\partial^2 U_{z,1}}{\partial t^2}\right] - \frac{r(1+\nu)}{3(1-2\nu)}\left(\tilde{M}^{-1}\left(\frac{\partial^2}{\partial t^2}\right) - 4\cdot\frac{\partial^2}{\partial z^2}\right)\frac{\partial T_0}{\partial z} -$$

$$-\frac{1}{r}\cdot\tilde{M}^{-1}\left[\frac{\partial^3 U_{r,1}}{\partial t^2 \partial z}\right] + \frac{r_i}{2}\cdot\frac{1+\nu}{3\nu}\cdot\tilde{M}^{-1}\left[\frac{\partial^3 U_{r,1}}{\partial t^2 \partial z}\right] = \frac{\rho}{\lambda\cdot\mu}\tilde{M}^{-2}\left(\frac{\partial^2 f_{rz}^{(i)}}{\partial t^2}\right); \qquad (i=1,\,2)$$

$$\left\{\tilde{M}^{-1}\left(\frac{\partial^2}{\partial t^2}\right) + \frac{r_i^2}{4}\tilde{M}^{-1}\left(\frac{\partial^2}{\partial t^2}\right)\left(a_0^2\frac{\partial}{\partial t} + a_{01}^2\cdot\tilde{M}^{-1}\left(\frac{\partial^2}{\partial t^2}\right) - \frac{\partial^2}{\partial z^2}\right)\right\}\cdot T_0+$$

$$+\left\{\tilde{M}^{-1}\left(\frac{\partial^2}{\partial t^2}\right) - \frac{r_i^2}{8}\tilde{M}^{-1}\left(\frac{\partial^2}{\partial t^2}\right)\left(a_0^2\frac{\partial}{\partial t} + a_{01}^2\cdot\tilde{M}^{-1}\left(\frac{\partial^2}{\partial t^2}\right) - \frac{\partial^2}{\partial z^2}\right)\right\}\cdot T_1 =$$

$$= \frac{\rho}{\lambda}\tilde{M}^{-1}\left(\frac{\partial^2 G_0^{(i)}}{\partial t^2}\right), \qquad (i=1,\,2)$$

$$(14)$$

where $\nu = const$ - Poisson's ratio, $a_0^2 = 1/c_0^2$; $a_{01}^2 = 1/c^2$; $U_{z,0}$, $U_{z,1}$-main parts of longitudinal displacement; $U_{r,0}$, $U_{r,1}$ - main parts of radial displacement, z - longitudinal coordinate; r - radial coordinate; t - time; T_0, T_1 - main parts of temperature T; α_0 - thermal expansion factor; $G_0^{(i)}$ - temperature condition on the shell surface; non-dimensional variables are introduced by formulas

$$z = z^*\xi; \quad bt = t^*\xi; \quad r = r^*\xi; \quad U_{r,1} = U_{r,1}^*\xi; \quad r_2 = r_2^*\xi;$$

$$U_{z,0} = U_{z,0}^*\xi; \quad U_{z,1} = U_{z,1}^*; \quad U_{r,0} = U_{r,0}^*; \quad r_1 = r_1^*\xi; \tag{15}$$

$$a_0^2 = a_0^{*2}/b \cdot \xi; \quad a_{01}^2 = a_{01}^{*2}/b; \quad \alpha_0 T_0 = T_0^*; \quad \xi \alpha_0 T_1 = T_1^*,$$

and stars (*) for convenience are omitted, here b is transversal wave propagation velocity in material of shell; ξ is radius of the intermediate surface. Algorithm allowing to calculate the fields of stresses, strains and temperature by the field of unknown function, is devised; the stressed-strained state and temperature pattern of the shell is defined by formulas

$$U_z = U_{z,0} + \frac{3-2\nu}{3\nu}\left(\frac{\partial^2}{\partial t^2}\right)^{-1}\frac{\partial T_0}{\partial z}; \qquad U_r = \frac{r}{2}U_{r,0} - \frac{1}{r}U_{r,1};$$

$$\tilde{M}^{-1}[\sigma_{rr}] = \mu\frac{2\nu}{1-2\nu}\left\{\frac{4\nu-1}{1-2\nu}U_{r,0} + \frac{2}{r_i^2}U_{r,1}+\right.$$

$$+\frac{2(1-4\nu)}{1-2\nu}\frac{\partial U_{z,0}}{\partial z} - \frac{1}{2}\left(\tilde{M}^{-1}\left[\frac{\partial^2 U_{r,1}}{\partial t^2}\right] - \frac{\partial^2 U_{r,1}}{\partial z^2}\right)+$$

$$\left.+\frac{3}{2}\frac{\partial U_{z,1}}{\partial z} - \tilde{M}\left(\frac{\partial^2}{\partial t^2}\right)^{-1}\left[\frac{\partial^3 U_{z,1}}{\partial z^3}\right] - \frac{1+\nu}{3(1-2\nu)}T_1\right\}; \tag{16}$$

$$\tilde{M}^{-1}[\sigma_{rz}] = \mu\frac{2\nu}{1-2\nu}\left\{\frac{1-4\nu}{1-2\nu}r\frac{\partial U_{r,0}}{\partial z} + \frac{r}{2}\cdot\left(\frac{4-6\nu}{1-2\nu}\cdot\frac{\partial^2 U_{z,0}}{\partial z^2} - \frac{8\nu}{1-2\nu}\cdot\frac{\partial^4 U_{z,0}}{\partial z^4}\right.\right.$$

$$\left.-\tilde{M}^{-1}\left[\frac{\partial^2 U_{z,0}}{\partial t^2}\right]\right) - \frac{r(1+\nu)}{3(1-2\nu)}\left(\frac{\partial T_0}{\partial z} - 4\cdot\tilde{M}\left(\frac{\partial^2}{\partial t^2}\right)^{-1}\left[\frac{\partial^3 T_0}{\partial z^3}\right]\right) - \frac{1}{r}\cdot\frac{\partial U_{r,1}}{\partial z}$$

$$\left.-\frac{1}{r}U_{z,1} + \frac{1+\nu}{3\nu}\cdot\frac{\partial T_1}{\partial z}\right\};$$

$$T = T_0 - \left(\ln(r) + \frac{1}{2}\right)T_1.$$

3 Longitudinal Impact on Viscoelastic Cylindrical Shell

As test problem the problem of longitudinal-radial vibrations of circular cylindrical viscoelastic shell is solved, which is excited by sudden applied impact load on the shell face on which rigid undeformed ring with small width is put on [5].

Boundary conditions of the problem are when $z = 0$

$$U_r = \frac{\partial}{\partial r}U_r = 0; \qquad \sigma_{zz} = -\frac{EH(t)}{1-\nu^2}, \qquad (17)$$

when $z \to \infty$

$$U_z = U_r = 0,$$

where $H(t)$ - unit function of Heveside.

Initial conditions are zero.

As basic resolving equations are taken the refined vibration equations (14). Problem is solved on base of equations of (14) ignoring the influence of temperature using Laplace transformation. Step of going from a transformation to the original time function is realized with using shift theorem and theorem of functions' convolution and it is got the following solution of problem:

$$U_{r,1} = \gamma_{41}\gamma_{21} \int_{z/c_0}^{t} e^{-c_1\tau/2} J_0\left(\gamma_2\sqrt{\tau^2 - z^2/c_0^2}\right) \left[\int_0^{t-\tau} f(\xi)\,d\xi\right] d\tau -$$

$$-\gamma_{41}\gamma_{18} \int_z^t G_1(t-\tau) e^{-c_1\tau/2} J_0\left(\gamma_1\sqrt{\tau^2 - z^2}\right) d\tau,$$

$$U_{z,1} = -\gamma_{41}\gamma_3\gamma_{18} \int_z^t e^{-c_1\tau/2} G_2(t-\tau) J_0\left(\gamma_1\sqrt{\tau^2 - z^2}\right) d\tau +$$

$$+\gamma_{41}\gamma_4\gamma_{21}c_0 \int_{z/c_0}^t e^{-c_1\tau/2} G_2(t-\tau) J_0\left(\gamma_2\sqrt{\tau^2 - z^2/c_0^2}\right) d\tau,$$

$$G_1(t) = f'(t) + \frac{3c_1}{2}f(t) + \frac{c_1^2}{2}\int_0^t f(\xi)\,d\xi, \quad G_2(t) = f(t) - c_1\int_0^t f(\xi)\,d\xi,$$

$$U_{z,0} = -\frac{\gamma_{41}}{c_0}\left\{c_0^2 \int_{z/c_0}^t e^{-c_1\tau/2} G_3(t-\tau) I_0\left(c_{01}\sqrt{\tau^2 - z^2/c_0^2}\right) d\tau +\right.$$

$$+\gamma_{18} \int_z^t e^{-c_1\tau/2} G_4(t-\tau) J_0\left(\gamma_1\sqrt{\tau^2 - z^2}\right) d\tau -$$

$$\left.- \gamma_{42} \int_{z/c_0}^t e^{-c_1\tau/2} G_5(t-\tau) J_0\left(\gamma_2\sqrt{\tau^2 - z^2/c_0^2}\right) d\tau\right\},$$

$$G_3(t) = \gamma_{54}f'(t) + \gamma_{55}f(t) + \gamma_{56}\int_0^t f(\xi)\,d\xi,$$

$$G_5(t) = \gamma_{44}f''(t) + 2c_1\gamma_{44}f'(t) + \gamma_{45}f(t),$$

$$G_4(t) = \gamma_{42}f''(t) + 2c_1\gamma_{42}f'(t) + \left(\gamma_{43} + \gamma_{42}c_1^2\right)f(t) + c_1\gamma_{43}\int_0^t f(\xi)\,d\xi,$$

$$U_{r,0} = -\nu\gamma_{41} \left\{ G_6 \left(t - \frac{z}{c_0} \right) e^{-\frac{c_1 z}{2c_0}} - \frac{c_{01} z}{c_0} \int\limits_{z/c_0}^{t} e^{-c_1 \tau/2} G_6 \left(t - \tau \right) \right.$$

$$\left. \times \frac{I_1 \left(c_{01} \sqrt{\tau^2 - z^2/c_0^2} \right)}{\sqrt{\tau^2 - z^2/c_0^2}} d\tau \right\} - \tag{18}$$

$$- \gamma_{12}\gamma_{41}\gamma_{51} \int\limits_{z}^{t} e^{-c_1\tau/2} G_7 \left(t - \tau \right) J_0 \left(\gamma_1 \sqrt{\tau^2 - z^2} \right) d\tau +$$

$$+ \gamma_{21}\gamma_{41}c_0 \int\limits_{z/c_0}^{t} e^{-c_1\tau/2} G_8 \left(t - \tau \right) J_0 \left(\gamma_2 \sqrt{\tau^2 - z^2/c_0^2} \right) d\tau,$$

$$G_6 \left(t \right) = \frac{\gamma_{46}}{(1-\nu)\omega_1} \left[f'' \left(t \right) + 2c_1 f' \left(t \right) \right] + \gamma_{49} f \left(t \right) + c_1 \gamma_{50} \int\limits_0^t f \left(\xi \right) d\xi,$$

$$G_7 \left(t \right) = f' \left(t \right) + \frac{3c_1}{2} f \left(t \right) + \frac{c_1^2}{2} \int\limits_0^t f \left(\xi \right) d\xi,$$

$$G_8 \left(t \right) = \gamma_{52} \left[f' \left(t \right) + \frac{3c_1}{2} f \left(t \right) \right] + \gamma_{53} \int\limits_0^t f \left(\xi \right) d\xi, \quad \gamma_{ij} = const.$$

Substituting last expressions into formulas (16) the stressed-strained state of arbitrary shell section is defined. Calculations are performed for viscoelastic material the physical characteristics of which are as follow:

$$\alpha_1 = 0,6; \quad \tau_1 = 0,16 \cdot 10^5 \, mks; \quad a = 2220 \, m/s; \quad b = 1040 \, m/s;$$
$$\nu = 0,36; \rho = 1,28 \cdot 10^{-6} \, kG \, s^2/sm^4; \quad \alpha_1 = 0,3; \quad \tau_2 = 0,14 \cdot 10^7 \, mks;$$
$$E = 3,6 \cdot 10^4 kG/sm^2; \quad \alpha_3 = 0,1; \quad \tau_1 = 0,11 \cdot 10^8 \, mks;$$
$$K \left(t \right) = \sum_{n=1}^{3} \alpha_n \tau_n^{-1} e^{-t/\tau_n}. \tag{19}$$

Herewith external load is given as unit function of Heveside: $f(t) = H(t)$. Displacements are calculated by formulas (16) ignoring the temperature. Results of calculations of displacement are given on Figs. 1 and 2, where U_z and U_r are longitudinal and radial displacement of meridian shell surface. From graph we can see that proposed theory gives satisfactory results nearly on all region $z \in \left(0, \sqrt{2 + 2\nu} \right)$, except only for points of shell, which are very closed to face, which is under external action, where boundary effect of Sen–Venan is very strong. The process of longitudinal deformation may be divided into three sections. In the first section, when $z \leq 2, 8$, boundary effects are sufficiently strong. The second section is enclosed in the zone from $z > 3$ till the second wave front $z \leq 10$. This section is characterizing by smooth decreasing of value of U_z, which indicates that in this section influent of viscosity is significant. Because of the influence of viscosity in the first section where $z \leq 2, 8$,

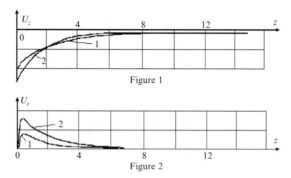

Figure 1

Figure 2

Figs. 1-2. Plots of longitudinal U_z and radial U_r displacements as functions of coordinate z for $\nu = 0,22$ (curve 1) and $\nu = 0,33$ (curve 2)

this section is run by wave for time $t \approx \frac{2,8}{\sqrt{2+2\nu}}$, which is too short. So in this section shell behaves as elastic. The longitudinal displacement is very small behind the first wave front wave till the second one and may be neglected. Radial displacement is also damping fast with the increase of distance from the face and for $z > 5$ we can neglect it.

4 Analysis of Thermo-Stressed State of Viscoelastic Circular Cylindrical Shell

Now let's consider the thermo-stressed state of semi-infinite circular cylindrical shell with inner r_1 and outer r_2 radius, excited by kinematical action on its rigid restrained face with heat conditions on the surfaces of the shell. Shell material is viscoelastic, isotropic and homogeneous. Surfaces of shell are free from mechanical effect. In this case the proposed vibration equations in non-dimensional variables (r, z, t) have the form:

$$
\left(\frac{2}{r_i^2} + \frac{1}{2} + \ln r_i \right) \tilde{M}^{-1} \left[\frac{\partial^2 U_{z,1}}{\partial t^2} \right] - \frac{4\nu - 1}{1 - 2\nu} \tilde{M}^{-1} \left[\frac{\partial^2 U_{r,0}}{\partial t^2} \right] +
$$

$$
+ \frac{2(1 - 3\nu)}{1 - 2\nu} \tilde{M}^{-1} \left[\frac{\partial^3 U_{z,0}}{\partial z \partial t^2} \right] + \left\{ \tilde{M}^{-1} \left(\frac{\partial^2}{\partial t^2} \right) \times \right.
$$

$$
\times \left(\frac{2}{r_i^2} + \ln r_i + \frac{1}{2} \right) + \frac{1 - 2\nu}{2\nu} \left[\ln r_i \tilde{M}^{-2} \left(\frac{\partial^4}{\partial t^4} \right) - \tilde{M}^{-1} \left(\frac{\partial^4}{\partial t^2 \partial z^2} \right) \right] \right\} U_{r,1} +
$$

$$
+ \left\{ \ln r_i \frac{1 - 4\nu}{2\nu} \tilde{M}^{-1} \left(\frac{\partial^2}{\partial t^2} \right) + \ln r_i \frac{\partial^2}{\partial z^2} \right\} \frac{\partial U_{z,t}}{\partial z} -
$$

$$
- \left\{ \left(\ln r_i + \frac{1}{2} \right) \frac{2(1 + \nu)}{3(1 - 2\nu)} - \ln r_i \frac{1 + \nu}{3\nu} \right\} \tilde{M}^{-1} \left[\frac{\partial^2 T_1}{\partial t^2} \right] = 0; (i = 1, \; 2)
$$

$$\frac{r_i(1-3\nu)}{1-2\nu} \cdot \tilde{M}^{-1}\left[\frac{\partial^3 U_{r,0}}{\partial t^2 \partial z}\right] +$$

$$+\frac{r_i}{2}\left\{\frac{4-6\nu}{1-2\nu} \cdot \tilde{M}^{-1}\left(\frac{\partial^4}{\partial t^2 \partial z^2}\right) - \frac{8\nu}{1-2\nu} \cdot \frac{\partial^4}{\partial z^4} - \tilde{M}^{-2}\left(\frac{\partial^4}{\partial t^4}\right)\right\}U_{z,0} -$$

$$-\frac{2r_i(1+\nu)}{3(1-2\nu)} \cdot \left[\tilde{M}^{-1}\left(\frac{\partial^3 T_0}{\partial t^2 \partial z}\right)\right] - \frac{1}{r_i}\tilde{M}^{-1}\left[\frac{\partial^3 U_{r,1}}{\partial z \partial t^2}\right] +$$

$$+\left\{\ln r_i\left[\frac{5\nu-1}{\nu}\tilde{M}^{-1}\left(\frac{\partial^4}{\partial z^2 \partial t^2}\right) + 4\frac{4\nu^2}{(1-2\nu)^2}\frac{\partial^4}{\partial z^4} - \frac{(1-2\nu)^2}{4\nu^2}\tilde{M}^{-2}\left(\frac{\partial^4}{\partial t^4}\right)\right] -$$

$$-\frac{1}{r_i} \cdot \tilde{M}^{-1}\left(\frac{\partial^2}{\partial t^2}\right)\right\}U_{z,1} + \frac{r_i}{2}\ln r_i\frac{1+\nu}{3\nu}\tilde{M}^{-1}\left[\frac{\partial^3 T_1}{\partial z \partial t^2}\right] = 0, \ (i=1,2)$$

$$\tilde{M}^{-1}\left[\frac{\partial^2 T_0}{\partial t^2}\right] - \tilde{M}^{-1}\left[\frac{\partial^2 T_1}{\partial t^2}\right]\left(\ln r_i + \frac{1}{r_i}\right) = \tilde{M}^{-1}\left[\frac{\partial^2 G_0^{(i)}}{\partial t^2}\right], \ (i=1,2)$$

$$(20)$$

where $\tilde{M}(\xi) = \xi(t) - \int_0^t \xi(\tau) K(t-\tau)d\tau$, $K(t-\tau)$ - arbitrary integrable core of viscoelastic operator, $\nu = const$ - Poisson's ratio; non-dimensional variables are introduced by formulas (15).

The stressed-strained of state and temperature pattern of the shell defined by formulas (16).

Assume that temperature condition is given as: on inner surface $r = r_1$:

$$G_0^{(1)} = a_1 e^{-(k_1 z + \omega_1 t)}, \tag{21}$$

on outer surface $r = r_2$:

$$G_0^{(2)} = a_2 e^{-(k_2 z + \omega_2 t)}. \tag{22}$$

Also proceed on the assumption that shell vibrations are excited by mechanical action applied on the face. In this case boundary conditions have form:

for $z = 0$

$$U_z = -f(t), \quad U_r = 0, \quad \sigma_{rr} = 0, \tag{23}$$

as $z \to \infty$

$$U_z = 0. \tag{24}$$

Initial conditions are as zero.

For solving problem the Laplace transformation on time is used with further using of method of constant's variation and method of exception. Step of going from a transformation to the original time function is realized with using shift theorem and theorem of functions' convolution. It is obtained the closed solution substitution of which into formulas (16) lets fully define temperature pattern and stressed-strained state of arbitrary shell section:

$$T_1(z,t) = \left[a_1 \cdot e^{-(k_1 z + \omega_1 t)} - a_2 \cdot e^{-(k_2 z + \omega_2 t)} \right] : \ln \frac{r_2}{r_1};$$

$$T_0(z,t) = \frac{\ln r_1 + 1/r}{\ln(r_2/r_1)} \cdot \left[a_1 \cdot e^{-(k_1 z + \omega_1 t)} - a_2 \cdot e^{-(k_2 z + \omega_2 t)} \right] + a_1 \cdot e^{-(k_1 z + \omega_1 t)};$$

$$U_{z,1}(z,t) = \frac{\gamma_{23}}{\gamma_{22}} e^{-c_1 t/2} \int_{\gamma_3 z}^{t} J_0 \left(\gamma_0 \sqrt{\tau^2 - \gamma_3^2 z^2} \right) d\tau +$$

$$+ \frac{\gamma_{24}}{\gamma_{22}} e^{-c_1 t/2} \int_{\gamma_4 z}^{t} J_0 \left(\gamma_0 \sqrt{\tau^2 - \gamma_4^2 z^2} \right) d\tau;$$

$$U_{r,1}(z,t) = \frac{\gamma_{23}\gamma_{48}}{\gamma_{22}\gamma_3} e^{-c_1 t/2} \left[J_0 \left(\gamma_0 \sqrt{t^2 - \gamma_3^2 z^2} \right) H(t - \gamma_3 z) \right] +$$

$$+ \frac{\delta_1}{\gamma_3} e^{-c_1 t/2} \int_{\gamma_3 z}^{t} J_0 \left(\gamma_0 \sqrt{\tau^2 - \gamma_3^2 z^2} \right) d\tau +$$

$$+ \frac{\gamma_{24}\gamma_{53}}{\gamma_{22}\gamma_4} e^{-c_1 t/2} \left[J_0 \left(\gamma_0 \sqrt{t^2 - \gamma_4^2 z^2} \right) H(t - \gamma_4 z) \right] +$$

$$+ \frac{\delta_2}{\gamma_4} e^{-c_1 t2} \int_{\gamma_4 z}^{t} J_0 \left(\gamma_0 \sqrt{\tau^2 - \gamma_4^2 z^2} \right) d\tau +$$

$$+ \delta_0 \left[a_1 e^{-k_1 z - \omega_1 t} - a_2 e^{-k_2 z - \omega_2 t} \right]; \tag{25}$$

$$U_{z,0}(z,t) = \int_{\gamma_5 z}^{t} F_1(t - \tau) e^{-c_1/t_2} J_0 \left(\gamma_0 \sqrt{\tau^2 - \gamma_5^2 z^2} \right) d\tau +$$

$$+ \delta_{82} \left[e^{-c_1 t/2} J_0 \left(\gamma_0 \sqrt{t^2 - \gamma_5^2 z^2} \right) H(t - \gamma_5 z) \right] +$$

$$+ \int_{\gamma_6 z}^{t} F_1(t - \tau) e^{-c_1 \tau/2} J_0 \left(\gamma_0 \sqrt{\tau^2 - \gamma_6^2 z^2} \right) d\tau +$$

$$+ \delta_{86} \left[e^{-c_1 t/2} J_0 \left(\gamma_0 \sqrt{t^2 - \gamma_6^2 z^2} \right) H(t - \gamma_6 z) \right] + \delta_{87} e^{-c_1 t/2} \times$$

$$\times \int_{\gamma_5 z}^{t} J_0 \left(\gamma_0 \sqrt{\tau^2 - \gamma_5^2 z^2} \right) d\tau + \delta_{88} e^{-c_1 t/2} \int_{\gamma_6 z}^{t} J_0 \left(\gamma_0 \sqrt{\tau^2 - \gamma_6^2 z^2} \right) d\tau +$$

$$+ e^{-k_1 z} \delta_{97} - \delta_{99} e^{-k_2 z} + \delta_{91} \left[e^{-c_1 t/2} J_0 \left(\gamma_0 \sqrt{t^2 - \gamma_3^2 z^2} \right) H(t - \gamma_3 z) \right] +$$

$$+\delta_6 e^{-c_1 t/2} \int_{\gamma_3 z}^{t} J_0 \left(\gamma_0 \sqrt{\tau^2 - \gamma_3^2 z^2} \right) d\tau +$$

$$+\delta_{91} \left[e^{-c_1 t/2} J_0 \left(\gamma_0 \sqrt{t^2 - \gamma_4^2 z^2} \right) H(t - \gamma_4 z) \right] +$$

$$+\delta_{95} e^{-c_1 t/2} \int_{\gamma_4 z}^{t} J_0 \left(\gamma_0 \sqrt{\tau^2 - \gamma_4^2 z^2} \right) d\tau ;$$

$$F_1(t) = -\frac{1}{\gamma_{22}} \left[\gamma_{25} f'(t) + \left(\gamma_{26} + \frac{c_1}{2} \gamma_{25} \right) f(t) \right] ;$$

$$F_2(t) = -\frac{1}{\gamma_{22}} \left[\gamma_{10} f'(t) + \left(\gamma_{11} + \frac{\gamma_{10} c_1}{2} \right) f(t) \right] ;$$

$$U_{r,0} = -\frac{q_1}{q_5} \left\{ \delta_{53} \left[e^{-c_1 t/2} J_0 \left(\gamma_0 \sqrt{t^2 - \gamma_3^2 z^2} \right) H(t - \gamma_3 z) \right] + \right.$$

$$+\delta_6 e^{-c_1 t 2} \int_{\gamma_3 z}^{t} J_0 \left(\gamma_0 \sqrt{\tau^2 - \gamma_3^2 z^2} \right) d\tau +$$

$$+\delta_{57} \left[e^{-c_1 t/2} J_0 \left(\gamma_0 \sqrt{t^2 - \gamma_4^2 z^2} \right) H(t - \gamma_4 z) \right] +$$

$$+\delta_{10} \delta_{14}^0 e^{-c_1 t/2} \int_{\gamma_4 z}^{t} J_0 \left(\gamma_0 \sqrt{\tau^2 - \gamma_4^2 z^2} \right) d\tau +$$

$$+ \int_{\gamma_5 z}^{t} F_3(t - \tau) e^{-c_1 \tau/2} J_0 \left(\gamma_0 \sqrt{\tau^2 - \gamma_5^2 z^2} \right) d\tau +$$

$$+\delta_{64} \left[e^{-c_1 t/2} J_0 \left(\gamma_0 \sqrt{t^2 - \gamma_5^2 z^2} \right) H(t - \gamma_5 z) \right] +$$

$$+ \int_{\gamma_5 z}^{t} F_4(t - \tau) e^{-c_1 \tau/2} J_0 \left(\gamma_0 \sqrt{\tau^2 - \gamma_6^2 z^2} \right) d\tau +$$

$$+\delta_{71} \left[e^{-c_1 t/2} J_0 \left(\gamma_0 \sqrt{t^2 - \gamma_6^2 z^2} \right) H(t - \gamma_6 z) \right] +$$

$$+e^{-k_1 z} a_1 \delta_{75} + \gamma_{43} a_1 e^{-(k_1 z + \omega_1 t)} - e^{-k_2 z} a_2 \delta_{79} - \delta_{48} a_2 e^{-(k_2 z + \omega_2 t)} ;$$

$$F_3(t) = \delta_{27} \left(\gamma_{25} f''(t) + \delta_{58} f'(t) + \delta_{59} f(t) \right) ;$$

$$F_4(t) = \delta_{28} \left(\gamma_{10} f''(t) + \delta_{65} f'(t) + \delta_{66} f(t) \right) .$$

Calculations are performed by formulas (16) for viscoelastic material the physical characteristics of which have form (19), herewith external load is given as function: $f(t) = g_0 \sin^2 [\pi t/t_1]$, $(t_1 = 20)$.

On Figs. 3–12 dependence of change in temperature, longitudinal and radial displacements and radial stress σ_{rr} of shell's external surface points with coordinate are presented for fixed value of time $t = 10$ and different values of non-dimensional parameters $\omega_1, \omega_2, a_1, a_2, k_1, k_2, r_2$, where k_1, k_2, ω_1 and ω_2 are correspondingly damping factors by coordinate and time.

On Figs. 3–6 this dependence is given at $\omega_1 = \omega_2 = k_1 = k_2 = 0,2$; $r_2 = 1,2$ for different values of temperature parameters a_1, a_2:0, 004; 0, 006; 0, 008; 0, 01. It can be seen from pictures that the change curves of temperature, displacement and stress are damping with the growth of distance. Herewith temperature changes according to given heat condition and has exponential damping character. Graphs of displacements and stresses have harmoniously damping form caused by action of external loading. Besides from Fig. 4 it can be seen that with the increase of the temperature parameters a_1, a_2 the value of longitudinal displacement also increases, while changes in radial

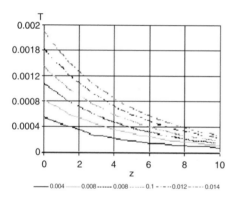

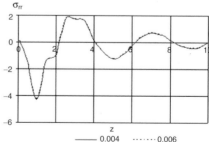

Fig. 3. The dependence of temperature change with coordinate z for different values of a_1 and a_2

Fig. 4. Plot of displacement U_z as function of coordinate z for $t = 10$

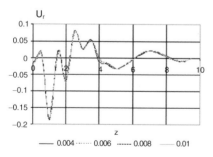

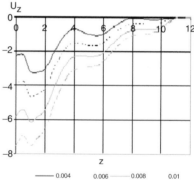

Fig. 5. Plot of displacement U_r as function of z

Fig. 6. Plot of stress σ_{rr} as function of z

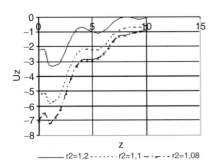

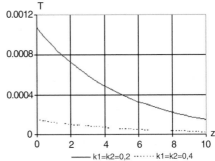

Fig. 7. Plot of displacement U_z as function of z at $h = 0,2$ (continuous line), $h = 0,1$ (dashed line) and $h = 0,08$ (dot-dashed line)

Fig. 8. The change of temperature with the distance

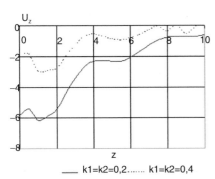

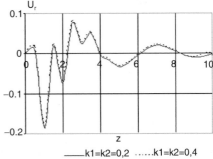

Fig. 9. The graphs of U_z displacement change with distance

Fig. 10. Plot of displacement U_r as function of z

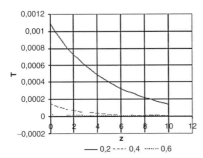

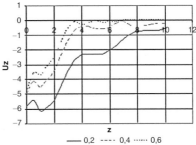

Fig. 11. Plot of temperature change as function of z at various values of factor k_1 and k_2

Fig. 12. Plot of displacement U_z as function of z $k_1, k_2 = 0, 2; 0, 4; 0, 6$

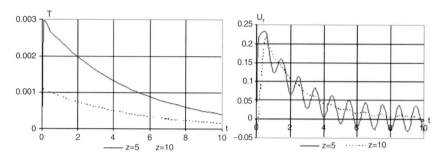

Fig. 13. Plot of temperature change as function of time in sections $z = 5$ and $z = 10$

Fig. 14. Plot of displacement U_z as function of time t in different sections

displacement and stress are insignificant (Figs. 5–6). On Fig. 7 the graphs of displacement change are presented at various values of thickness $h = 0, 2$ (corresponding to $r_1 = 1, 0$; $r_2 = 1, 2$); $h = 0, 1$ ($r_1 = 1, 0$; $r_2 = 1, 1$), $h = 0, 08$ ($r_1 = 1, 0$; $r_2 = 1, 08$), at the following values of parameters: $a_1 = a_2 = 0, 004$; $\omega_1 = \omega_2 = k_1 = k_2 = 0, 2$. With reduction of shell thickness, here the increase in values of displacement of shell surface points is observed. Changes of thickness of a layer ($h = 0, 2$) or shell ($h = 0, 1$; $h = 0, 08$) lead to essential change of longitudinal displacements in the sections bounded with coordinates in an interval $0 \leq z \leq 12$. In the further, with growth of coordinate z the displacement disturbance is not observed.

On Figs. 8–10 the results of temperature T and displacements change for value of parameters of temperature $a_1 = a_2 = 0, 008$ and factors $k_1 = k_2 = 0, 2$, at values of factors $\omega_1 = \omega_2 = 0, 2$ (continuous line) and $\omega_1 = \omega_2 = 0, 4$ (dashed line) depending on longitudinal coordinate are presented.

On Figs. 11–12 this dependence is presented for $a_1 = a_2 = 0, 008$ and $\omega_1 = \omega_2 = 2$ at various values of factors k_1, k_2, equal to 0,2; 0,4; 0,6. Comparison of the dates resulted on these graphs shows, that reduction of temperature and displacement values is observed with increase in values of factors k_1, k_2 and ω_1, ω_2. Herewith from Figs. 9 and 10 it can be seen, that in sections situating in immediate proximity to the face at increasing the values ω_1, ω_2, displacements decrease 2–3 times whereas at increasing the values k_1, k_2 the difference of displacement values makes 15–20.

On Figs. 13–14 the dependence of values change of temperature T and radial displacement of points of external surface in time in the fixed shell sections: $z = 5$ and $z = 10$ are represented at the following values of: $a_1 = a_2 = 0, 008$, $\omega_1 = \omega_2 = 0, 2$, $k_1 = k_2 = 0, 2$. Here the drop of displacements and temperature values in time is observed. Thus in section $z = 5$ displacement change has oscillatory character with the damping amplitude tending to constant value. In section $z = 10$ such vibrations occur with small amplitudes and graphs of displacement change have smooth form. In both cases starting at $t \geq 0$, U_r displacement disturbance tends to constant value eventually.

In conclusion we can say that the proposed vibration theory of termo-viscoelastic cylindrical shell gives satisfactory results and has enough broad applicability area even in case of suddenly applied load. If loading and temperature changes smoothly enough, they allow to define a stress and temperature actually at all sections of shell and displacements of its arbitrary points.

References

1. F.A. AMIRKULOVA (2006): Calculation of thermostressed state of circular cylindrical shell. Proceedings of Int. Conf. "Differential equations and their application", Samarkand, pp. 66–69.

2. F.A. AMIRKULOVA (2005): Vibrations of semi-infinite thermoviscoelastic cylindrical layer with restrained face. Proceedings of the int. conf. "Modern problems of Mathematical Physics and informational technology", Tashkent, pp. 116–120.

3. F.A. AMIRKULOVA (2003): Numerical-analytical analysis of thermo-stressed state of viscoelastic cylindrical shell. Compilation of abstracts of the Second M.I.T. conference on Computational Fluid and Solid Mechanics, June 17–20, Cambridge, MA, USA, p. 1.

4. F.A. AMIRKULOVA (2002): Nonstationary longitudinal-radial vibrations of circular thermoelastic shells. Proceedings of the 7th International conference on Shell Structures: Theory and applications, Gdansk-Jurata, Poland, pp. 45–46.

5. KH. KHUDOYNAZAROV, F.A. AMIRKULOVA (2002): Longitudinal-radial vibrations of circular cylindrical termoviscoelastic shells. Messenger of Russian People's Friendship University, Special issue: Geometry and design, Moscow, Vol. 1, pp. 66–70.

6. F.A. AMIRKULOVA, R.I. HOLMURODOV (2002): Thermo-stressing condition of the cylindrical shells with the nonstationary heating on the boundary. Messenger of Russian People's Friendship University, Special issue: Geometry and design, Moscow, Vol. 1, pp. 62–65.

7. A.E. ARMENAKAS, D.C. GAZIS, G. HERRMANN (1969): Free vibration of circular cylindrical shell. New York, Pergamon.

8. I.G. FILIPPOV, O.A. EGORICHEV (1983): Wave processes in linear viscoelastic medium. Moscow, Mashinostroeniye.

9. W. FLUGGE (1973): Axially symmetric motions of a two-layer Timoshenko type cylindrical shells. Berlin Heidelberg New York, Springer, p. 526.

10. Z.P. JONES, Z.S. WHITTIER (1969): Axially symmetric motions of a two-layer Timoshenko type cylindrical shells. AIAA J., Vol. 7(2), pp. 244–250.

11. G. HERMANN, I. MIRSKY (1956): Three-dimensional and shell-theory analysis of axially symmetric motions of cylinders. J. Appl. Mech., Vol. 23(4), pp. 563–568.

12. MARKUS STEFAN (1988): The mechanics of vibtrations of cylindrical shells. Amsterdam, Elsevier, p. 195.

13. V. NOVACSKIY (1975): The theory of elasticity. Moscow, Mir, p. 872. (in Russian).

14. S.Y. RAO: Vibrations of layered shells with transversal shear and rotary inertia effects. J. Sound Vib., Vol. 1, pp. 147–170.

Mathematical Vibration Modelling of the Pre-Stressed Viscoelastic Thick-Walled Cylindrical Shell

Feruza Abdukadirovna Amirkulova

Summary. In the paper the mathematical model of the pre-stressed viscoelastic thick walled cylindrical shell's axisymmetrical vibrations, based on the deduction of the integro-differential vibration equations of such shell, is developed. The deduction of the general and approximate vibrations equations of the thick-walled shell, which in limiting cases can proceed in the vibration equations of column and thin-walled cylindrical shell, is based on the use of exact solutions of the three-dimensional problem of the elasticity theory in transformations concerning potentials of longitudinal and transversal waves. The received equations belong to hyperbolic type and describe the waves distribution concerning dispersion. Herewith proposed approach allows different particular cases and generalizations. Alongside with the vibration equations, the algorithm, allowing by results of the solution of the vibration equations uniquely define the stress–strain state of considered system in its any section at the arbitrary time moment and correctly formulate initial and boundary conditions at the formulation of applied mechanics problems, is developed. Using Laplace transformation and the constants' variation method at the various boundary and initial conditions, a number of applied problems are solved. Reverse transition in the area of originals has been realized by means of the reversion table using shift theorem and theorem of folding function. Plots show how the change in the value of initial displacements affects on the strain–stress distribution.

1 Introduction

Initial stresses in structure elements can result from various physical and chemical processes and technological operations. It is also arises in earth's under action of different geostatic and geodynamic influence, in composites, rocks, blood vessels and etc. Besides many design elements are influenced by pre-stresses caused by imperfections of technology at their manufacturing and mistakes at their installation and other reasons. At the same time, the preliminary stressed state can be created also purposefully proceeding from the certain constructive reason, therefore it is necessary to consider the prestressed-strained state of structures at the calculation of their stress–strain state under action of non-stationary loadings.

Grin [13], Trusdell [21], Guz [14–16] and others have made the essential contribution in creation of modern linearized dynamic theory in initially stressed bodies. At present investigations on specified problem are conducted not only in solid mechanics but also in rock mechanics, acoustics, seismology, solid physics, geophysics and a number of other areas of science and technology. Results of these investigations were reflected in numerous publications [10–12, 18–20, 22] where elastic body's motion was described by 2D applied theories created by applying Kirchhoff–Love hypotheses which simplify equations and at the same time lead to essential lacks and mistakes.

Despite of a plenty of researches and basic works there is a significant circle of not enough investigated questions keeping the urgency: creation of the exact and effective approached analytical and numerical methods of studying the influence of initial stresses on the stressed–strained state of structure in view of variety of boundary conditions and complexities of external influence. Questions concerns to this circle of problems are studied in the represented work where axisymmetric vibrations of initially stressed viscoelastic circular cylindrical shell are studied on the base of proposed refined vibration equations of initially stressed cylindrical shell [1–9, 17]. These equations take into account the effect of deformation on transversal shear and the rotary inertia and are more generally than equations of Timoshenko type and Herrmann-Mirsky.

2 The Basic Relations

The circular cylindrical viscoelastic shell of constant thickness with internal r_1 and external r_2 radiuses is considered. It is supposed, that the material of a layer is homogeneous, isotropic and preliminary strained, and pre-stressed state is homogeneous, and dependences between stresses and deformations are linear. In this case dependence between tensor components of stress and deformation have form

$$\sigma_{ii} = L(\varepsilon) + 2M(\varepsilon_{ii}); \quad \sigma_{ij} = M(\varepsilon_{ij}), \quad (i \neq j), \tag{1}$$

where ε - cubic strain; L, M - viscoelastic operators

$$(L, M)\varsigma = (\lambda, \mu) \left[\varsigma(t) - \int_0^t [K_1(t - \tau), K_2(t - \tau)] \right] \varsigma(\tau)d\tau;$$

λ, μ - Lame coefficients; it is assumed that operators L, M are reversible and cores $K_1(t)$ and $K_2(t)$ are arbitrary.

Let's admit, that pre-stressed state is homogeneous. Generally, when thickness of shell is changeable initial displacements have form

$$U_z^{(0)} = a_0 z + b\theta + c_0 r; \quad U_\theta^{(0)} = a_1 z + b_1 \theta + c_1 r, \quad U_r^{(0)} = a_2 z + b_2 \theta + c_2 z, \tag{2}$$

where a_i, b_i, c_2 - are generally constant and are not small and c_2 is a function of a_i, b_i.

Let's note some special cases when dependence (2) takes simple form:

Case 1^0. Let's admit thickness of a considered cylindrical shell is constant, then $a_2 = b_2 = c_0 = c_1 = 0$ in (2) and initial displacements are equal to

$$U_z^{(0)} = a_0 z + b_0 \theta; \quad U_\theta^{(0)} = a_1 z + b_1 \theta; \quad U_r^{(0)} = c_2 r. \tag{3}$$

Let's designate through U_z, U_r, U_θ accordingly disturbed longitudinal, radial and torsional displacements and consider them as small quantities. Then displacements subject to initial displacement (2) will become as

$$\bar{U}_z = (1 + a_0)U_z + a_1 U_\theta; \quad \bar{U}_\theta = (1 + b_1)U_\theta + b_0 U_z; \quad \bar{U}_r = (1 + c_2)U_r, \tag{4}$$

and in the further for convenience of writing we shall neglect "hyphens" above displacements.

In this case stresses depending on small disturbed and homogeneous finite deformations have form

$$\sigma_{rr} = (1 + c_2)(L + 2M)\left(\frac{\partial U_r}{\partial r}\right) + (1 + a_0)L\left(\frac{\partial U_z}{\partial z}\right) + a_1 L(U_\theta) + a_1 L\left(\frac{\partial U_\theta}{\partial z}\right) +$$

$$+ \frac{1}{r}\left[(1 + b_1)L\left(\frac{\partial U_\theta}{\partial \theta}\right) + b_0 L\left(\frac{\partial U_z}{\partial \theta}\right) + (1 + c_2)L(U_r)\right];$$

$$\sigma_{zz} = (1 + c_2)L\left(\frac{\partial U_r}{\partial r}\right) + (1 + a_0)(L + 2M)\left(\frac{\partial U_z}{\partial z}\right) + a_1(L + 2M)U_\theta +$$

$$+ \frac{1}{r}\left[(1 + b_1)L\left(\frac{\partial U_\theta}{\partial \theta}\right) + b_0 L\left(\frac{\partial U_z}{\partial \theta}\right) + (1 + c_2)L(U_r)\right] + a_1 L\left(\frac{\partial U_\theta}{\partial z}\right);$$

$$\sigma_{\theta\theta} = (1 + c_2)L\left(\frac{\partial U_r}{\partial r}\right) + (1 + a_0)L\left(\frac{\partial U_z}{\partial z}\right) + a_1 L(U_\theta) + \frac{1 + b_1}{r}(L + 2M)\times$$

$$\times \left(\frac{\partial U_\theta}{\partial \theta}\right) + \frac{b_0}{r}(2M + L)\left(\frac{\partial U_z}{\partial \theta}\right) + \frac{1 + c_2}{r}(L + 2M)(U_r) + a_1 L\left(\frac{\partial U_\theta}{\partial z}\right); \tag{5}$$

Similarly density of initially stressed material of shell is distinguished from density of setting out material, i.e. its density ρ and is equal to

$$\rho_1 = \frac{\rho}{(1 + a_0)(1 + c_2)^2}.$$

Case1_a^0. For longitudinal-radial vibrations $U_\theta = 0$, $b_0 = b_1 = 0$ and initial longitudinal and radial displacement are equal to

$$U_z^{(0)} = a_0 \cdot z; \quad U_r^{(0)} = c_2 \cdot r. \tag{6}$$

Displacements subject to initial displacements have form

$$\bar{U}_z = (1 + a_0)U_z; \quad \bar{U}_r = (1 + c_2)U_r; \tag{7}$$

stresses depending on deformation

$$\sigma_{rr} = (1 + c_2)(L + 2M)\left(\frac{\partial U_r}{\partial r}\right) + (1 + a_0)L\left(\frac{\partial U_z}{\partial z}\right) + \frac{1}{r}(1 + c_2)L(U_r);$$

$$\sigma_{zz} = (1 + c_2)L\left(\frac{\partial U_r}{\partial r}\right) + (1 + a_0)(L + 2M)\left(\frac{\partial U_z}{\partial z}\right) + \frac{1}{r}(1 + c_2)L(U_r); \quad (8)$$

$$\sigma_{\theta\theta} = (1 + c_2)L\left(\frac{\partial U_r}{\partial r}\right) + (1 + a_0)L\left(\frac{\partial U_z}{\partial z}\right) + \frac{1}{r}(L + 2M)(U_r);$$

$$\sigma_{rz} = (1 + a_0)M\left(\frac{\partial U_z}{\partial r}\right) + (1 + c_2)M\left(\frac{\partial U_r}{\partial z}\right).$$

Case 1_b^0. For torsional vibrations $U_z = U_r = 0$, $a_1 = 0$, displacements subject to initial displacements (4) have form

$$\bar{U}_\theta = (1 + b_1)U_\theta \quad (9)$$

and $\sigma \div \varepsilon$ dependence subject to initial displacement takes following form

$$\sigma_{r\theta} = (1 + b_1)M\left(\frac{\partial U_\theta}{\partial r}\right) - \frac{1 + b_1}{r}M(U_\theta);$$

$$\sigma_{z\theta} = (1 + b_1)M\left(\frac{\partial U_\theta}{\partial z}\right). \quad (10)$$

3 Longitudinal-Radial Vibrations of Pre-Stressed Circular Cylindrical Shells

3.1 Derivation of General Equations of Longitudinal-Radial Vibrations Equations of Initially Stressed Isotropic Circular Cylindrical Shells

The longitudinal-radial vibrations of cylindrical shell are excited by stresses on its surfaces $r = r_i$ $(i = 1, 2)$

$$\sigma_{rr} = f_{rr}^{(i)}(z, t), \quad \sigma_{rz} = f_{rz}^{(i)}(z, t). \quad (11)$$

Shell motion is described by equations

$$\frac{\partial \sigma_{rr}}{\partial r} + \frac{\partial \sigma_{zr}}{\partial z} + \frac{\sigma_{rr} - \sigma_{\theta\theta}}{r} = \rho\frac{\partial^2 U_r}{\partial t^2}$$

$$\frac{\partial \sigma_{rz}}{\partial r} + \frac{\partial \sigma_{zz}}{\partial z} + \frac{\sigma_{rz}}{r} = \rho\frac{\partial^2 U_r}{\partial t^2}.$$

Subject to relationship (8) between stresses and displacements the equations of shell motion will take form

$$(1 + c_2) \left\{ (L + 2M) \left(\frac{\partial^2 U_r}{\partial r^2} + \frac{1}{r} \frac{\partial U_r}{\partial r} - \frac{1}{r^2} U_r \right) + M \left(\frac{\partial^2 U_r}{\partial z^2} \right) \right\} +$$

$$+ (1 + a_0) \left(\frac{\partial^2 U_r}{\partial r \partial z} \right) = \rho \frac{\partial^2 U_r}{\partial t^2},$$

$$(1 + c_2)(L + M) \left(\frac{\partial}{\partial r} + \frac{1}{r} \right) \frac{\partial U_r}{\partial z} +$$

$$+ (1 + a_0) \left\{ M \left(\frac{\partial^2 U_z}{\partial r^2} + \frac{1}{r} \frac{\partial U_z}{\partial r} \right) + (L + 2M) \left(\frac{\partial^2 U_z}{\partial z^2} \right) \right\} = \rho \frac{\partial^2 U_z}{\partial t^2}. \quad (12)$$

Representing displacements as

$$U_r = \int_0^\infty \left. \begin{matrix} \sin kz \\ -\cos kz \end{matrix} \right\} dk \int_{(l)} \bar{U}_r e^{pt} dp,$$

$$U_z = \int_0^\infty \left. \begin{matrix} \cos kz \\ \sin kz \end{matrix} \right\} dk \int_{(l)} \bar{U}_z e^{pt} dp \quad (13)$$

and substituting into motion equations (12), we can obtain

$$\Delta_0 \bar{U}_r - \left[\bar{M} \left(\bar{L} + 2\bar{M} \right)^{-1} k^2 + \frac{1}{1 + c_2} \left(\bar{L} + 2\bar{M} \right)^{-1} \rho p^2 \right] \bar{U}_r -$$

$$- \frac{1 + a_0}{1 + c_2} k \bar{G}_1 \frac{d\bar{U}_z}{dr} = 0,$$

$$\Delta_0 \frac{d\bar{U}_z}{dr} - \left[\bar{M}^{-1} \left(\bar{L} + 2\bar{M} \right) k^2 + \frac{1}{1 + a_0} \bar{M}^{-1} \rho p^2 \right] \frac{d\bar{U}_z}{dr} +$$

$$+ k \frac{1 + c_2}{1 + a_0} \bar{M}^{-1} \left(\bar{L} + 2\bar{M} \right) \Delta_0 \bar{U}_r = 0, \quad (14)$$

where

$$\bar{G}_1 = \left(\bar{L} + \bar{M} \right) \left(\bar{L} + 2\bar{M} \right)^{-1}; \quad \Delta_0 = \frac{d^2}{dr^2} + \frac{1}{r} \frac{d}{dr} - \frac{1}{r^2};$$

$$\bar{L} \div L; \quad \bar{M} \div M; \quad \bar{U}_r \div U_r; \quad \bar{U}_z \div U_z.$$

The general solution of (14) takes form

$$\bar{U}_r = C_1 I_1 (\alpha_1 r) + D_1 K_1 (\alpha_1 r) + C_2 I_1 (\alpha_2 r) + D_2 K_1 (\alpha_2 r);$$

$$\bar{U}_z = \frac{1 + c_2}{1 + a_0} \bar{G}_1 \frac{\alpha_1^2 - \alpha^2}{k\alpha_1} [C_1 I_0 (\alpha_1 r) - D_1 K_0 (\alpha_1 r)] +$$

$$+ \frac{\alpha_2^2 - \alpha^2}{k\alpha_2} [C_2 I_0 (\alpha_2 r) - D_2 K_0 (\alpha_2 r)], \quad (15)$$

where

$$\alpha^2 = \frac{1}{1+c_2}\left(\bar{L}+2\bar{M}\right)^{-1}\rho p^2 + \bar{M}\left(\bar{L}+2\bar{M}\right)^{-1}k^2,$$

$$\alpha_1^2 + \alpha_2^2 = \bar{N}_2 p^2 + \bar{N}_3 k^2, \qquad \alpha_1^2 \alpha_2^2 = \bar{N}_1 p^4 + \bar{N}_2 p^2 k^2 + k^4,$$

$$\bar{N}_1 = \frac{\rho^2}{(1+a_0)(1+c_2)}\left(\bar{L}+2\bar{M}\right)^{-1}\bar{M}^{-1};$$

$$\bar{N}_2 = \frac{\rho}{(1+a_0)}\bar{M}^{-1} + \frac{1}{(1+c_2)}\left(\bar{L}+2\bar{M}\right)^{-1};$$

$$\bar{N}_3 = \left(\bar{L}+2\bar{M}\right)\bar{M}^{-1} + \left(\bar{L}+2\bar{M}\right)^{-1}\bar{M} - \left(\bar{L}+\bar{M}\right)^2\left(\bar{L}+2\bar{M}\right)^{-1}\bar{M}^{-1}.$$

Let's transform into a power series of radial coordinate r expressions (15) and enter auxiliary functions $\bar{U}_{r,0}, \bar{U}_{z,0}, \bar{U}_{r,1}, \bar{U}_{z,1}$ by formulas

$$\bar{U}_{r,0} = \alpha_1 C_{10} + \alpha_2 C_{20}, \quad \bar{U}_{r,1} = \xi^{-1}\left(\alpha_1^{-1}D_1 + \alpha_2^{-1}D_2\right),$$

$$\bar{U}_{z,0} = \frac{\alpha_1^2 - \alpha^2}{k\alpha_1 \bar{G}_1}C_{10} + \frac{\alpha_2^2 - \alpha^2}{k\alpha_2 \overset{(0)}{B}_1}C_{20},$$

$$\xi\bar{U}_{z,1} = \frac{\alpha_1^2 - \alpha^2}{k\alpha_1 \bar{G}_1}D_1 + \frac{\alpha_2^2 - \alpha^2}{k\alpha_2 \bar{G}_1}D_2, \tag{16}$$

where

$$[C_{10}, C_{20}] = [C_1, C_2] + [D_1, D_2]\cdot\left[\ln\frac{(\alpha_1,\alpha_2)\xi}{2} - \psi(1)\right].$$

For unknowns $\bar{U}_r, \bar{U}_z$ using (16) we can get

$$\bar{U}_z = \frac{1+c_2}{1+_0}\frac{1}{\alpha^2 k\bar{G}_1}\sum_{n=0}^{\infty} k\,\bar{G}_1\alpha_1^2\alpha_2^2\left(\bar{P}_n - \alpha^2\bar{P}_{n-1}\right)\bar{U}_{z,0}\frac{(r/2)^{2n}}{(n!)^2}$$

$$-\frac{1+c_2}{1+_0}\left\{\frac{1}{\alpha^2 k\bar{G}_1}\sum_{n=0}^{\infty}\left(\alpha_1^2 - \alpha^2\right)\left(\alpha_2^2 - \alpha^2\right)\bar{P}_n\bar{U}_{r,0}\right\}\frac{(r/2)^{2n}}{(n!)^2}$$

$$-\frac{\xi}{k\,\bar{G}_1}\sum_{n=0}^{\infty}\eta_{6,n}\,(r)\left[k\,\bar{G}_1\left(\bar{P}_{n+1} - \alpha^2\,\bar{P}_n\right)\bar{U}_{z,1}\right.$$

$$\left. - \left(\alpha_1^2 - \alpha^2\right)\left(\alpha_2^2 - \alpha^2\right)\bar{P}_n\,\bar{U}_{r,1}\right]\frac{(r/2)^{2n}}{(n!)^2}\right\},$$

$$\bar{U}_r = \frac{\xi}{r}\bar{U}_{r,1} + \frac{1}{\alpha^2}\sum_{n=0}^{\infty}\left[\left(\alpha^2\,\underset{n+1}{\bar{P}} - \alpha_1^2\alpha_2^2\,\underset{n}{\bar{P}}\right)\bar{U}_{r,0} + k\bar{G}_1\alpha_1^2\alpha_2^2\bar{U}_{z,0}\right]\frac{(r/2)^{2n}}{(n!)\,(n+1)!} +$$

$$+\xi\sum_{n=0}^{\infty}\eta_{7,n}\,(r)\left[\left(\alpha^2\,\underset{n+1}{\bar{P}} - \alpha_1^2\alpha_2^2\,\underset{n}{\bar{P}}\right)\bar{U}_{r,1} + k\bar{G}_1\,\underset{n+1}{\bar{P}}\,\bar{U}_{z,1}\right]\frac{(r/2)^{2n}}{n!\,(n+1)!},$$

$$\tag{17}$$

where

$$\bar{\bar{P}}_n = \sum_{i=0}^{n-1} \alpha_2^{2(n-i-1)} \alpha_1^{2i}; \quad \bar{\bar{P}}_0 \equiv 0; \quad \bar{\bar{P}}_1 \equiv 1; \quad \bar{\bar{P}}_2 = \alpha_1^2 + \alpha_2^2; \quad \alpha_1^2 \alpha_2^2 \underset{-1}{\bar{\bar{P}}} = -1;$$

$$\eta_{6,n}(r) = \ell n \frac{r}{\xi} - \sum_{k=1}^{n} \frac{1}{k}; \quad \eta_{7,n}(r) = \eta_{6,n}(r) - \frac{1}{2(n+1)};$$

ξ is a radius of the intermediate surface of cylindrical shell defined as

$$\xi = \frac{r_1}{2} \left(\chi - \frac{r_1}{r_2} \right), \tag{18}$$

where constant χ fulfils with inequality

$$2 + \frac{r_1}{r_2} \leq \chi \leq 2\frac{r_2}{r_1} + \frac{r_1}{r_2}.$$

Substituting general solutions into transformed by Laplace and Fur'e boundary conditions we get

$$\left(\alpha_1^2 - \alpha^2 + k^2 \bar{G}_1 \right) [C_1 I_1(\alpha_1 r_i) + D_1 K_1(\alpha_1 r_i)] +$$
$$+ \left(\alpha_2^2 - \alpha^2 k^2 \bar{G}_1 \right) [C_2 I_1(\alpha_2 r_i) + D_2 K_1(\alpha_2 r_i)] +$$
$$= \frac{1}{1+c_2} \bar{G}_1 \bar{M}^{-1} \left[k f_{rz}^{(i0)}(k,p) \right], \quad (i = 1,2) \tag{19}$$

$$\left[\alpha_1 \bar{G}_1 (\bar{L} + 2\bar{M}) - \frac{\alpha_1^2 - \alpha^2}{\alpha_1} \bar{L} \right] [C_1 I_0(\alpha_1 r_i)] +$$
$$+ \left[\alpha_2 \bar{G}_1 (\bar{L} + 2\bar{M}) - \frac{\alpha_2^2 - \alpha^2}{\alpha_2} \bar{L} \right] [C_2 I_0(\alpha_2 r_i) - D_2 K_0(\alpha_2 r)] -$$
$$+ \frac{2\bar{M}\bar{G}_1}{r} [C_1 I_1(\alpha_1 r_i) + D_1 K_1(\alpha_1 r_i) + C_2 I_1(\alpha_2 r_i) + D_2 K_1(\alpha_2 r_i)]$$
$$= \frac{1}{1+c_2} \bar{G}_1 \left[\bar{f}_r^{(i)}(k,p) \right], \quad (i = 1,2).$$

Using standard expansion of Bessel's functions into a power series of radiuses r_1 and r_2 in the derived equations we can get the system of four algebraic equations relatively to entered functions $\bar{U}_{r,0}$, $\bar{U}_{r,1}$, $\bar{U}_{z,0}$, $\bar{U}_{z,1}$. Step of going from Laplace and Fur'e transforms to the original time and space functions is realized, and it is obtained the system of vibration equations taking form

$$a_{1i} U_{z,0} + a_{2i} U_{r,0} + \xi(a_{3i} U_{z,1} + a_{4i} U_{r,1}) = \frac{1}{1+c_2} G_1 M^{-1} \lambda_1 \left[\frac{\partial}{\partial z} f_{rz}^{(i)}(z,t) \right],$$

$$b_{1i} U_{z,0} + b_{2i} U_{r,0} + \xi(b_{3i} U_{z,1} + b_{4i} U_{r,1}) = \frac{1}{1+c_2} G_1 \lambda_1 \left[f_{rz}^{(i)}(z,t) \right]. \tag{20}$$

Here operators a_{ij}, b_{ij} $(i = 1, 2;\ j = \overline{1, 4})$ are defined by formulas

$$a_{1i} = \sum_{n=0}^{\infty} G_1 \lambda_2^2 \frac{\partial}{\partial z} \left[P_{n+1} - \left(\lambda_1 + \frac{\partial^2}{\partial z^2} B_1 \right) P_n \right] \frac{(r_i/2)^{2n+1}}{n!\,(n+1)!};$$

$$a_{2i} = -\sum_{n=0}^{\infty} \left[\lambda_1 P_{n+2} - \left(\lambda_1^2 + \lambda_2^2 + \lambda_1 \frac{\partial^2}{\partial z^2} G_1 \right) P_{n+1} + \left(\lambda_1 + \frac{\partial^2}{\partial z^2} G_1 \right) P_n \lambda_2^2 \right]$$
$$\times \frac{(r_i/2)^{2n+1}}{n!(n+1)!};$$

$$a_{3i} = \sum_{n=0}^{\infty} \eta_{7,n}\,(r_i)\,G_1 \lambda_2^2 \frac{\partial}{\partial z} \left[P_{n+2} - \left(\lambda_1 + \frac{\partial^2}{\partial z^2} G_1 \right) P_{n+1} \right] \frac{(r_i/2)^{2n+1}}{n!\,(n+1)!} + \frac{1}{r_i} G_1 \frac{\partial}{\partial z} \lambda_1;$$

$$a_{4i} = -\sum_{n=0}^{\infty} \eta_{7,n}\,(r_i)\,\lambda_1 \left[\lambda_1 P_{n+2} - \left(\lambda_1^2 + \lambda_2^2 + \lambda_1 \frac{\partial^2}{\partial z^2} G_1 \right) P_{n+1} \right.$$
$$\left. + \left(\lambda_1 + \frac{\partial^2}{\partial z^2} B_1 \right) P_n \lambda_2^2 \right] \frac{(r_i/2)^{2n+1}}{n!\,(n+1)!} + \frac{1}{r_i} G_1 \frac{\partial^2}{\partial z^2} \lambda_1;$$

$$b_{1i} = -\sum_{n=0}^{\infty} G_1 \lambda_2^2 \frac{\partial}{\partial z} \left[(G_1(L+2M) - L)\,P_n + \lambda_1 L P_{n-1} - \frac{G_1 M}{(n+1)} P_n \right] \frac{(r_i/2)^{2n}}{(n!)^2};$$

$$b_{2i} = \sum_{n=0}^{\infty} \left[G_1(L+2M)\,(\lambda_1 P_{n+1} - P_{n+2}) + L\left(\lambda_1^2 - \lambda_1 \lambda_3 + \lambda_2^2 \right) P_n - \right.$$
$$\left. - \frac{G_1 M}{(n+1)}\,(\lambda_1 P_{n+1} - \lambda_2^2 P_n) \right] \frac{(r_i/2)^{2n}}{(n!)^2};$$

$$b_{3i} = -\sum_{n=0}^{\infty} \lambda_1 \left\{ \eta_{6,n}\,(r_i)\,\frac{\partial}{\partial z} \left[G_1(L+2M) - G_1 L\,(P_{n+\bar{1}} \lambda_1 P_n) \right] - \right.$$
$$\left. -\eta_{7,n}\,(r_i) \cdot \frac{\partial}{\partial z} \cdot \frac{M G_1^2}{(n+1)} P_{n+1} \right\} \frac{(r_i/2)^{2n}}{(n!)^2} + \frac{1}{r_i} \lambda_1;$$

$$b_{4i} = \sum_{n=0}^{\infty} \lambda_1 \eta_{6,n}\,(r_i) \left[G_1(L+2M)\,(\lambda_1 P_{n+\bar{1}} - \lambda_2^2 P_n) + L\left(\lambda_1^2 - \lambda_1 \lambda_3 + \lambda_2^2 \right) P_n \right]$$
$$- \eta_{7,n}\,(r_i) \cdot \frac{M G_1}{(n+1)}\,(\lambda_1 P_{n+1} - \lambda_2^2 P_n) \frac{(r_i/2)^{2n}}{(n!)^2}.$$

Herewith operators λ_1^n, λ_2^n, λ_3^n, P_n are equal to

$$\lambda_1^n = (L+2M)^{-n} \left[\left(\frac{\rho}{1+c_2}\frac{2}{t^2} - M\frac{2}{z^2} \right)^n \right];$$

$$\lambda_3^n = \left\{ N_2 \left(\frac{2}{t^2} \right) - N_3 \left(\frac{2}{z^2} \right) \right\}^n ; \qquad \lambda_2^2 P_{-1} = -1;$$

$$\lambda_2^{2n} = A_{11}^{-n} \left[N_1 \left(\frac{4}{t^4} \right) - \left[\frac{\rho}{1 + c_2} M^{-1} + \frac{\rho}{1 + a_0} (L + 2M)^{-1} \right] \left(\frac{4}{t^2 z^2} \right) + \frac{4}{z^4} \right]^n ;$$

$$N_1 \div \bar{N}_1; \quad N_2 \div \bar{N}_2; \quad N_3 \div \bar{N}_3;$$

$$P_0 \equiv 0; \quad P_1 \equiv 1; \quad P_2 = \lambda_3; \quad P_3 = \lambda_3^2 - \lambda_2^2;$$

$$P_4 = \lambda_5 \left(\lambda_3^2 - 2\lambda_2^2 \right); \quad P_5 = \lambda_3^4 - 3\lambda_2^2 \lambda_3^3 + \lambda_2^4; \dots$$

Equations (20) are general equations of longitudinal-radial vibrations of cylindrical shell with initial stresses. Here we can get different approximate and refined vibration equation and consider various limited cases. According to form of operators λ_i^n, a_{ij}, b_{ij} system (20) contains the nth time and directional derivatives of functions $U_{r,0}$, $U_{r,1}$, $U_{z,0}$, $U_{z,1}$. Therefore it is reasonable to restrict the number of terms in equations for using them in applied problems.

3.2 Formulas for Definition of Displacements and Stresses in Shell Sections

Having inverted expressions (17) on z and t formulas for displacements are deduced

$$U_z = \frac{1 + c_2}{1 + a_0} \cdot \left\{ (G_1 \cdot \lambda_1)^{-1} \cdot \sum_{n=0}^{\infty} [\lambda_4 \cdot G_1 \cdot (P_{.n} - \lambda_1 \cdot P_{n-1}) U_{z,0} + \right.$$

$$+ \left(\frac{\partial}{\partial z} \right)^{-1} \cdot [\lambda_4 - \lambda_1 \cdot \lambda_3 + \lambda_1^2] Q_n U_{r,0}] \frac{(r\backslash 2)^{2n}}{(n!)} +$$

$$+ \sum_{n=0}^{\infty} \{ (P_{.n+1} - \lambda_1 \cdot P_n) U_{z,1} +$$

$$+ G_1^{-1} \cdot \left(\frac{\partial}{\partial z} \right)^{-1} [\lambda_4 - \lambda_1 \cdot \lambda_3 + \lambda_1^2] P_n U_{r,1} \frac{(r\backslash 2)^{2n}}{(n!)^2},$$

$$U_r = \lambda_1^{-1} \cdot \sum_{n=0}^{\infty} \left[\lambda_1 (P_{.n+1} - \lambda_4 \cdot P_n) \cdot U_{r,0} - G_1 \cdot \lambda_4 \cdot \left(\frac{\partial u_{z,0}}{\partial z} \right) \right] \frac{(r\backslash 2)^{2n}}{n!(n+1)!} +$$

$$+ \sum_{n=0}^{\infty} \eta_{z,n}(r) \cdot \left[(\lambda_1 P_{n+1} - \lambda_4 P_n) U_{r,1} - P_{n+1} \left(\frac{\partial u_{z,1}}{\partial z} \right) \right] \frac{(r\backslash 2)^{2n}}{n!(n+1)!} + \frac{1}{r} \cdot U_{r,1}.$$

$$(21)$$

Being limited to zero approximation from (21) we shall have

$$U_z = \frac{1 + c_2}{1 + a_0} [U_{z,0} + \ln U_{z,1}] ;$$

$$U_r = \frac{1}{r} U_{r,1} + \frac{r}{2} \left[U_{r,0} - G_1 \lambda_4 \lambda_1^{-1} \frac{\partial U_{z,0}}{\partial z} \right] + \left(\ln r - \frac{1}{2} \right) \left[\lambda_1 U_{r,1} - \frac{\partial}{\partial z} G_1 U_{z,1} \right] \frac{r}{2} .$$

$$(22)$$

Similar formulas take place for components of stress tensor

$$\sigma_{rr} = \lambda_1^{-1} B_1^{-1} \left[b_1 U_{z,0} + b_2 U_{r,0} + \xi \left(b_3 U_{z,1} + b_4 U_{r,1} \right) \right],$$

$$\sigma_{rz} = B_1^{-1} \lambda^{-1} \left[a_1 U_{z,0} + a_2 U_{r,0} + \xi \left(a_3 U_{z,1} + a_4 U_{r,1} \right) \right], \qquad (23)$$

where

$$a_i = a_{ij}\big|_{r_i=r}, \quad b_i = b_{ij}\big|_{r_i=r}, \quad B_1 = G_1 M^{-1} \frac{\partial}{\partial z}.$$

3.3 Limiting Cases and Particular Forms of the Longitudinal-Radial Vibrations Equations of the Preliminary-Strained Shell

The derived vibration equations (20) and formulas for stress and displacements suppose the following limiting and special cases:

1. If shell material is elastic, i.e. $M = \mu$, $L = \lambda$, (20) will pass in the corresponding equations for an elastic shell.

2. If $r_1 = 0$, then $\xi = 0$ and from (20) we can get longitudinal vibrations equations of preliminary-strained circle cylinder

$$a_1 U_{z,0} + a_2 U_{r,0} = \frac{1}{1+c_2} G_1 M^{-1} \lambda_1 \left[\frac{\partial f_{rz}^{(2)}(z,t)}{\partial z} \right],$$

$$b_1 U_{z,0} + b_2 U_{r,0} = \frac{1}{1+c_2} G_1 M^{-1} \lambda_1 \left[\frac{\partial f_r^{(2)}(z,t)}{\partial z} \right], \qquad (24)$$

where $a_i = a_{ij}\big|_{r_i=r}$, $b_i = b_{ij}\big|_{r_i=r}$.

If $r_2 = r_1(1 + \varepsilon)$ and $\varepsilon > 0$ is small quantity, then from (20) will follow vibration equations for initially stressed thin walled cylindrical shell.

At zero approach and a constancy of Poisson factor these equations in dimensionless variables (r, z, t) will become as

$$\tilde{M}^{-1} \left(\frac{\partial^3 U_{z,0}}{\partial t^2 \partial z} \right) - \frac{2(1-\nu)}{1-2\nu} \frac{\partial^3 U_{z,0}}{\partial z^3} - q_1 \frac{\partial^2 U_{r,0}}{\partial z^2} + \frac{1}{r_i^2} \frac{\partial U_{z,1}}{\partial z} -$$

$$- \frac{1}{2} \left(\tilde{M}^{-1} \left(\frac{\partial^3 U_{z,1}}{\partial z \partial t^2} \right) - \frac{\partial^3 U_{z,1}}{\partial z^3} \right) + q_1 d_2 \frac{1-2\nu}{4(1-\nu)} \times$$

$$\times \left(\tilde{M}^{-1} \left(\frac{\partial^4 U_{r,1}}{\partial t^2 \partial z^2} \right) - \frac{\partial^4 U_{r,1}}{\partial z^4} \right) + \frac{1}{r_i^2} \frac{\partial^2 U_{r,1}}{\partial z^2} = 0; \qquad (i = 1, 2) \qquad (25)$$

$$\frac{2\nu}{1-2\nu} \cdot \frac{\partial U_{z,0}}{\partial z} + \frac{1}{1-2\nu} \cdot U_{r,0} - \frac{1}{2} \frac{\partial U_{z,1}}{\partial z} - \frac{2 U_{r,1}}{r_i^2} +$$

$$\frac{1-2\nu}{4(1-\nu)} \left(d_2 \tilde{M}^{-1} \left(\frac{\partial^2 U_{r,1}}{\partial t^2} \right) - \frac{\partial^2 U_{r,1}}{\partial z^2} \right) = 0, \qquad (i = 1, 2)$$

where ν - Poisson factor, dimensionless variables are entered by formulas

$$z = z^*\xi; \quad bt = t^*\xi; \quad r_1 = r_2^*\xi; \quad r_2 = r_2^*\xi;$$
$$U_{r,1} = U_{r,1}^*\xi; \quad U_{z,0} = U_{z,0}^*\xi; \quad U_{z,1} = U_{z,1}^*; \quad U_{r,0} = U_{r,0}^*, \tag{26}$$

b - velocity of transversal wave propagation in the shell material.

4. If $a_0 = 0$ and $c_2 = 0$, then (20) coincide with the refined equations of a cylindrical layer without taking into account the pre-stress.

4 The Torsional Vibration Equations of Circular Cylindrical Shell with Initial Displacements

Torsional vibrations of cylindrical shell are excited on its surface by stresses

$$\sigma_{r\theta} = f_{r\theta}^{(i)}(r, t), \quad r = r_i \quad (i = 1, 2). \tag{27}$$

Shell motion is described by equation

$$\frac{\partial \sigma_{r\theta}}{\partial r} + \frac{\partial \sigma_{z\theta}}{\partial z} + \frac{2}{r}\sigma_{r\theta} = \rho\frac{\partial^2 U_\theta}{\partial t^2}. \tag{28}$$

At torsional vibrations displacements in view of initial displacements have form (9). Dependence between stress and small disturbed and initial displacements have form (10). Substitution of the last in the motion equations (28) leads to next the equation

$$(1 + b_1)\left\{\frac{\partial^2}{\partial r^2} - \frac{1}{r^2} + \frac{1}{r}\frac{\partial}{\partial r} + \frac{\partial^2}{\partial z^2}\right\}U_\theta = \rho M^{-1}\frac{\partial^2 U_\theta}{\partial t^2}. \tag{29}$$

Let's assume

$$U_\theta = \int_0^\infty \left\{\begin{matrix}\sin kz \\ -\cos kz\end{matrix}\right\}dk\int_l \bar{U}_\theta e^{pt}dp, \tag{30}$$

then from (29) we can get

$$\frac{d^2\bar{U}_\theta}{dr^2} + \frac{1}{r}\frac{\partial\bar{U}_\theta}{\partial r} - \left(\beta^2 + \frac{1}{r^2}\right)\bar{U}_\theta = 0, \tag{31}$$

where

$$\beta^2 = k^2 + (1 + b_1)^{-1}\rho p^2 \bar{M}^{-1}, \quad \bar{M} = \mu\left[1 - \bar{f}(p)\right].$$

General solution of (31) is equal to

$$\bar{U}_\theta = A_1 I_1(\beta r) + A_2 K_1(\beta r). \tag{32}$$

Having presented external acting stresses also in the form of

$$f_{r\theta}^{(i)}(z, t) = \int\limits_0^\infty \left\{ \begin{matrix} \sin kz \\ -\cos kz \end{matrix} \right\} dk \int\limits_l \bar{f}_{r\theta}^{(i)} e^{pt} dp, \quad r = r_i \ (i = 1, 2) \qquad (33)$$

and substituting them into boundary conditions (27) we will get equations

$$\frac{\partial \bar{U}_\theta}{\partial r} - \frac{1}{r}\bar{U}_\theta = \frac{1}{1+b_1} \bar{M}^{-1}\left[\bar{f}_{r\theta}^{(i)}\right], \quad r = r_i \ (i = 1, 2). \qquad (34)$$

Having substituted general solution (32) into (34) we get

$$\beta\left[A_1 I_1(\beta r_i) - A_2 K_2(\beta r_i)\right] = \frac{1}{1+b_1}\bar{M}^{-1}\left[\bar{f}_{r\theta}^{(i)}\right], \quad (i = 1, 2). \qquad (35)$$

Let's expand solution (32) into a power series of radius r

$$\bar{U}_\theta = \frac{A_2}{\beta^2} + \sum_{n=0}^\infty \left\{ A_1 + A_2\left[\ln\frac{\beta r}{2} - \frac{1}{2}\psi(n+1) - \frac{1}{2}\psi(n+2)\right] \right\} \frac{(\beta r/2)^{2n+1}}{n!(n+1)!}$$

and enter main parts of transformed torsional displacement $\bar{U}_\theta$ of intermediate surface of shell.

Assume

$$A_{10} = \frac{2}{\beta}\bar{U}_{\theta,0}; \quad A_2 = \beta\xi\bar{U}_{\theta,1}, \qquad (36)$$

where $A_{10} = A_1 + A_2\left[\ln\frac{\beta\xi}{2} - \frac{1}{2}\psi(1) - \frac{1}{2}\psi(2)\right]$.

Substituting last values of integration constant A_i into transformed torsional displacement we can get

$$\bar{U}_\theta = \left[\frac{r}{\xi} + \xi\sum_{n=0}^\infty \eta_{4,n}\frac{(r/2)^{2n+1}}{n!(n+1)!}\beta^{2n+2}\right]\bar{U}_{\theta,1} + 2\sum_{n=0}^\infty \frac{(r/2)^{2n+1}}{n!(n+1)!}\beta^{2n}\bar{U}_{\theta,0}, \quad (37)$$

where

$$\eta_{4,n}(r) = \ln\frac{r}{\xi} - \frac{1}{2}\sum_{k=1}^n \frac{2k+1}{k(k+1)}. \qquad (38)$$

Similarly, having expanded boundary conditions (35) into a power series of r_i we can get

$$\frac{1}{2}\left(1 - \frac{4}{\beta^2 r_i^2}\right)\beta A_2 + \sum_{n=0}^\infty \left[A_{10} + \eta_{5,n}(r_i)A_2\right]\frac{\beta^{2n+3}(r/2)^{2n+2}}{n!(n+2)!} =$$

$$= \frac{1}{1+b_1}\bar{M}^{-1}\left[\bar{f}_{r\theta}^{(i)}(k, p)\right]; \quad (i = 1, 2), \qquad (39)$$

where

$$\eta_{5,n}(r_i) = \frac{1}{n+2}.$$

Substituting values of expressions (36) in (39) we have

$$\xi \left(\frac{\beta^2}{2} - \frac{2}{r_i^2} \right) \bar{U}_{\theta,1} + 2 \sum_{n=0}^{\infty} \frac{(r/2)^{2n+2}}{n!(n+2)!} \beta^{2n+2} \bar{U}_{\theta,0} +$$

$$+ \xi \sum_{n=0}^{\infty} \eta_{5,n} \frac{(r/2)^{2n+2}}{n!(n+2)!} \beta^{2n+4} \bar{U}_{\theta,1} =$$

$$= \frac{1}{1+b_1} \bar{M}^{-1} \left[\bar{f}_{r\theta}^{(i)}(k, p) \right], \quad (i = 1, 2). \tag{40}$$

Having converted last equations we get

$$c_{1i} U_{\theta,0} + \xi c_{2i} U_{\theta,1} = \frac{1}{1+b_1} \bar{M}^{-1} \left[\bar{f}_{r\theta}^{(i)}(k, p) \right], \quad (i = 1, 2) \tag{41}$$

where

$$c_{1i} = 2 \sum_{n=0}^{\infty} \frac{(r_i/2)^{2n+2}}{n!(n+2)!} \lambda_1^{n+1};$$

$$c_{2i} = \frac{1}{2} \left(\lambda_1 - \frac{4}{r_i^2} \right) + \sum_{n=0}^{\infty} \eta_{5,n}(r_i) \frac{(r_i/2)^{2n+2}}{n!(n+2)!} \lambda_1^{n+1};$$

$$\lambda_1^n = \left\{ \frac{\rho}{1+b_1} M^{-1} \left[\frac{\partial^2}{\partial t^2} \right] - \frac{\partial^2}{\partial z^2} \right\}^n.$$

Equations (41) are general equations of torsional vibrations of isotropic viscoelastic cylindrical shell subject to initial stresses. At zero approach system (41) will take form

$$\frac{r_1^2}{4} \lambda_1 U_{\theta,0} + \xi \left\{ \frac{1}{2} \left(\lambda_1 - \frac{2}{r_1^2} \right) + \left(\ln \frac{r_1}{\xi} - \frac{1}{2} \right) \frac{r_1^2}{8} \lambda_1^2 \right\} U_{\theta,1} =$$

$$= \frac{1}{1+b_1} \bar{M}^{-1} \left[\bar{f}_{r\theta}^{(1)}(k, p) \right],$$

$$\frac{r_2^2}{4} \lambda_1 U_{\theta,0} + \xi \left\{ \frac{1}{2} \left(\lambda_1 - \frac{2}{r_2^2} \right) + \left(\ln \frac{r_2}{\xi} - \frac{1}{2} \right) \frac{r_2^2}{8} \lambda_1^2 \right\} U_{\theta,1} =$$

$$= \frac{1}{1+b_1} \bar{M}^{-1} \left[\bar{f}_{r\theta}^{(2)}(k, p) \right], \tag{42}$$

where

$$\lambda_1 = \frac{\rho}{1+b_1} M^{-1} \left[\frac{\partial^2}{\partial t^2} \right] - \frac{\partial^2}{\partial z^2}.$$

Further we shall result formulas for determination of stressed–strained state of shell. Having converted on k and p expression (37) we get formula for displacement

$$U_\theta(r, z, t) = g_1 U_{\theta,0} + g_2 U_{\theta,1}, \tag{43}$$

where

$$g_1 = 2\sum_{n=0}^{\infty} \frac{(r/2)^{2n+1}}{n!(n+1)!}\lambda^n, \quad g_2 = \frac{1}{r} + \sum_{n=0}^{\infty} \eta_{4,n}(r)\frac{(r/2)^{2n+1}}{n!(n+1)!}\lambda^{n+1}.$$

Using the above-stated analogy it is easy to derive formulas for component of stress tensor through unknown functions

$$\sigma_{r\theta} = (1+b_1)M\left[c_1\,U_{\theta,0} + \xi\,c_2\,U_{\theta,1}\right],$$

$$\sigma_{z\theta} = (1+b_1)M\left[g_1\frac{\partial U_{\theta,0}}{\partial z} + \xi\,g_2\,\frac{\partial U_{\theta,1}}{\partial z}\right], \tag{44}$$

where $c_1 = c_{1i}\,|_{r_i=r}\,;\quad c_2 = c_{2i}\,|_{r_i=r}\,.$

Formulas (43) and (44) allow to determine values of displacement and stress in arbitrary section of shell.

5 Application of Vibration Equations for Solving Problems

For approbation of the derived vibration equations as test problems are solved the problems of axisymmetrical vibrations of pre-stressed semi-infinite circular cylindrical thick shell at kinematical excitement on the end face.

5.1 Longitudinal-Radial Vibrations of Pre-Stressed Cylindrical Shell with Rigid Restrained Face

Let's consider the problem of longitudinal-radial vibrations of semi-infinite cylindrical shell with rigid restrained face, taking into account pre-stress and viscous parameters of material. The problem is solved taking the Laplace transform with further using method of variation of constant on the base of refined equations. The stressed–strained state of shell subject to initial displacements is defined.

Vibrations equations (20) ignoring external loading will take form

$$\tilde{M}^{-1}\left(\frac{\partial^3 U_{z,0}}{\partial t^2 \partial z}\right) - \frac{2(1-\nu)}{1-2\nu}\frac{\partial^3 U_{z,0}}{\partial z^3} - q_1\frac{\partial^2 U_{r,0}}{\partial z^2} + \frac{1}{r_i^2}\frac{\partial U_{z,1}}{\partial z} -$$

$$-q_1 d_2\left(\ln r_i - \frac{1}{2}\right)\frac{1-2\nu}{2(1-\nu)} \times \left(\tilde{M}^{-1}\left(\frac{\partial^4 U_{r,1}}{\partial t^2 \partial z^2}\right) - \frac{\partial^4 U_{r,1}}{\partial z^4}\right) +$$

$$+\frac{1}{r_i^2}\frac{\partial^2 U_{r,1}}{\partial z^2} + \left(\ln r_i - \frac{1}{2}\right)\left(\tilde{M}^{-1}\left(\frac{\partial^3 U_{z,1}}{\partial z \partial t^2}\right) - \frac{\partial^3 U_{z,1}}{\partial z^3}\right) = 0; \quad (i = 1,\,2)$$

$$\frac{2\nu}{1-2\nu} \cdot \frac{\partial U_{z,0}}{\partial z} + \frac{1}{1-2\nu} \cdot U_{r,0} - \left(\ln r_i + \frac{1}{2}\right)\frac{\partial U_{z,1}}{\partial z} - \frac{2U_{r,1}}{r_i^2} + \left(\frac{\ln r_i}{1-2\nu} + \frac{1}{2}\right) \times$$

$$\times \frac{1-2\nu}{2(1-\nu)}\left(d_2\tilde{M}^{-1}\left(\frac{\partial^2 U_{r,1}}{\partial t^2}\right) - \frac{\partial^2 U_{r,1}}{\partial z^2}\right) = 0, \quad (i = 1,\, 2) \qquad (45)$$

ν - Poisson factor;

$$q_1 = \frac{1-2v}{2v} - \frac{4(1-v)^2}{1-2v}; \quad d_1 = \frac{1}{(1+c_2)^2(1+a_0)^2}; \quad d_2 = \frac{1}{(1+c_2)^3(1+a_0)}$$

Initial condition are zero. Boundary conditions at the rigid restrained face have form:

when $z = 0$

$$U_z = -f(t), \quad U_r = 0, \qquad (46)$$

as $z \to \infty$

$$U_{z,0} = U_{z,1} = 0, U_{r,0} = U_{r,1} = 0.$$

Stressed–strained state of shell is defined by formulas

$$U_z = \frac{1+c_2}{1+a_0}\left[U_{z,0} + \ln r U_{z,1}\right];$$

$$U_r = \frac{1}{r}U_{r,1} + \frac{r}{2}U_{r,0} + \left(\ln r - \frac{1}{2}\right)\left[\lambda_1 U_{r,1} - \frac{\partial}{\partial z}G_1 U_{z,1}\right]\frac{r}{2};$$

$$\sigma_{rz} = \tilde{M}(1+c_2)\left\{\frac{r}{2}\cdot\left[(1+c_2)^2\tilde{M}^{-1}\left(\frac{\partial^2 U_{z,0}}{\partial t^2}\right) - \frac{2(1-v)}{1-2v}\cdot\frac{\partial^2 U_{z,0}}{\partial z^2}\right] - \right.$$

$$-\frac{r}{2}q_1\cdot\frac{\partial U_{r,0}}{\partial z} + \frac{r}{2}\left(\ln r - \frac{1}{2}\right)\left[(1+c_2)^2\cdot\tilde{M}^{-1}\left(\frac{\partial^2 U_{z,1}}{\partial t^2}\right) - q_2\frac{\partial^2 U_{z,1}}{\partial z^2}\right] +$$

$$-\left[(1+c_2)(1+a_0)\frac{1-2v}{2(1-v)}\cdot\tilde{M}^{-1}\left(\frac{\partial^3 U_{r,1}}{\partial t^2\partial z}\right) - \frac{1-2v}{2(1-v)}\cdot\frac{\partial^3 U_{r,1}}{\partial z^3}\right] \times$$

$$\left.\times\frac{r}{2}(\ln r - \frac{1}{2})q_1 + \frac{1}{r}\cdot\frac{\partial U_{r,1}}{\partial z} + \frac{1}{r}U_{z,1}\right\}; \qquad (47)$$

$$\sigma_{rr} = \tilde{M}(1+c_2)\frac{2v}{1-2v}\cdot\frac{\partial U_{z,0}}{\partial z} + \frac{1}{1-2v}\cdot U_{r,0} -$$

$$-\left(\ln r + \frac{1}{2}\right)\frac{\partial U_{z,1}}{\partial z} - \frac{2}{r^2}\cdot U_{r,1} + \left(\frac{\ln r}{1-2v} + \frac{1}{2}\right)\times$$

$$\times\frac{1-2\nu}{2(1-\nu)}\left[(1+c_2)(1+a_0)\tilde{M}^{-1}\left(\frac{\partial^2 U_{r,1}}{\partial t^2}\right) - \frac{\partial^2 U_{r,1}}{\partial z^2}\right],$$

where

$$\lambda_1 = \tilde{M}^{-1}(1+c_2)(1+a_0)\cdot\frac{1-2v}{2(1-v)}\cdot\frac{\partial^2}{\partial t^2} - \frac{1-2v}{2(1-v)}\cdot\frac{\partial^2}{\partial z^2}.$$

Having applied Laplace transformation on time to system (45), with the subsequent using of method of constants variation the solution in transformations is received. Step of going from Laplace and Fur'e transforms to the original time and space function is realized using shift theorem and theorem of functions' convolution. It is got closed solution which substitution into formulas (47) lets fully define stressed–strained state of arbitrary shell section

$$U_{z,1} = \frac{\gamma_{51}}{\gamma_7} \int\limits_{\gamma_7 z}^{t} e^{-\frac{c_1}{2}\tau} F_1(t-\tau) I_0 \left(\gamma_9 \sqrt{\tau^2 - \gamma_7^2 z^2} \right) d\tau -$$

$$- \frac{\gamma_{50}}{\gamma_5} \int\limits_{\gamma_5 z}^{t} e^{-\frac{c_1}{2}\tau} F_2(t-\tau) I_0 \left(\gamma_8 \sqrt{\tau^2 - \gamma_5^2 z^2} \right) d\tau;$$

$$F_1(t) = \gamma_{80} \int\limits_{0}^{t} f(\xi) d\xi + \gamma_{81} \left(f'(t) - \frac{c_1}{2} f(t) \right);$$

$$F_2(t) = \gamma_{82} \int\limits_{0}^{t} f(\xi) d\xi + \gamma_{84} \left(f'(t) + \frac{c_1}{2} f(t) \right);$$

$$U_{r,1} = \gamma_{50} \int\limits_{\gamma_5 z}^{t} f(t-\tau) e^{-\frac{c_1}{2}\tau} I_0 \left(\gamma_8 \sqrt{\tau^2 - \gamma_5^2 z^2} \right) d\tau -$$

$$- \gamma_{51} \int\limits_{\gamma_7 z}^{t} f(t-\tau) e^{-\frac{c_1}{2}\tau} I_0 \left(\gamma_9 \sqrt{\tau^2 - \gamma_7^2 z^2} \right) d\tau;$$

$$U_{z,0} = \frac{1}{\gamma_{10}} \int\limits_{\gamma_{10} z}^{t} \left[\gamma_{60} F_0(t-\tau) + \gamma_{61} \int\limits_{0}^{t-\tau} f(\xi) d\xi \right] \times$$

$$\times e^{-\frac{c_1}{2}\tau} I_0 \left(\gamma_0 \sqrt{\tau^2 - \gamma_{10}^2 z^2} \right) d\tau -$$

$$- q_4 \left\{ \frac{1}{\gamma_5} \int\limits_{\gamma_5 z}^{t} e^{-\frac{c_1}{2}\tau} I_0 \left(\gamma_8 \sqrt{\tau^2 - \gamma_5^2 z^2} \right) \times \right.$$

$$\times \left[\gamma_{62} F_0(t-\tau) + \gamma_{63} \int\limits_{0}^{t-\tau} f(\xi) d\xi \right] d\tau + \tag{48}$$

$$+ \frac{1}{\gamma_7} \int\limits_{\gamma_7 z}^{t} \left[\gamma_{64} F_0(t-\tau) + \gamma_{65} \int\limits_{0}^{t-\tau} f(\xi) d\xi \right] \times$$

$$\times e^{-\frac{c_1}{2}\tau} I_0 \left(\gamma_9 \sqrt{\tau^2 - \gamma_7^2 z^2} \right) d\tau;$$

$$F_0 = f'(t) + \frac{c_1}{2} f(t);$$

$$U_{r,0} = \frac{1}{\gamma_{10}} \int_0^t \left[\gamma_{70} f'(t-\tau) + \gamma_{71} f'(t-\tau) + \gamma_{72} f(t-\tau) \right] \times$$

$$\times I_0 \left(\gamma_0 \sqrt{\tau^2 - \gamma_{10}^2 z^2} \right) e^{-\frac{c_1}{2}\tau} d\tau + \frac{\gamma_{50}}{\gamma_5} \times$$

$$\times \int_0^t \left[\gamma_{73} f''(t-\tau) + \gamma_{73} c_1 f'(t-\tau) + \gamma_{75} f(t-\tau) \right]$$

$$I_0 \left(\gamma_8 \sqrt{\tau^2 - \gamma_5^2 z^2} \right) e^{-\frac{c_1}{2}\tau} d\tau -$$

$$-\frac{\gamma_{51}}{\gamma_{10}} \int_{\gamma_7 z}^t \left[\gamma_{76} f''(t-\tau) + \gamma_{76} c_1 f'(t-\tau) + \gamma_{78} \ f(t-\tau) \right]$$

$$I_0 \left(\gamma_9 \sqrt{\tau^2 - \gamma_7^2 z^2} \right) e^{-\frac{c_1}{2}} d\tau.$$

Calculations are performed by formulas (47) for viscoelastic material the physical characteristics areas follow:

$$\alpha_1 = 0,78; \alpha_2 = 0,18; \alpha_1 = 0,1;$$

$$\tau_1 = 80mks\,; \tau_2 = 0,14 \cdot 10^6 mks; \ \tau_3 = 0,11 \cdot 10^7;$$

$$K(t) = \sum_{n=1}^3 \alpha_n \tau_n^{-1} e^{-t/\tau_n}.$$

Results are represented on Figs. 1–6.

On Figs. 1 and 2 dependence of change in longitudinal displacement and radial stress σ_{rr} with coordinate are presented for fixed value of time $t = 10$ when external load is given as function: where $H(t)$ - unit function of Heaveside, $A_0 = const$. From the resulted graphs it is visible, that in the sections close to end face $(0 < z < 4)$ the influence of factor of preliminary intensity (a_0) is strong. Depending on the value of factor a_0 in points close to the end face displacement can increase two $(a_0 = -0,68)$ - three $(a_0 = -0,7)$ time. The similar picture is observed or stress σ_{rr} which can accept big values in sections close to the face depending on a_0, and in sections removed from the end face on distance $z = 6$ and more the full drop of displacement and stress is observed. From here follows, that at $z > 6$ for loadings as $H(t)$-type the influence of factor of preliminary intensity can be neglected.

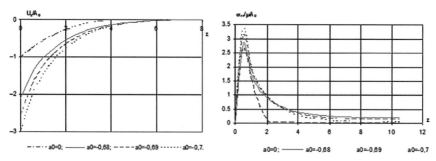

Fig. 1. Plot of displacement U_z/A_0 as functions of coordinate z for $t = 10$

Fig. 2. The dependence of change in stress with coordinate z

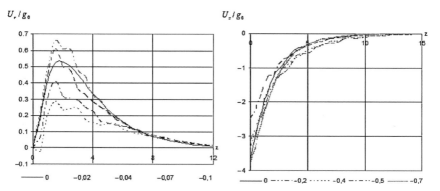

Fig. 3. Plot of displacement U_r against coordinate z

Fig. 4. The change of displacement U_z with coordinate for various values of a_0

On Figs. 3 and 4 graphs of displacements and U_r against coordinate z are presented when external loading is has form: $f(t) = g_0 \sin^2 [\pi t/t_1]$ for $t = 20$ and various fixed values of a_0. From these graphs follows, that general picture of changes of displacements and U_r looks like previous ones, but unlike here displacement U_r at $a_0 \geq -0,04$ will be bigger, and at $a_0 \leq -0,07$ will be less than one at $a_0 = 0$. On Figs. 5 and 6 graph of change of radial displacement in time are presented, here also loading is given in the form of smooth harmonic function. On Fig. 5 results in section $z = 5$ are presented in view of initial displacements ($a_0 = -0,01; -0,02; -0,03; -0,04$) and ignoring them ($a_0 = 0$). Apparently from graphs it can be seen at initial moments of action of external loading the sharp increase in value of displacement which eventually at $t = 0$ passes in the stable state is observed, moreover with increase of initial displacements the value of radial displacement increases. On Fig. 6 values of change of radial displacement in various shell sections ($z = 5, 10, 15$) at $a_0 = 0$ are resulted. With growth of distance from the face in shell sections reduction of displacement values is observed.

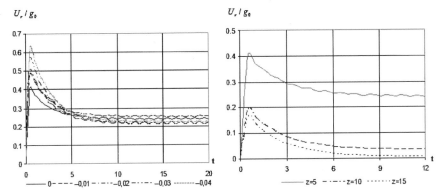

Fig. 5. The change of displacement U_r in time in section $z = 5$ for different values of a_0

Fig. 6. Plot of displacement U_r as function of time in various shell sections at $a_0 = 0$

5.2 Torsional Waves in Viscoelastic Cylindrical Shell with Initial Stresses

Let's consider vibration process in semi-infinite pre-stressed viscoelastic circular cylindrical shell, at kinematical excitation at the face. Mathematical formulation of problem is based on using of refined vibration equations, which at the absence of external loadings in dimensionless variables (r, z, t) take form

$$\frac{r_i^2}{4}\left(\frac{1}{1+b_1}\tilde{M}^{-1}\left[\frac{\partial^2 U_{\theta,0}}{\partial t^2}\right] - \frac{\partial^2 U_{\theta,0}}{\partial z^2}\right) + \frac{1}{2}\left(\left(\frac{1}{1+b_1}\tilde{M}^{-1}\left[\frac{\partial^2 U_{\theta,1}}{\partial t^2}\right] - \frac{\partial^2 U_{\theta,1}}{\partial z^2}\right) - \right.$$

$$\left. - \frac{4}{r_i^2}U_{\theta,1}\right) + \left(\ln r_i - \frac{1}{2}\right)\frac{r_i^2}{8}\times$$

$$\times\left(\frac{1}{(1+b_1)^2}\tilde{M}^{-2}\left[\frac{\partial^4 U_{\theta,1}}{\partial t^4}\right] - \frac{2}{1+b_1}\tilde{M}^{-1}\left[\frac{\partial^4 U_{\theta,1}}{\partial t^2\partial z^2}\right] - \frac{\partial^4 U_{\theta,1}}{\partial z^4}\right) = 0, \; (i = 1, 2) \tag{49}$$

where $\tilde{M} = 1 - \int_0^t K(t-\tau)d\tau$, $K(t)$ - core of viscoelastic operator dimensionless variables are introduced by formulas

$$z = z^*\xi; \quad bt = t^*\xi; \quad r_1 = r_1^*\xi; \quad r_2 = r_2^*\xi; \quad U_{\theta,1} = U_{\theta,1}^*\xi; \quad U_{\theta,0} = U_{\theta,0}^*,$$

ξ - radius of intermediate surface, $U_{\theta,0}$, $U_{\theta,1}$ - main parts of torsional displacement U_θ, b - velocity of transversal wave propagation in material of shell.

Initial torsional displacement has form

$$\bar{U}_\theta = (1+b_1)U_\theta \quad (b_1 = const).$$

Boundary conditions of the problem: when $z = 0$:

$$U_\theta = -f(t); \quad \sigma_{z\theta} = 0; \quad \sigma_{r\theta} = 0;$$

as $z \to \infty$

$$U_\theta = 0. \tag{50}$$

Initial conditions are zero.

Stressed–strained state of shell is defined by formulas

$$U_\theta = \left(\frac{1}{r} + \frac{r}{2} \ln r \cdot \lambda_1 \right) U_{\theta,1} + r U_\theta;$$

$$\sigma_{z\theta} = (1 + b_1) \mu \tilde{M} \left\{ r \frac{\partial U_{\theta,0}}{\partial z} + \left[\frac{1}{2} + \ln r \cdot \lambda_1 \right] \frac{\partial U_{\theta,1}}{\partial z} \right\}; \tag{51}$$

$$\sigma_{r\theta} = (1 + b_1) \mu \tilde{M} \left\{ \frac{r^2}{4} \lambda_1 U_{\theta,0} + \left[\frac{1}{2} \left(\lambda_1 - \frac{4}{r^2} \right) + \left(\ln r - \frac{1}{2} \right) \frac{r^2}{8} \lambda_1^2 \right] U_{\theta,1} \right\},$$

where

$$\lambda_1 = \frac{1}{1 + b_1} \tilde{M}^{-1} \left[\frac{\partial^2}{\partial t^2} \right] - \frac{\partial^2}{\partial z^2}.$$

The problem is solved using Laplace transformation on time. Step of going from Laplace and Fourier transforms to the original time and space function is realized using shift theorem and theorem of functions' convolution.

$$U_{\theta,1} = \exp^{-c_1 t} \left\{ \left[I_0 \left(\gamma_0 \sqrt{t^2 - \gamma_1^2 z^2} \right) \delta(t - \gamma_1 z) + \frac{\partial}{\partial t} I_0 \left(\gamma_0 \sqrt{t^2 - \gamma_1^2 z^2} \right) \times \right. \right.$$

$$\times H(t - \gamma_1 z) - \frac{c_1}{2} I_0 \left(\gamma_0 \sqrt{t^2 - \gamma_1^2 z^2} \right) H(t - \gamma_1 z) \right] + \frac{1}{\gamma_1} \left(\gamma_{90} + \frac{c_1}{2} \gamma_{91} \right) \times$$

$$\times I_0 \left(\gamma_0 \sqrt{t^2 - \gamma_1^2 z^2} \right) H(t - \gamma_1 z) + \frac{\gamma_{92}}{\gamma_1} \int_{\gamma_1 z}^{t} I_0 \left(\gamma_0 \sqrt{t^2 - \gamma_1^2 z^2} \right) d\tau +$$

$$+ \frac{\gamma_{93}}{\gamma_2} \left[I_0 \left(\gamma_3 \sqrt{t^2 - \gamma_2^2 z^2} \right) \delta(t - \gamma_2 z) \right] \left[\gamma_{100} + \gamma_{84} \frac{\partial^3}{\partial t^3} + \gamma_{98} \frac{\partial^2}{\partial t^2} + \gamma_{99} \frac{\partial}{\partial z} \right] \times$$

$$\times \left(\left(e^{-\frac{c_1}{2} t} \right) \cdot I_0 \left(\gamma_0 \sqrt{t^2 - q_0^2 z^2} \right) + H(t - q_0 z) \right) + \frac{1}{\gamma_1} \times$$

$$+ \frac{\partial}{\partial t} \left(I_0 \left(\gamma_3 \sqrt{t^2 - \gamma_2^2 z^2} \right) \right) H(t - \gamma_2 z) - \frac{c_1}{2} I_0 \left(\gamma_0 \sqrt{t^2 - \gamma_2^2 z^2} \right) H(t - \gamma_2 z) \right] +$$

$$+ \frac{1}{\gamma_2} \left(\gamma_{94} + \frac{c_1}{2} \gamma_{93} \right) I_0 \left(\gamma_3 \sqrt{t^2 - \gamma_2^2 z^2} \right) H(t - \gamma_2 z) +$$

$$+ \frac{\gamma_{95}}{\gamma_2} \int_{\gamma_2 z}^{t} I_0 \left(\gamma_3 \sqrt{t^2 - \gamma_2^2 z^2} \right) d\tau \right\} + \gamma_{96} e^{-k_1 z};$$

$$U_{\theta,0} = -\frac{\gamma_{97}}{q_0} \int\limits_{q_0 z}^{t} \int\limits_{0}^{t-\tau} f(\xi) d\xi \, e^{-c_1 \tau} I_0 \left(\gamma_0 \sqrt{t^2 - \gamma_1^2 z^2} \right) d\tau +$$

$$+ \frac{e^{-k_1 z}}{q_0^2} \left\{ -q_{11} \left(\frac{\partial^2}{\partial t^2} + c_1 \frac{\partial}{\partial t} + \frac{c_1^2}{4} \right) + \gamma_{82} \right\} +$$

$$+ e^{-c_1^* t} \left(\frac{1}{q_0^*} \right) \gamma_{100} I_0 \left(\gamma_0 \sqrt{t^2 - q_0^2 z^2} \right) H(t - q_0 z) + \gamma_{88} \int\limits_{q_0 z}^{t} I_0 \left(\gamma_0 \sqrt{t^2 - q_0^2 z^2} \right) d\tau +$$

$$+ \frac{1}{\gamma_1} \left[\gamma_{102} I_0 \left(\gamma_0 \sqrt{t^2 - \gamma_1^2 z^2} \right) H(t - \gamma_1 z) + \gamma_{77} \gamma_{74} \int\limits_{\gamma_1 z}^{t} I_0 \left(\gamma_0 \sqrt{t^2 - \gamma_1^2 z^2} \right) d\tau \right] +$$

$$+ \frac{1}{\gamma_2} \left[\gamma_{104} I_0 \left(\gamma_3 \sqrt{t^2 - \gamma_2^2 z^2} \right) H(t - \gamma_2 z) + \gamma_{68} \gamma_{71} \int\limits_{\gamma_2 z}^{t} I_0 \left(\gamma_0 \sqrt{t^2 - \gamma_2^2 z^2} \right) d\tau \right] \right\}.$$

$$(52)$$

Substituting last expressions into formulas (51), stress tensor component $\sigma_{r\theta}$ and displacement U_θ as function of coordinate and time are calculated for viscoelastic material parameters of which was given above. External loading again is given as smooth function

$$f(t) = g_0 \, \sin^2 \left[\pi t / t_1 \right].$$

The results are presented on Figs. 7–12. On Figs. 7 and 8 graphs of stress change in time for fixed section $z = 5$ and different values of parameter b_1 $(0; 0, 05; 0, 1; 0, 15; 0, 2)$ are presented, herewith case $b_1 = 0$ is corresponded to case without initial displacements. Here we can see the influence of factor of initial displacement on change of stress which is essential at initial moments

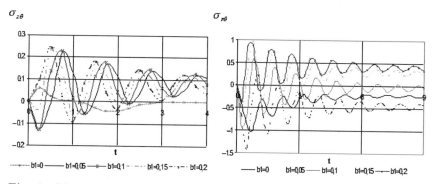

Fig. 7. Plot of $\sigma_{z\theta}$ stress change in time for $z = 5$

Fig. 8. Plot of stress $\sigma_{r\theta}$ as function of time for $z = 5$

of interaction. By these graphs it is possible to draw a conclusion, that in the beginning of interaction the sharp increase in stress values is observed, and eventually oscillatory process is stabilized and passes in the established condition.

On Figs. 9 and 10 graphs of displacement and stresses are presented as function of coordinate, at the fixed values of time t and factor b_1. On Figs. 9 and 10 accordingly graphs of stresses $\sigma_{r\theta}$ and are presented at $t = 1$ and various values of factor b_1. From graphs it is visible, that with growth of distance from an end face the stress amplitude falls and on distance approximately 6–7 radiuses from an end face it can be neglected, and this is observed at any value of factor b_1.

On Figs. 11 and 12 graphs of displacement change and stress in view of ($b_1 = 0, 2$-dashed line) and without taking into account ($b_1 = 0$-continuous line) factor of initial displacement are presented for the various fixed values of time $t = 1, 5, 10, 15$. Here eventually reduction of values of displacement and stress is observed. Herewith graphs of torsional displacement presented on

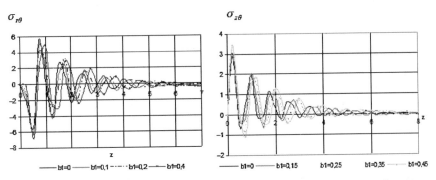

Fig. 9. Plot of $\sigma_{r\theta}$ stress change in co-ordinate for $t = 1$

Fig. 10. Plot of stress $\sigma_{z\theta}$ as function of coordinate z for $t = 1$

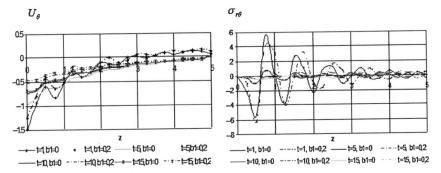

Fig. 11. Plot of displacement U_θ as function of z for fixed values of time $t = 1, 5, 10, 15$

Fig. 12. Plot of $\sigma_{r\theta}$ stress as function of coordinate for $t = 1, 5, 10, 15$

Fig. 11 at $t = 1$ have harmonious character, and in other cases more smooth form.

On Fig. 12 in sections close to the face the sharp increase in values of stress $\sigma_{r\theta}$ is observed, which in the process of removal from an end face gradually decreases. From the presented graphs it is visible, that in the shell sections equal to 5–6 radiuses the values of displacement and stresses might be neglected.

References

1. AMIRKULOVA F.A.: The influence of pre-stress on axisymmetrical vibrations of cylindrical layer. Proceedings of International Conference on Computational Solid Mechanics, Moscow, Russia, 2006, pp. 425–428.

2. AMIRKULOVA F.A.: Torsional waves in initially stressed thick cylindrical shell wish non-zero initial conditions. Proceedings of the 8th International Conference on Shell Structures: Theory and Applications, Gdansk-Jurata, Poland, 2005, Balkema, London, pp. 285–289.

3. AMIRKULOVA F.A.: Torsional waves in viscoelastic cylindrical layer with initial displacements. Problems of Mechanics, Tashkent, 2005, No. 6, pp. 6–11.

4. AMIRKULOVA F.A.: Stress-strain state analysis of cylindrical layer with initial displacement. Proceedings of International conference on problems of mechanics and seismic dynamics of structures, Tashkent, Uzbekistan, 2004, pp. 23–26.

5. AMIRKULOVA F.A., UMIROV A.: Torsion vibrations of pre-stressed viscoelastic circular cylindrical layer. Proceedings of the conference on Actual problem of modern science and technology, Jizzax, Uzbekistan, 2004, pp. 37–41.

6. AMIRKULOVA F.A.: Refined vibrations equations of anisotropic pre-stressed cylindrical layer. Absracts of the International conference on Differential equations, Almaty, 24–26 September, 2003, pp. 76–77.

7. AMIRKULOVA F.A.: The refined equation of longitudinal-radial vibrations of prestressed viscoelastic cylindrical layer. Uzbek Journal Problems of Mechanics, Tashkent, 2002, Vol. 6, pp. 27–32.

8. AMIRKULOVA F.A.: Nonstationary axisymmetric vibration of transversal-isotropic circular cylindrical layer with initial stress. Reports of Academy Sciences R. Uzbekistan, Tashkent, 2002, Vol. 5, pp. 16–21.

9. AMIRKULOVA F.A., JAVLIYEV B.K.: The torsional vibrations of prestressed viscoelastic transversal-isotropic circular cylindrical layer. Uzbek Journal Problems of Mechanics, Tashkent, 2002, Vol. 6, pp. 18–22.

10. BABICH S.YU., GUZ A.N.: Complex potentials in a plane dynamic problem for initially stressed compressible elastic bodies. Prikladnaya Mekhanika. 1981. 17(7): 75–83.

11. BANGO A.M., GUZ A.N.: Elastic waves in initially stressed bodies interacting with liquid (review). Prikladnaya Mekhanika. 1997. 33(6): 3–39.

12. DEMIRAY H., SUHUBI E.S.: Small torsional oscillations in initially twisted circular subbed cylinder. Int. J. Eng. Sci. 1970. 8(1): 13.

13. GRIN A.E.: Torsional vibrations of prestressed circular cylinder. In "Problems of continuum mechanics". Moscow. 1953.

14. GUZ A.N.: Elastic waves in bodies with initial (residual) stresses. International Applied Mechanics. 2002. 38(1): 23–59.

15. GUZ A.N.: Fundamentals of the three dimentional theory of stability of deformable bodies. Berlin: Springer, 1999.
16. GUZ A.N.: Propogation laws, Vol. 2 of the two-volume series elastic waves in initially stressed bodies. Kiev: Naukova Dumka, 1986.
17. KHUDOYNAZAROV KH.: Nonstationary interaction of cylindrical shells and bars with deformable medium. Tashkent, 2003.
18. RAMAKANTH J.: Longitudinal vibrations of pre-stressed circular cylinder. Bull. Acad. Pol. Sci. Ser. Tech. 1964. 12(11): 495–505.
19. SUHUBI E.S.: Small longitudinal vibration of an initially stressed circular cylinder. Int. J. Eng. Sci. 1965. 2(5): 509–515.
20. SYTENOK N.A.: Torsional waves in laminated composites with initial stresses. Prikl. Mekh. 1984. 20(8): 100–103. 21.
21. TRUSDELL K.: Initial course of rational continuum mechanics. Moscow: Mir, 1975.
22. WAGH D.K.: Torsional waves in an elastic cylinder with Cauchy's initial stress. Gerlands Beitr. Geophys. 1972. 81(6): 489–493.

Viscoelastic Fluids in a Thin Domain: A Mathematical Study for a Non-Newtonian Lubrication Problem

G. Bayada, L. Chupin, and S. Martin

Summary. After describing the process allowing obtaining classical Reynolds equation describing thin flow model for Newtonian fluids, we give a short description of the present state of art in the modelling of various non Newtonian thin flows. At last, we give some recent results concerning visco-elastic flows for which it is not possible to gain a generalized Reynolds equation. At the contrary, the thin flow corresponding model relies primary on the computation of the velocity field.

1 Mathematical Aspect of Lubrication Problems

In many fields of engineering and applied sciences, different technologies reduce friction and wear between relative moving surfaces. Lubrication is part of a larger discipline, the tribology, which included also the study of friction and wear [1]. Lubrication is mainly concerned with the presence of a lubricant used to prevent contact between close surfaces in relative motion [2].

In some sense it is primary a fluid mechanics problem whose originality deals with the thin gap in which the fluid stays. Moreover, contrary to the classical problems in fluid mechanics, the velocity is not the primary unknown. People in the field of lubrication are mainly interested by the knowledge of the pressure which, due the thin gap, can reach more than one Giga Pascal.

Let us consider the classical Navier–Stokes system defined in $\Omega = \omega \times (\mathrm{O},\ \mathrm{H}(x_1, x_2))$ in which ω is a smooth domain of the (x_1, x_2) plane:

$$\rho\, \mathbf{U}.\,\mathrm{Grad}(\mathbf{U}) = -\mathrm{grad}(p) + \eta\,\Delta\mathbf{U} + \mathbf{F},$$
$$\mathrm{div}(\mathbf{U}) = 0. \qquad (1)$$

Assuming that H is small with respect of the dimensions of ω, so that the pressure can be considered as a function of (x_1, x_2) only and, after suitable dimensionless procedure, retaining leading terms in the preceding system, we get:

$$\eta\frac{\partial^2 u_i}{\partial x_3} = \frac{\partial p}{\partial x_i} \quad i = 1, 2 \qquad \frac{\partial p}{\partial x_3} = 0.$$

It is now possible to integrate these equations in the x_3-direction, using velocity boundary conditions on the lover and upper part of Ω, namely:

$$\mathbf{U}(x_1, x_2, \mathrm{H}(x_1, x_2)) = 0, \quad \mathbf{U}(x_1, x_2, 0) = (s, 0, 0),$$

so obtaining the velocity as a function of the pressure.

Putting this last expression in the divergence free equation and integrating it also in the x_3 direction allows us to eliminate the velocity and to obtain an equation in pressure only: the so-called Reynolds equation which dates back to 1886.

$$\frac{\partial}{\partial x_1}\left(H^3 \frac{1}{12\,n}\frac{\partial p}{\partial x_1}\right) + \frac{\partial}{\partial x_2}\left(H^3 \frac{1}{12\,n}\frac{\partial p}{\partial x_2}\right) = s/2 \frac{\partial H}{\partial x_1}. \tag{2}$$

Such procedure can be rigorously described by way of asymptotic expansion with respect of the small parameter $\varepsilon = H/L$ in which L is the characteristic length of ω ([3], [4]).

In some sense Reynolds equation is for the Navier–Stokes system similar to the plate equation for the full 3-D elasticity system.

Such Reynolds equation is a classical elliptic one. However in most of the lubricated devices, a biphasic phenomena occurs: the cavitation (the occurrence of a bubbles of air into the fluid) and a free boundary formulation has to be introduced.

The classical one is a variational inequality (VI) of the first kind which takes full account of the fact that pressure into the fluid cannot fall below the saturation pressure (taken as zero for convenience):

Let $K = (\varphi, \varphi \in H^1(\omega), \varphi \geq 0)$, then p is searched as the solution in K of the VI:

$$\iint_\omega H^3 \frac{1}{12\,n}\left(\frac{\partial p}{\partial x_1}\frac{\partial(\varphi - p)}{\partial x_1} + \frac{\partial p}{\partial x_2}\frac{\partial(\varphi - p)}{\partial x_2}\right) dx_1 dx_2$$

$$\geq -\iint_\omega s/2 \frac{\partial H}{\partial x_1}(\varphi - p)dx_1 dx_2 \qquad \forall \varphi, \varphi \in K.$$

Numerous studies exist about existence, uniqueness and various properties of solution of this inequality [5].

As this VI model is not a conservative one, another more complex model has been proposed in the literature: the pressure saturation model. In this approach, the pressure is no longer the only unknown and a saturation function θ is introduced. The problems read now as:

Find p and θ such that: $p \geq 0$, $0 \leq \theta \leq 1$, $p(1 - \theta) = 0$

$$\iint_\omega H^3 \frac{1}{12\,n}\left(\frac{\partial p}{\partial x_1}\frac{\partial\varphi}{\partial x_1} + \frac{\partial p}{\partial x_2}\frac{\partial\varphi}{\partial x_2}\right) dx_1 dx_2$$

$$= \iint_\omega s/2 \frac{\partial\varphi}{\partial x_1} H\theta \, dx_1 dx_2 \qquad \forall \varphi, \varphi \in H^1(\omega). \tag{3}$$

Mathematical results about this elliptic-hyperbolic free boundary problem can be found in ([6, 7]).

Remark. Clearly, the procedure is not a rigorous one as if the way to passed from (1) to (2) can be justified, there does not exist so far a rigorous way to pass to a 3-D diphasic model, to be defined, to (3). This is still an open question.

2 Thin Film Non-Newtonian Fluid

In the preceding section, the fluid has been considered as a Newtonian one. However it is well known that numerous biological fluids, blood or physiological secretions like tears or synovial fluids present non-Newtonian characteristic. In engineering applications, people are interested to control the characteristic of the flows in order to suit various requirements such as maintaining its qualities in a wide range of temperature and stresses. Commercial lubricants are then modified with different additives to be able to protect engines both in winter and in summer with the same product. This addition leads also a non-Newtonian behaviour of the actual lubricant: the usual assumption of a linear relation ship of shear stress and shear strain may failed due to the additives contained in the lubricant and to the very severe operating conditions.

Moreover, it is to be noticed that, in most of the practical applications, the geometry of the flow to be considered is anisotropic. This is the case in lubrication studies which are mainly devoted to thin film flows, in the study of the spreading of tears or in the description of polymers through thin dies.

The idea is then to try to obtain generalized Reynolds equations describing non-Newtonian effects and easier to manage than the full 3-D non-Newtonian Navier–Stokes system.

Near all models for non-Newtonian flows are associated to a thin film asymptotic lubrication model in a somewhat heuristic way. These models are often discussed in the engineering literature. The main reason is the non linearity of the basic non-Newtonian 3-D equation which induces some difficulty by passing to the limit making ε tends to zero. Moreover, these non-Newtonian 3-D models are most often depending of some rheological parameters whose dependence with respect to ε is almost always an important feature in the formal or rigorous asymptotic process to get a Reynolds type equation.

Micropolar fluids are perhaps the closest to Newtonian one by its mathematical aspect. They are used to model flow in which there are a lot of small rigid particles. A new field (the micro rotation) together with a supplementary equation is added to the Stokes system [8]. Additional parameter, the characteristic length of the micro-rotation effects is introduced and has to be compared with the gap, so leading to define a critical ratio involving these two parameters. One of the first studies concerning micropolar lubrication appears in [9] in which a modified Reynolds equation is introduced. The related mathematical proof has been given in [10]. From practical use, the difficulty lies in the knowledge of the boundary condition to be satisfied by the micro rotation at the fluid-solid interface. Assuming homogeneous boundary conditions is of common use. However a more physical situation assuming a "boundary viscosity effect" condition can be introduce [11] and leads naturally to a slip conditions on this interface [12]. Various situations have been considered in [13].

Another class of problems for which the asymptotic procedures are mathematically well described included quasi Newtonian fluids (Carreau's law, power law, Willamson's law, in which various stresses velocity relations are chosen) ([14]–[16]). For the Carreau law, for example the asymptotic law strongly depends on the values of the Reynolds number Re written as $R_e = \varepsilon^\gamma$ and of the coefficients r in the power law. In some cases a Reynolds–Carreau equation can be obtained while for other cases it is not possible to decoupled velocity and pressure equations. In [17] a general approach is described with zero stress boundary conditions on the lateral boundaries which induce additional difficulties. Neither the less it has been possible to find an equation for the pressure only. This is not the case for Bingham fluid [18] in which the limit inequality appears in term of both pressure and velocity. Near all these papers however are devoted to problems with zero velocity at the boundary. They are not representative of lubrication boundaries conditions which are characterized by relative velocity between upper and lower surfaces of the devices. So, application of these results in lubrication need to revisit the proof especially in terms of dependence of the rheological coefficients with respect to ε.

The second order flow, including several rate of strain terms in the constitutive equation has also been considered for lubrication in several mechanical papers ([19], [20]) and corresponding references. Due to the non linearity various assumptions are made to gain a limit problem and resulting equations are not the same, although numerical results are close one together.

3 Thin Film Viscoelastic Fluids

The case of visco elastic fluid is less clear from the mathematical aspect. Introducing viscoelasticity behaviour is primary described by the Deborah number De which is associated to the relaxation time. One of the family of laws used to describe this phenomena is the Olroyd-B model based upon a constitutive equation which is an interpolation between purely elastic and purely viscous behaviour, thus introducing a supplementary parameter r describing the relative proportion of both behaviours (solvent to solute ratio) [21].

$$\rho\, \mathbf{U}.\operatorname{Grad}(\mathbf{U}) - \eta(1-r)\Delta\mathbf{U} + \operatorname{grad}(p) - \operatorname{div}(\boldsymbol{\sigma})\ = \mathbf{F},$$
$$\operatorname{div}(\mathbf{U}) = 0,$$
$$\lambda\left(\mathbf{U}.\operatorname{Grad}(\boldsymbol{\sigma}) + \mathbf{U}.\nabla\boldsymbol{\sigma} + g_a(\nabla\mathbf{U}, \boldsymbol{\sigma})\right) + \mathbf{f}(\boldsymbol{\sigma})\boldsymbol{\sigma} = +2\eta\, r\mathbf{D}(\mathbf{U}).$$

The parameter λ is proportional to $1/\text{De}$ and the bilinear application g_a is defined up to $\mathbf{D(U)}$ and $\mathbf{W(u)}$, the symmetric and skew symmetric parts of the velocity gradient by

$$g_a(\nabla U, \sigma) = \sigma.W(U) - W(U).\sigma - a(\sigma.D(U) + D(U).\sigma) \qquad 0 \leq a \leq 1.$$

As a particular case, this model retains the generalized Maxwell model ($r = 1$, $f = $ Identity) and Phan-Thein Tanner model for example. From mathematical aspects, few results exist concerning these models and the way how to obtain thin film approximation is highly heuristic.

Using Maxwell equation as starting point, a way has been to take the ε parameter as a leading small parameter and then to use the Deborah number as a perturbation parameter [22]. Another approach has been to choose the D_e of the same order of magnitude than ε [23]. This assumption allows balancing the order of Newtonian and non-Newtonian contribution (see also [24] for mechanical comments). Boundary conditions are chosen in order to be applied to usual lubrication problems. In all these cases a Reynolds-type equation is gained.

With the same assumption and starting with the full Olroy-B model leads to a limit problem which has both the velocity field and the pressure as unknowns. Let us describe briefly the procedure:

After scaling both equations and stress tensor in an adequate way, an asymptotic 2D problem is obtained [25] from the OLROYD-B system:

$$-(1-r)\frac{\partial^2 u_i}{\partial x_3} - r\frac{\partial}{\partial x_3}\left(B\left(\frac{\partial u_i}{\partial x_3}\right)\right) = -\frac{\partial p}{\partial x_i} + F_i \qquad i = 1, 2. \qquad (4)$$

$$\text{div}(\mathbf{U}) = 0$$

In which the non linear operator B is defined by:

$$B(t) = \frac{t}{1 + D_e^2(1 - a^2)\, t^2}.$$

This result generalizes the work of ([23], [24]), concerning not only the rheological model but also the dimension (2D instead of 1D for the pressure asymptotic problem). Obtaining the asymptotic problem is partly the result of an heuristic process, so that the solvability of this problem has to be rigorously proved.

As already mentioned, one feature of this system is that it is not possible to eliminate the velocity so obtaining an equation with respect to the pressure. The basic idea is to make exactly the contrary. The pressure has to be first eliminated by working in a weak sense in a divergence free space. Then a non linear equation with unknowns (u_1, u_2) is posed in that kind of space which is defined by:

Let $KI = (\varphi, \; \varphi \in (L^2(\Omega))^2, \; \frac{\partial \varphi}{\partial x_3}(s,0) \in (L^2(\Omega))^2, \; \varphi(x_1, x_2, H(x_1, x_2)) = 0,$

$$\varphi(x_1, x_2, 0) = (s, 0)$$

such that $\quad \forall \theta, \theta \in D(\overline{\omega}), \; \iint\limits_{\omega} (\nabla_x \theta . \int\limits_0^{H(x)} \varphi(x_3) dx_3) dx_1 dx_2 = 0.$

So that the problem becomes:

Find **U** in KI such that for any φ in KI, we have:

$$(1-r) \iiint\limits_\Omega \frac{\partial(\mathbf{U})}{\partial x_3} \frac{\partial(\varphi - \mathbf{U})}{\partial x_3} dX + r \iiint\limits_\Omega \mathbf{B}(\frac{\partial \mathbf{U}}{\partial x_3}) \frac{\partial(\varphi - \mathbf{U})}{\partial x_3} dX$$

$$\geq \iiint\limits_\Omega \mathbf{F}.(\varphi - \mathbf{U}) dX. \tag{5}$$

It is easy to prove that the left hand side of this inequality is a bounded monotone and coercive operator for $0 < r < 8/9$. Following [25], this gives the existence and uniqueness of the solution of this inequality. The pressure as a function of $(x_1, \; x_2)$ is easily recovered using De Rham theorem.

Interestingly this range of parameter for which the problem can be solved is exactly the same for which the 3-D Olroyld initial problem has a solution.

A new algorithm related to the Uzawa one is presented and convergence theorems can be found in [26].

Remark: Taking diphasic phenomena as the cavitation in a visco elastic fluid is not clear at all. Even if the constraint can be written on term of pressure as for a Newtonian fluid, it is not possible to include it in (5) which relies only on the velocity.

References

1. P. Jost, *The tasks of tribology societies in a changing world*, Proceedings of Second World Tribology Congress, Viena, 2001.
2. J. Frene, D. Nicolas, B. Degueurce, D. Berthe, M. Godet, *Hydrodynamic lubrication: bearingsand thrust bearings*, Elsevier Science, 1997.
3. G. Bayada, M. Chambat, *The transition between the Stokes equation and the Reynolds equation: a mathematical proof*, Appl. Math. Optimization, 14, 1986, 73, 94.
4. S.A. Nazarov, *Asymptotic solution of Navier–Stokes problem on the flow of a thin layer of fluid.* Sibirsk Math. Zh, 31, 1986, 131,144.
5. G. Cimatti, On *a problem of the theory of lubrication governed by a variational inequality*, Appl. Math. Optimization, 3, 1977, 227, 242.
6. G. Bayada, M. Chambat *Sur quelques modélisations de la zone de cavitation en lubrification hydrodynamique*, J. of Theor. Appl. Mech., 5, 1986, 703–729.

7. S. Alvarez, J. Carillo, *A free boundary problem in the theory of lubrication*, Com. Part. Diff. Equat., 19, 1994, 1743, 1761.
8. A.C. Eringen, *Theory for micropolar fluid*, J. Math. Mech., 16, 1966, 1–16.
9. J. Prakash, P. Sinha, *Lubrication theory for micropolar fluid and its application to a journal bearing*, Int. J. Eng. Sci., 13, 1975, 217–232.
10. G. Bayada, G. Lukaswecizw, *On micropolar fluid in the theory of lubrication*, Int. J. Eng. Sci., 34, 1996, 1477–1490.
11. N.M. Bessonov, *A new generalization of the Reynolds condition for a micropolar fluid and its application*, Trib. Int., 27, 1994 105–108.
12. G. Bayada, N. Benhaboucha, M. Chambat, *New models in micropolar fluid and their application to lubrication*, Math. Mod. Methods Appl. Sciences, 15, 2005, 343–374.
13. G. Bayada, M. Chambat, R. Gamouana, *Micropolar effects in the coupling of a thin film past a porous medium*, Asympt. Anal., 30, 2002, 187–216.
14. A. Bourgeat, A. Mikelic, R. Tapiero, *Dérivation des équations moyennées décrivant un écoulement non newtonien dans un domaine de faible épaisseur*, C.R. Acad. Sci., Paris, I, 316, 1993, 965–970.
15. F. Boughanin, R. Tapiero, *Derivation of the two-dimensional Carreau law for a quasi-newtonian fluid flow through a thin slab*, Appl. Anal. 57, 1995, 243–269.
16. R. Bunoiu, J. Saint Jean Paulin, *Fluide à viscosité non linéaire dans un domaine de faible épaisseur dans le cas de lubrification.* C. R. Acad. Sci. Paris I. 323, 1996, 1097–1102.
17. J.M. Sac Epee, K. Taous, *On a wide class of non linear models for non Newtonian fluids with mixed boundary conditions in thin film domain*, Asympt. Anal., 44, 2005, 151–171.
18. R. Bunoiu, S. Kesavan, *Asymptotic behaviour of a Bingham fluid in thin layer*, J. Math. Anal. Appl., 293, 2004, 405–418.
19. W.G. Sawyer, J.A. Tichy, *Non Newtonian lubrication with the second order fluid*, J. Tribology, 120, 1998, 622–628.
20. P. Huang Zhi Heng Li, Y.G Meng, S.Z.W. Wen, *Study of thin film lubrication with second order flow*, ASME J. of Tribology 2002, 547–552.
21. G. Guillopé, J.C. Saut, *Existence results for the flow of visco elastic fluids with a differential constitutive law*, Non linear analysis, 15, 1990, 9, 849–869.
22. J. Tichy, *Non Newtonian lubricatin with the convected Maxwell model*, ASME J. Tribology, 118, 1996, 344–349.
23. R. Zhang, X. Kai Li, *Non Newtonian effects on lubricant film flow*, J. Eng. Math., 51, 2005, 1–13.
24. F. Talay Akyildiz, H. Bellout, *Viscoelastic lubrication with Phan-Thein-Tanner fluid* (P.T.T), ASME J. Tribology, 126, 2004, 288–291.
25. J.L. Lions, *Quelques méthodes de résolution de problèmes aux limites non linéaires*, Dunod Ed., 1969.
26. G. Bayada, L. Chupin, S. Martin, *Viscoelastic fluids in thin domain*, Quart. J. Appl. Math, To appear 2007. 20.

From the African *sona* Tradition to New Types of Designs and Matrices

Paulus Gerdes

Introduction: Mathematics in African History

From the earliest times onwards humans in Africa have created and developed mathematical ideas. For an annotated bibliography of mathematics in African history and cultures, see Gerdes and Djebbar (2004, 2007).[1]

Among the earliest "mathematical artifacts" known worldwide, several are from Africa. A small piece of the fibula of a baboon, marked with 29 notches, was found in a cave in the Lebombo mountains between South Africa and Swaziland. The bone has been dated to approximately 35,000 BC. Well known and widely discussed is another bone found at Ishango (Congo), dated at 20,000 BC.

One of the oldest mathematical texts from Ancient Egypt is a collection of problems composed by the scribe Ahmose (ca. 1,650 BC), probably copied from a text about 200 years older. It contains various methods of approximate solution. For instance, the length of the side of the square that has approximately the same area as that of the circle is determined as 8/9 of the diameter of the circle, which implies 3.1605 as a close approximation for the value of π. The "pinnacle of achievement" of mathematics in Ancient Egypt is the exact result for the volume of a truncated pyramid with square base. In his paper "Africa, the cradle of world mathematics?" (1985), presented in Nairobi for the University of the United Nations, Henri Hogbe-Nlend, the first president (1976–1986) of the African Mathematical Union, stated that mathematics in Pharaonic Africa was intuitive, demonstrative and rational. Furthermore, that Africa is the mother of geometry. In his book "Egyptian Geometry: Contribution of Ancient Africa to World Mathematics", the Congolese linguist and Egyptologist Théophile Obenga underlines that the title

---

[1] This bibliography is the outcome of work done by AMUCHMA, the Commission on the History of Mathematics in Africa, created in 1986 by the African Mathematical Union (AMUCHMA webpage: http://www.math.buffalo.edu/mad/AMU/amuchma_online.html).

of the Ahmose papyrus presents the oldest known description or definition of what mathematics is about: "Correct method of investigation of nature in order to understand all that exists, each mystery, all secrets" (Obenga, 1995, p. 290).

During the Hellenistic period and its aftermath, famous mathematicians worked in Alexandria, like the geometers Euclid (ca. 365–300 BC), Heron (ca. 100 AD), Claudius Ptolemeus (second century AD), and the number theorist Diophantus (third century AD). Hypathia of Alexandria (ca. 370–415 AD) is the first female mathematician known in history. In the same period, several other mathematicians are known from the Maghreb, like Theodorus (ca. 465–398 BC), Eratosthenes (ca. 276–194 BC), and Nicotelese (ca. 250 BC), all of Cyrene, Theodoses (second century BC) of Tripoli and Apuleius of Madaura (ca. 124–170 AD).

North Africa played an important role in the genesis of algebra in Islamic culture (for an overview, see Djebbar 2005). North African mathematicians from Egypt to the Maghreb made their contributions, like Abu Kamil (d. 930), Abu Bakr al-Hassar (twelfth century), Samaw'al (d. 1175), Ibn al-Yasamin (d. 1204), Ibn Rashiq (c. 1275), Ibn al-Banna (1256–1321), Uqbani (1320–1408), Ibn Qunfudh (1339–1407), Ibn al-Ha'im (1352–1412), Ibn Haydur (d. 1413), Ibn al-Majdi (1365–1447), Qatrawani (fifteenth century), Sibt al-Maradini (1423–1506), Ibn Ghazi (1437–1513). Important mathematicians born outside Africa worked for many years in North Africa, like Ibn al-Haytham (965–1041), Al-Qurashi (d. 1184), and Al-Qalasadi (1412–1485). Ibn Mun^c im (d. 1228) of Andalusian origin settled in Marrakech where he laid the foundations of combinatorial analysis, including a presentation of the so-called *triangle of Pascal*, more than four centuries before Blaise Pascal (1623–1662). Several mathematical notations used today had been conceived in the Maghreb. In recent years many mathematical manuscripts from the medieval Maghreb have been discovered, analysed and edited, underscoring the mathematical heritage of North Africa.

Mathematical ideas from Ancient Egypt to Islamic Egypt and from the Maghreb during the 'Middle Ages' found their way to Europe and have contributed substantially to the international development of mathematics (Djebbar 2001, 2005). Leonardo de Pisa (Fibonacci) had studied in Algeria before he wrote his famous *Liber Abaci* (1202).

Hundreds of mathematical manuscripts – written in Arabic and in various African languages – from Timbuktu in today's Mali remain to be analysed to lift the veil from some of the mathematical connections between Africa South of the Sahara and the North of the continent. Only one manuscript from Timbuktu, written by al-Arwani (probably sixteenth century) has been partially analysed so far. The astronomer-mathematician Muhammed ibn Muhammed al Katsinawi (ca. 1740) from Katsina in today's Nigeria was well known in Egypt and the Middle East.

Thomas Fuller (1710–1790), brought from West Africa as a slave to North America in 1724, became famous in the "New World" for his mental

 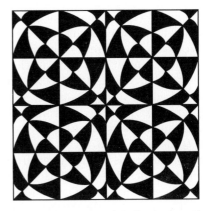

Fig. 1. Examples of one-colour and two-colour *litema* patterns (cf. Gerdes, 1998)

calculations (see Fauvel and Gerdes, 1990). As an old man in 1788, he was able to calculate the answer to questions like $7 \times 7 \times 7 \times 7 \times 7 \times 7 \times 7 = ?$ and "How many seconds lived a man 70 years, 17 days and 12 hours old" in an instant. "Discovered" by the anti-slavery militants, he turned into one of their important examples that the "black man is not mentally inferior to the white man" (cf. Fauvel and Gerdes, 1990). His example shows also that there were strong traditions of mental calculation in the African region from which he originated, probably the Gulf of Benin.

Our Cameroonian colleague Edward Njock stressed in 1985, that geometry is omnipresent in African culture: "Pure mathematics is the art of creating and imagining. In this sense black art is mathematics." As an illustration the *litema* wall decoration by women in Lesotho may serve (cf. Gerdes, 1998, 87–170). Figure 1 presents examples of one-colour and two-colour *litema* patterns.

For an overview of geometrical and other mathematical ideas from Africa south of the Sahara, see Zaslavsky's classic study (1973, 1999), Eglash (1999), and Gerdes (1998, 1999a). Geometrical ideas in basket and mat weaving are analysed in several studies by Gerdes and Bulafo (1994) and Gerdes (2000a,b, 2003a,b, 2004).

Due to the slave trade, wars of conquest and colonial occupation, many African scientific and technological traditions entered into decline. Substantial knowledge was lost.

With the end of the colonial period and the conquest of national independence a period of mathematical revival started on the African continent (cf. Sica 2005). Only a few African mathematicians had received a doctorate before independence, like the Egyptians Ali Mostafa Mosharafa (Ph.D. 1923) and Mohamed Mursy-Ahmed (Ph.D. 1931) (Independence Egypt 1937), the Nigerians Chike Obi (1950), Adegoke Olubummo (1955) and James Ezeilo (1959) (Independence Nigeria 1960), and the Sierra Leonen Awadagin

Williams (1958) (Independence Sierra Leone 1961). Since Independence, African mathematicians were awarded doctoral degrees by universities in (at least) 53 different countries, mostly situated in three continents: Africa (44%), Europe (36%) and America (20%). During the second half of the twentieth century more than 3,000 Africans earned a doctorate in mathematics. Hundreds have been working as researchers in Europe and North America (Gerdes 2007a). The percentage of female doctorate holders is still only 11%. African mathematicians have organized themselves in national and regional associations. In 1976 the African Mathematical Union was created. African mathematicians are doing research in various fields of pure and applied mathematics, including applications to urgent problems the continent is facing, like desertification, malaria and HIV/AIDS.

One aspect of the African mathematical renaissance is the reconstruction of partially lost knowledge and the exploration of its potential both in mathematics education and research. The case of *sona* geometry seems to me a particular powerful example in this respect, as will be shown.

Sona Geometry

In southern central Africa the *sona* drawing and story telling tradition incorporated various geometrical ideas (cf. Gerdes 1995, 1997a, 2006a). This tradition of drawings in the sand called *sona* (sing. *lusona*) was developed among the Cokwe of northeast Angola and related peoples. Each boy learnt the meaning and execution of the easier *sona* during the initiation rites. Drawing specialists, called *akwa kuta sona*, transmitted the more complicated *sona* to their male descendants. These drawing experts were at the same time the story tellers who used the *sona* as illustrations, referring to proverbs, fables, games, riddles, animals. Figure 2 presents a *lusona* and tells the corresponding story of 'The hunter and the dog.'

The *sona* drawings were executed in the following way: After cleaning and smoothing the ground, the drawing experts first set out with their fingertips a net of equidistant points and then they draw a line figure that embraces the points of the network. The experts execute the drawings swiftly. Once drawn, the designs are generally immediately wiped out.

Slave trade, colonial penetration and occupation provoked a cultural decline and the loss of a great deal of knowledge about *sona*. On the basis of an analysis of *sona* reported by missionaries, colonial administrators and ethnographers, it was possible to reconstruct some mathematical elements in the *sona* tradition.

As the examples in Fig. 3 suggest, symmetry and monolinearity played an important role as cultural values: most Cokwe *sona* are symmetrical or monolinear. Monolinear means composed of only one (smooth) line; a part of the line may cross another part of the line, but never a part of the line may touch another part.

The Hunter and the Dog

An old storyteller said that a certain hunter, named Tshipinda, went on a hunt, taking the dog Kawa, and caught a wild goat. Upon returning to the village, the hunter divided the meat with Calala, the owner of the dog. Kawa was left with the bones.

After some time, Tshipinda again asked for the services of the dog, but the latter refused to help him. He told the hunter to take Calala since it was with him that he was accustomed to dividing the meat.

Fig. 2. In this drawing, the isolated point in the centre represents the hunter, and the isolated point on the left side, the dog.

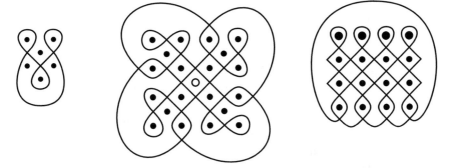

Fig. 3. Examples of symmetrical, monolinear *sona* (cf. Gerdes, 2006a–c, p. 70, 43, 87)

The drawing experts developed a whole series of geometric algorithms for the construction of monolinear, symmetrical designs. Figure 4 displays two monolinear *sona* belonging to the same class in the sense that, although the dimensions of the underlying grids are different, both *sona* are drawn applying the same geometric algorithm.

The drawing experts also invented various rules for the building up monolinear *sona*. The following presents a first example. Figure 5 shows three monolinear *sona*. They are similar to each other: each presents a basic design of triangular form. Figure 6 presents another example: each hand draws simultaneously one half!

The drawing experts who invented these *sona* probably began with triangular patterns and transformed them into monolinear patterns with the help of one or more loops (see the example in Fig. 7). The monolinear patterns

Fig. 4. Two *sona* drawn with the same geometric algorithm (cf. Gerdes, 2006a–c, p. 71, 72)

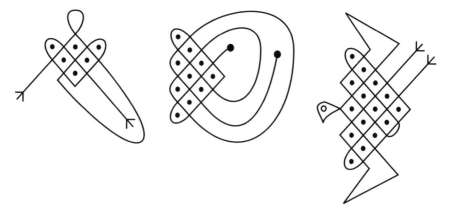

Fig. 5. (a) An eagle carrying a chicken **(b)** A person dead to fatigue **(c)** A *thimunga* bird in flight (cf. Gerdes, 2006a–c, p. 115, 116)

so obtained were adapted topologically (maybe later by other drawing specialists) so that they could express the ideas the drawers wanted to transmit through them.

Sona experts also discovered various rules for chaining monolinear *sona* to form bigger monolinear *sona*. Figure 8 displays an example of the use of a chain rule: it indicates how the appearance of the monolinear drawing in Fig. 8c may be explained on the basis of the monolinearity of the two patterns in Fig. 8a and the way they have been chained together (see Fig. 8b).

When analysing and reconstructing mathematical elements of the *sona* tradition, the author found that there are reported *sona*, which clearly do not conform to the cultural values of symmetry and monolinearity. We seem to be dealing with "mistakes". Figure 9a gives an example of a reported *lusona* with mistakes; and Fig. 9b the reconstructed drawing without mistakes.

Fig. 6. Trunks of the *kajana* tree (cf. Gerdes, 2006a–c, cover design)

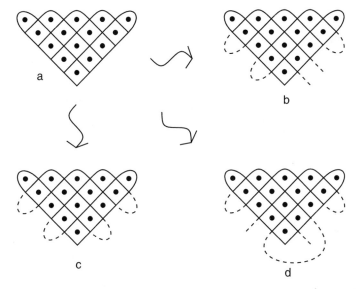

Fig. 7. Transformation of a triangular design (**a**) into monolinear designs (**b, c, d**) (cf. Gerdes, 2006a–c, p. 118)

The drawing experts may have committed consciously some of these "mistakes" to deceive some reporters – the "white man", associated with slave trade, colonial administration and Christianity–, and so to protect their knowledge. The consistency, however, of the *sona* tradition as a whole makes it possible to reconstruct part of it, like the example in Fig. 9b, and to analyse mathematical rules and theorems underlying the building up of monolinear or symmetrical *sona* (see the books Gerdes, 1995, 1997a, 2006a–c).

In the books (Gerdes, 1997b; 1999a–c; 2007c) and in several papers examples are presented of how the reconstructed *sona* tradition may be explored in mathematics education to present stimulating, challenging problems to the

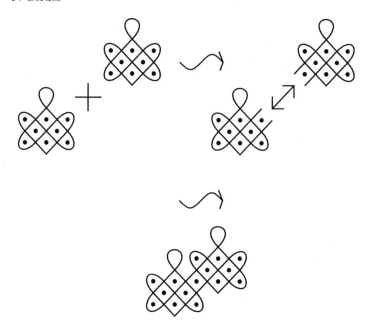

Fig. 8. Example of the application of a chain rule (cf. Gerdes, 2006a–c, p. 42)

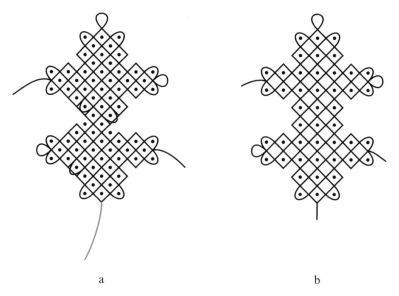

a   b

Fig. 9. (a) Reported *lusona*, representing a lioness with her two cubs. The drawing is neither symmetrical nor monolinear. (b) Reconstructed symmetrical and monolinear line drawing (the tails are added at the end) (cf. Gerdes, 2006a–c, p. 166, 167)

pupils and students, and to motivate, in particular, young Africans to take proud in their scientific heritage. The study of mathematical aspects of the *sona* tradition led to the discovery of new mathematical ideas, like Lunda-designs and various classes of matrices, like cycle matrices, as will be shown in the next section.

Mathematical Research Inspired by the Reconstructed Sona Tradition

Wolfgang Jaritz of the University of Graz (Austria) may have been the first to do mathematical research inspired by the *sona* tradition. Informed about *sona* by the anthropologist Gerhard Kubik, Jaritz studied the properties of a particular class of what we called plaited-mat *sona* and compared these line drawings to the paths of a ball at a billiard table (Jaritz, 1983). In 1990 I published a first paper on a larger class *sona* that includes the plaited-mat designs, wherein the concept of *mirror curves* is proposed and where *Lunda-designs* are presented for the first time (Gerdes, 1990). Inspired by this research, Slavik Jablan (University of Belgrade, Serbia) has studied mirror curves and their relationship with mathematical knot theory (Jablan, 1995, 2001). In the early 1990s Robert Lange (Brandeis University MA, USA) developed *sona tiles*. Franco Favilli and his students at the University of Pisa (Italy) have been developing software for the construction of mirror curves and Lunda-designs (Favilli et al. 2002; Vitturi and Favilli, 2006). Mark Schlatter (Centenary College of Louisiana, USA) has been studying mirror curves and permutations (Schlatter, 2000, 2001, 2004, 2005; cf. Peterson, 2001). Nils Rossing of the University of Science and Technology (Trondheim, Norway) and Christoph Kirfel of the University of Bergen (Norway) applied methods of *sona* analysis by mirror curves to the mathematical analysis of a class of traditional Norwegian rope mats (Rossing and Kirfel, 2003). Myself, I advanced with the study of Lunda-designs (cf. Gerdes, 1999b [Chap. 4]; Gerdes, 1996, 1997a, 1999a–c, 2000b, 2002a,h, 2005) and a sub-class called *Liki-designs* (Gerdes, 2002b,c). I found several interesting classes of matrices, like *cycle* (Gerdes, 2002b, 2006b,c, 2007b), *helix* (Gerdes, 2002e), *cylinder* (Gerdes, 2002f) and *chessboard matrices* (Gerdes, 2002g). Several of these papers were published in *Visual Mathematics* (*) and other on-line journals. Earlier links between Lunda-designs, determinants and magic squares were established (Gerdes, 2000b). The newness and the multiple relationships of mathematical ideas arising from the analysis of the *sona* tradition with other areas of mathematics reflects the profoundness and the mathematical fertility of the ideas of the Cokwe master drawers.

In this section I will briefly show how the study of the *sona* tradition from Angola led me to discover and analyse successively mirror curves, Lunda- and Liki-designs, and cycle matrices.

(a) Mirror curves

The "chased-chicken-path" *lusona* (Fig. 4a) may be considered as a mirror curve, that is:

* * It is the smooth version of the polygonal path described by a light ray emitted from the starting place S at an angle of 45° to the rows of the grid (see Fig. 10);
* * As the ray travels through the grid it is reflected by the sides of the rectangle and by the "double-sided mirrors" it encounters on its path. The mirrors are placed horizontally and vertically, midway, between two neighbouring grid points, as in Fig. 11.

Figure 12 presents the position of the mirrors in the case of the "chased-chicken-path" design. Once defined the concept of mirror curve in general, I started to look for the properties of mirror curves.

To facilitate the execution of mirror curves, one may draw them on squared paper with a distance of two units between two successive grid points. In this way, a monolinear drawing such as the "chased-chicken" path passes exactly once through each of the unit squares inside the rectangle surrounding the grid (see Fig. 13).

This gives the possibility of enumerating the small squares modulo 2, being 1 the number attributed to the unit square where one starts the line, and 0 the number of the second unit square through which the curve passes, and so on successively 101010... until the closed curve is complete. In this way a {0, 1}-matrix is produced. Colouring the unit squares with number 1 black, and the ones with number 0 white, a black-and-white design is obtained.

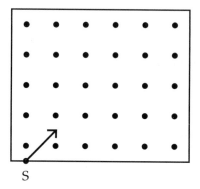

Fig. 10. Emission of the light ray

Fig. 11. Possible mirror positions

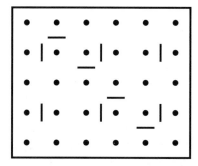

Fig. 12. Mirror-design of the "chased-chicken" path

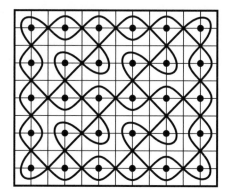

Fig. 13. "Chased-chicken" path drawn on squared paper

(b) Lunda-designs and matrices

As this type of black-and-white design generated by mirror curves was discovered in the context of analysing *sona* from the Cokwe, who predominantly inhabit the Lunda region of Angola, I gave them the name of *Lunda-designs*. Figure 14 presents two examples of Lunda-designs.

Searching for the common characteristics of Lunda-designs, the following symmetry properties may be observed and proved:

(i) In each row there are as many black unit squares as there are white unit squares;

(ii) In each column there are as many black unit squares as there are white unit squares;

(iii) Along the border each grid point always has one black unit square and one white unit square associated with it (see Fig. 15);

(iv) Of the four unit squares between two arbitrary (vertical or horizontal) neighboring grid points, two are always black (see Fig. 16).

Conversely, it holds that any rectangular black-and-white design that satisfies the properties (i), (ii), (iii), and (iv) is a Lunda-design. In other words,

Fig. 14. Two examples of Lunda-designs

Fig. 15. Border situation

Fig. 16. Possible situations between vertically or horizontally, neighbouring grid points

for any rectangular black-and-white design that satisfies the properties (i), (ii), (iii), and (iv) a mirror curve that generates it may be constructed (cf. Gerdes 1996).

The characteristics (i), (ii), (iii), and (iv) may be used to define Lunda-designs (of dimensions mxn). Moreover, the local symmetry characteristics (iii) and (iv) are sufficient for this definition, as they imply the global symmetry properties (i) and (ii). The particular symmetry characteristics of Lunda-designs turn them often aesthetically attractive (cf. Gerdes 2005).

Lunda-designs may be generalized in several ways. Circular and hexagonal Lunda-designs are some interesting possibilities (cf. Gerdes 2002a). Figure 17 presents an example of a hexagonal grid and a hexagonal Lunda-design (cf. Gerdes 1996, 1999a). Instead of enumerating the unit squares through which a mirror curve passes modulo 2, they can be enumerated *modulo t*, if t is a divisor of 4mn, where m and n are the dimensions of the rectangle. In this way t-valued matrices and t-Lunda-designs are created. Figure 18 gives examples of 3 and 4-Lunda-designs.

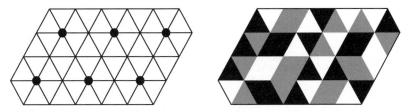

Fig. 17. Example of a hexagonal grid and hexagonal Lunda-design

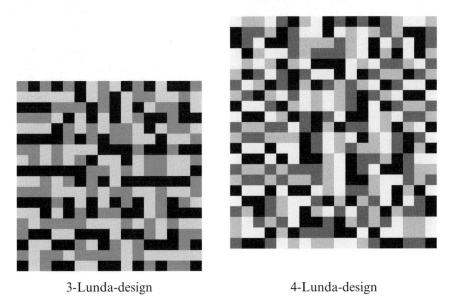

3-Lunda-design   4-Lunda-design

Fig. 18. Examples of a 3- and a 4-Lunda-design

Fig. 19. Stronger condition

(c) Liki-designs

It was on the eve of the fourth anniversary of my daughter Likilisa that I started to analyse a particular class of 2-Lunda-designs. As these designs turned out to have some interesting properties I gave them the name of Liki-designs.

In the case of Liki-designs, the fourth property is substituted by the following stronger condition:

(iv') Of the four unit squares between two arbitrary (vertical or horizontal) neighbouring grid points, two adjacent unit squares are always black, while the other two are white (see Fig. 19).

Fig. 20. Diagonally opposed unit square always have different colours

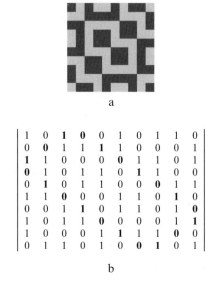

a

$$\begin{vmatrix} 1 & 0 & \mathbf{1} & \mathbf{0} & 0 & 1 & 0 & 1 & 1 & 0 \\ 0 & \mathbf{0} & 1 & \mathbf{1} & 1 & 0 & 0 & 0 & 1 \\ \mathbf{1} & 1 & 0 & 0 & 0 & \mathbf{0} & 1 & 1 & 0 & 1 \\ \mathbf{0} & 1 & 0 & 1 & 1 & 0 & \mathbf{1} & 1 & 0 & 0 \\ 0 & \mathbf{1} & 0 & 1 & 1 & 0 & 0 & \mathbf{0} & 1 & 1 \\ 1 & 1 & \mathbf{0} & 0 & 0 & 1 & 1 & 0 & \mathbf{1} & 0 \\ 0 & 0 & 1 & \mathbf{1} & 0 & 1 & 1 & 0 & 1 & \mathbf{0} \\ 1 & 0 & 1 & 1 & \mathbf{0} & 0 & 0 & 0 & 1 & \mathbf{1} \\ 1 & 0 & 0 & 0 & 1 & \mathbf{1} & 1 & 1 & \mathbf{0} & 0 \\ 0 & 1 & 1 & 0 & 1 & 0 & \mathbf{0} & 1 & 0 & 1 \end{vmatrix}$$

b

Fig. 21. Example of a square Liki-design and associated Liki-matrix

The new condition (iv') may be described as follows. Consider the four unit squares between two vertically or horizontally neighbouring grid points. Two of them that belong to different rows and different columns always have different colours (Fig. 20).

The two properties (i) and (iv') imply that a square Liki-design and its associated Liki-matrix are composed of cycles of alternating black and white unit squares and of cycles of alternating 1s and 0s, respectively.

Figure 21 presents an example of a square Liki-design and its corresponding Liki-matrix. The matrix has five $\{0,1\}$ – cycles; one cycle is represented in bold (Fig. 21b). A question that naturally emerges is what will happen with the powers of Liki-matrices.

Figure 22 displays the first powers of Liki-matrix A. The third power has the same cycle structure as the first power: the first cycle of the third power is composed of alternating 16s and 9s, the second cycle of alternating 15s and 10s, etc. The even powers do not have the same cycle structure. Their diagonals are constant and they present other cycles, like the cycle of 2s of the second power. Figure 23 compares the cycle structures of the odd and even powers of the Liki-matrix A. A cycle structure of the first type I call a first order cycle structure. A cycle structure of the second type I call a second order cycle structure.

$$\begin{vmatrix} 5 & 2 & 2 & 1 & 1 & 3 & 3 & 3 & 3 & 2 \\ 2 & 5 & 1 & 2 & 3 & 1 & 3 & 3 & 2 & 3 \\ 2 & 1 & 5 & 3 & 2 & 3 & 1 & 2 & 3 & 3 \\ 1 & 2 & 3 & 5 & 3 & 2 & 2 & 1 & 3 & 3 \\ 1 & 3 & 2 & 3 & 5 & 2 & 2 & 3 & 1 & 3 \\ 3 & 1 & 3 & 2 & 2 & 5 & 3 & 2 & 3 & 1 \\ 3 & 3 & 1 & 2 & 2 & 3 & 5 & 3 & 2 & 1 \\ 3 & 3 & 2 & 1 & 3 & 2 & 3 & 5 & 1 & 2 \\ 3 & 2 & 3 & 3 & 1 & 3 & 2 & 1 & 5 & 2 \\ 2 & 3 & 3 & 3 & 3 & 1 & 1 & 2 & 2 & 5 \end{vmatrix}$$

$$A^2$$

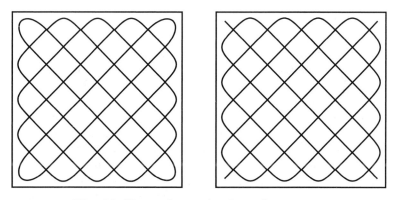

$$A^3 \qquad\qquad\qquad A^4$$

Fig. 22. The first powers of the Liki-matrix A

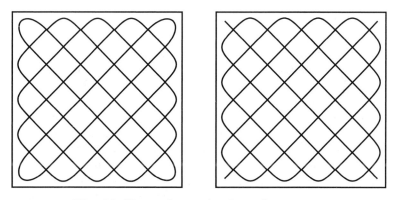

Fig. 23. First and second order cycle structure

The powers of a Liki-matrix, like the matrices A^2, A^3, etc., are themselves not Liki-matrices, but they display cycle structures. Let us call them *cycle matrices*. As the numbers on the cycles on the odd powers are alternating, we may say that these cycle matrices have period 2. As the numbers on the cycles on the even powers are constant, we say that these cycle matrices have a period 1.

(d) Cycle matrices

In this way we may introduce the concept of a cycle matrix of period 2, independent of the context of Liki-designs in which the concept was discovered. Figure 24 displays two cycle matrices of period 2. Both have a first order cycle structure whereas their products AB and BA have a second order cycle structure. The cycles of AB and BA have a difference in phase of one unit, and A + B is a cycle matrix of period 1.

The following table holds for the multiplication of cycle matrices of period 2 (Table 1).

As this multiplication table is similar to the multiplication table of negative and positive numbers, we may call matrices that have a first order cycle structure *negative cycle matrices*, and matrices that have a second order cycle structure *positive cycle matrices*.

The study of cycle matrices of period 2 led me to the study of cycle matrices of any period. Figure 25 displays two (positive) cycle matrices E and F of period 3. For instance, the first cycle of E is composed of repeating {−2, 1, 4}'s. The product EF is a positive cycle matrix of period 3.

The same multiplication table is true for even and odd cycle matrices of any period p. Figure 26 presents an example of two negative cycle matrices

$$
A = \begin{vmatrix} -1 & 3 & -4 & 6 & -2 & 0 \\ 3 & 6 & -1 & 0 & -4 & -2 \\ -4 & -1 & -2 & 3 & 0 & 6 \\ 6 & 0 & 3 & -2 & -1 & -4 \\ -2 & -4 & 0 & -1 & 6 & 3 \\ 0 & -2 & 6 & -4 & 3 & -1 \end{vmatrix}
\qquad
B = \begin{vmatrix} 4 & -2 & 3 & 5 & -3 & 2 \\ -2 & 5 & 4 & 2 & 3 & -3 \\ 3 & 4 & -3 & -2 & 2 & 5 \\ 5 & 2 & -2 & -3 & 4 & 3 \\ -3 & 3 & 2 & 4 & 5 & -2 \\ 2 & -3 & 5 & 3 & -2 & 4 \end{vmatrix}
$$

$$
AB = \begin{vmatrix} 14 & 7 & 5 & -17 & 18 & -9 \\ 5 & 14 & 18 & 7 & -9 & -17 \\ 7 & -17 & 14 & -9 & 5 & 18 \\ 18 & 5 & -9 & 14 & -17 & 7 \\ -17 & -9 & 7 & 18 & 14 & 5 \\ -9 & 18 & -17 & 5 & 7 & 14 \end{vmatrix}
\qquad
BA = \begin{vmatrix} 14 & 5 & 7 & 18 & -17 & -9 \\ 7 & 14 & -17 & 5 & -9 & 18 \\ 5 & 18 & 14 & -9 & 7 & -17 \\ -17 & 7 & -9 & 14 & 18 & 5 \\ 18 & -9 & 5 & -17 & 14 & 7 \\ -9 & -17 & 18 & 7 & 5 & 14 \end{vmatrix}
$$

Fig. 24. Two cycle matrices of period 2 and their products

Table 1. Multiplication table of cycle matrices of period 2

A	B	AB
First order	First order	Second order
First order	Second order	First order
Second order	First order	First order
Second order	Second order	Second order

$$\begin{vmatrix} -2 & 1 & 3 & -3 & 5 & 0 \\ -4 & 5 & -4 & 2 & 5 & 2 \\ -3 & 1 & 0 & -2 & 5 & 3 \\ 3 & 5 & -2 & 0 & 1 & -3 \\ 2 & 5 & 2 & -4 & 5 & -4 \\ 0 & 5 & -3 & 3 & 1 & -2 \end{vmatrix}$$

E

$$\begin{vmatrix} 5 & -1 & 2 & -3 & 4 & -2 \\ 3 & 4 & 3 & 0 & 4 & 0 \\ -3 & -1 & -2 & 5 & 4 & 2 \\ 2 & 4 & 5 & -2 & -1 & -3 \\ 0 & 4 & 0 & 3 & 4 & 3 \\ -2 & 4 & -3 & 2 & -1 & 5 \end{vmatrix}$$

F

$$\begin{vmatrix} -22 & 11 & -22 & 42 & 31 & 34 \\ 7 & 64 & 19 & 7 & 4 & 19 \\ -22 & 31 & -22 & 34 & 11 & 42 \\ 42 & 11 & 34 & -22 & 31 & -22 \\ 19 & 4 & 7 & 19 & 64 & 7 \\ 34 & 31 & 42 & -22 & 11 & -22 \end{vmatrix}$$

EF

Fig. 25. Two positive cycle matrices of period 3 and their product

$$\begin{vmatrix} 1 & 2 & -1 & -3 & 4 & 5 & 0 & 3 & 2 \\ 0 & 4 & 3 & -3 & 6 & 4 & 1 & 3 & 5 \\ 8 & 5 & 6 & 4 & 2 & 7 & 1 & 0 & 1 \\ 7 & 0 & 4 & 1 & 5 & 1 & 8 & 2 & 6 \\ 1 & 6 & 5 & 3 & 3 & 0 & 4 & 4 & -3 \\ 5 & 3 & -3 & 2 & 2 & 0 & 1 & 4 & -1 \\ 0 & 4 & 2 & -1 & 3 & 1 & 5 & 2 & -3 \\ 4 & 3 & -3 & 5 & 4 & 1 & 0 & 6 & 3 \\ 1 & 2 & 1 & 6 & 0 & 8 & 7 & 5 & 4 \end{vmatrix}$$

G

$$\begin{vmatrix} 2 & 1 & -2 & -1 & -2 & -1 & 2 & 0 & 1 \\ 0 & 4 & 2 & 3 & 6 & 2 & -3 & 0 & 1 \\ 7 & 5 & -3 & 4 & 4 & 5 & -4 & 3 & 2 \\ 5 & 3 & 4 & 2 & 5 & -4 & 7 & 4 & -3 \\ -3 & 6 & 1 & 2 & 0 & 0 & 2 & 4 & 3 \\ -1 & 0 & -1 & 1 & 1 & 2 & 2 & -2 & -2 \\ 2 & -2 & 1 & -2 & 0 & 2 & -1 & 1 & -1 \\ 2 & 0 & 3 & 1 & 4 & -3 & 0 & 6 & 2 \\ -4 & 4 & 2 & -3 & 3 & 7 & 5 & 5 & 4 \end{vmatrix}$$

H

$$\begin{vmatrix} -39 & 27 & 5 & 5 & 14 & 25 & 7 & 19 & 26 \\ -28 & 76 & 9 & 20 & 52 & 71 & -1 & 57 & 54 \\ 63 & 84 & -10 & 45 & 68 & 39 & 27 & 34 & 8 \\ 27 & 68 & 8 & -10 & 34 & 63 & 39 & 84 & 45 \\ 71 & 57 & 20 & 54 & 76 & -1 & -28 & 52 & 9 \\ 7 & 14 & 26 & 5 & 19 & -39 & 25 & 27 & 5 \\ 25 & 19 & 5 & 26 & 27 & 7 & -39 & 14 & 5 \\ -1 & 52 & 54 & 9 & 57 & -28 & 71 & 76 & 20 \\ 39 & 34 & 45 & 8 & 84 & 27 & 63 & 68 & -10 \end{vmatrix}$$

GH

Fig. 26. The negative cycle matrices of period 6 and their product

G and H of dimensions 9×9 of period 6. The product GH is a positive cycle matrix. Figure 27 displays the cycle structure of negative and positive cycle matrices of dimensions 9×9.

(e) Final comments

Mathematical research may appear as an unending story of discovering new concepts, new relationships, new theorems, and new applications. The example of the discovery of mirror curves, Lunda-designs and cycle matrices shows how an old African cultural practice may inspire and stimulate mathematical research. This line of research does not stop with cycle matrices. Other concepts like helix and cylinder matrices (Gerdes, 2002 e,f) were discovered and surely many more will follow. The book (Gerdes, 2007b) presents an introduction the attractive, visual properties of cycle matrices.

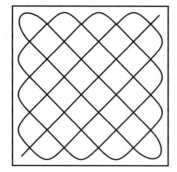 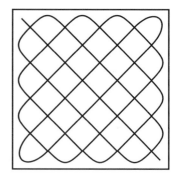

Fig. 27. Cycle structure of negative and positive cycle matrices of dimensions 9×9

Acknowledgements

Abridged version of a paper presented at the International Workshop on Mathematical Modelling, Simulation, Visualization and e-learning, Rockefeller Foundation, Bellagio, Italy, November 20–25, 2006.

References

Djebbar, Ahmed (2001), *Une histoire de la science arabe.* Editions du Seuil, Paris.

Djebbar, Ahmed (2003), A Panorama of Research on the History of Mathematics in al-Andalus and the Maghreb between the Ninth and the Sixteenth Century, in: Jan P. Hogendijk & A. Sabra (Eds.), *The Enterprise of Science in Islam, New perspectives*, MIT Press, Cambridge MA, 309–350.

Djebbar, Ahmed (2005), *L'algèbre arabe: Naissance d'un art*, Vuibert, Paris.

Eglash, Ron (1999), *African Fractals. Modern Computing and Indigenous Design*, Rutgers University Press, New Brunswick NJ.

Fauvel, John & Gerdes, Paulus (1990), African Slave and Calculating Prodigy: Bicentenary of the Death of Thomas Fuller, *Historia Mathematica*, New York, Vol. 17, 141–151.

Favilli, Franco, Maffei, Laura & Venturi, Irene (2002): SONA Drawings: From the Sand to the Silicon, in: Sebastiani Ferreira E. (Ed.), *Proceedings of the II International Congress on Ethnomathematics, Summary Booklet*, Ouro Preto, p. 35.

Gerdes, Paulus & Bulafo, Gildo (1994), *Sipatsi: Technology, Art and Geometry in Inhambane*, Universidade Pedagógica, Maputo, 112 pp.

Gerdes, Paulus & Djebbar, A. (2004), *Mathematics in African History and Cultures. An Annotated Bibiography*, African Mathematical Union, Cape Town. New edition: Lulu.com, 2007, 340 pp. [ISBN 978-1-4303-1537-7].

Gerdes, Paulus & Djebbar, A. (2007), *Les mathématiques dans l'histoire et les cultures africaines. Une bibliographie annotée*, Université de Lille, 332 pp.

Gerdes, Paulus (1990), On ethnomathematical research and symmetry, *Symmetry: Culture and Science*, Budapest, 1(2), 154–170.

Gerdes, Paulus (1995), *Une tradition géométrique en Afrique – Les dessins sur le sable*, L'Harmattan, Paris (3 volumes) [ISBN 2-7384-3552-8, 2-7384-3553-6, 2-7384-3654-4].

Gerdes, Paulus (1996), *Lunda Geometry — Designs, Polyominoes, Patterns, Symmetries*, Universidade Pedagógica, Maputo, New expanded edition, 2007, lulu.com, Morrisville NC, 152 pp.

Gerdes, Paulus (1997a), *Ethnomathematik dargestellt am Beispiel der Sona Geometrie*, Spektrum Verlag, Berlin/Heidelberg/Oxford, 1997, 436 pp. [ISBN 3-8274-0201-8].

Gerdes, Paulus (1997b), *Recréations géométriques d'Afrique – Lusona – Geometrical recreations of Africa*, L'Harmattan, Paris, 127 pp. [ISBN 2-7384-5168-3].

Gerdes, Paulus (1998), *Women, Art and Geometry in Southern Africa*, Africa World Press, Trenton NJ, 200 pp. [ISBN 0-86543-601-0 (hardback), ISBN 0-86543-602-9 (paperback)].

Gerdes, Paulus (1999a), *Geometry from Africa: Mathematical and Educational Explorations*, The Mathematical Association of America, Washington DC, 210 pp. [ISBN 0-88385-715-4].

Gerdes, Paulus (1999b) On Lunda-designs and some of their symmetries, *Visual Mathematics*, Belgrade, 1(1).[*2]

Gerdes, Paulus (1999c), On the geometry of Celtic knots and their Lunda-designs, *Mathematics in School*, Leicester, 28(3), 29–33.

Gerdes, Paulus (2000a), *Le cercle et le carré: Créativité géométrique, artistique, et symbolique de vannières et vanniers d'Afrique, d'Amérique, d'Asie et d'Océanie* [The circle and the square: Geometric, artistic and symbolic creativity of basket weavers from Africa, America, Asia and Oceania], L'Harmattan, Paris, 301 pp. [ISBN 2-7384-9235-5].

Gerdes, Paulus (2000b), On Lunda-designs and the construction of associated magic squares of order 4p, *The College Mathematics Journal*, Washington DC, 31(3), 182–188.

Gerdes, Paulus (2002a), Symmetrical explorations inspired by the study of African cultural activities, in: István Hargittai & Torvand Laurent (Eds.), *Symmetry 2000*, Portland Press, London, 75–89.

Gerdes, Paulus (2002b), From Liki-designs to cycle matrices: The discovery of attractive new symmetries, *Visual Mathematics*, 4(1).[*]

Gerdes, Paulus (2002c), New designs from Africa, *Plus Magazine*, 19 (http://plus.maths.org/issue19/features/liki/index.html)

Gerdes, Paulus (2002d), m-Canonic mirror curves, *Visual Mathematics*, 4(1).[*]

Gerdes, Paulus (2002e), Helix matrices, *Visual Mathematics*, 4(2).[*]

Gerdes, Paulus (2002f), Cylinder matrices, *Visual Mathematics*, 4(2).[*]

Gerdes, Paulus (2002g), A note on chessboard matrices, *Visual Mathematics*, 4(3).[*]

Gerdes, Paulus (2002h), Variazioni sui disegni Lunda, in: Michele Emmer (Ed.), *Matematica e Cultura 2002*, Springer, Milan, 135–146.

Gerdes, Paulus (2003a), *Awakening of Geometrical Thought in Early Culture*, MEP Press, Minneapolis MN, 184 pp. [ISBN 0-930656-75-X].

Gerdes, Paulus (2003b), *Sipatsi: Cestaria e Geometria na Cultura Tonga de Inhambane* [Sipatsi: Basketry and Geometry in the Tonga culture of Inhambane], Moçambique Editora, Maputo, 176 pp. [ISBN 902-47-9908-2].

Gerdes, Paulus (2004), *Basketry, Geometry, and Symmetry in Africa and the Americas*, E-book, Visual Mathematics, Belgrade, 2004 [on-line available at: www.mi.sanu.ac.yu/vismath/]

---

[2] The papers marked by[*] are available at: http://members.tripod.com/vismath/pap.htm or http://www.mi.sanu.ac.yu/vismath/pap.htm.

Gerdes, Paulus (2005), Lunda Symmetry where Geometry meets Art, in: Michele Emmer (Ed.), *The Visual Mind II*, MIT Press, Boston, 335–348.

Gerdes, Paulus (2006a), *Sona geometry from Angola: Mathematics of an African tradition*, Polimetrica, Monza, 232 pp. [ISBN 978-88-7699-055-7].

Gerdes, Paulus (2006b), Symmetries of alternating cycle matrices, *Visual Mathematics*, 8(2).*

Gerdes, Paulus (2006c), On the representation and multiplication of basic alternating cycle matrices, *Visual Mathematics*, 8(2).*

Gerdes, Paulus (2007a), *African Doctorates in Mathematics: A Catalogue*, Lulu.com, London, etc., 384 pp. [ISBN 978-1-4303-1867-5].

Gerdes, Paulus (2007b), *Adventures in the World of Matrices*, Nova Science Publishers, New York (in press).

Gerdes, Paulus (2007c), *Drawings from Angola: Living Mathematics*, Lulu.com, Morrisville NC, 72 pp. [ISBN 978-1-4303-2313-6].

Jablan, Slavik (1995), Mirror generated curves, *Symmetry: Culture and Science*, Budapest, 6(2), 275–278.

Jablan, Slavik (2001), Mirror curves, in: Sarhangi, R. & Jablan, S. (Eds.), *Bridges: Mathematical Connections in Art, Music, and Science Conference Proceedings*, Southwestern College, Winfield [reproduced in: *Visual Mathematics*, 3(2)*].

Jaritz, Wolfgang (1983), Über Bahnen auf Billardtischen – oder: Eine mathematische Untersuchung von Ideogrammen Angolanischer Herkunft, *Berichte der mathematisch-statistischen Sektion im Forschungszentrum Graz*, Graz, 207, 1–22.

Njock, Georges Edward (1985), Mathématiques et environnement socio-culturel en Afrique noire, *Presence Africaine*, Paris, No. 135, 3–21.

Obenga, Théophile (1985), *La géométrie égyptienne. Contribution de l'Afrique antique à la Mathématique mondiale*, L'Harmattan, Paris.

Peterson, Ivars (2001), Sand Drawings and Mirror Curves, *Science News*, Washington DC (available at: http://www.sciencenews.org/20010922/mathtrek.asp).

Rossing, Nils & Kirfel, Christoph (2003), *Matematisk beskrivelse av taumatter* [Mathematical description of rope mats], NTNU, Trondheim.

Schlatter, Mark (2000), *Mirror Curves and Permutations* (available at: http://personal.centenary.edu/~mschlat/sonaarticle.pdf).

Schlatter, Mark (2001), Sona sand drawings and permutation groups, in: R. Sarhangi & S. Jablan (Eds.), *Bridges: Mathematical Connections in Art, Music, and Science Conference Proceedings*, Southwestern College, Winfield (USA) [reproduced in: *Visual Mathematics*, 3(2)*].

Schlatter, Mark (2004), Permutations in the sand, *Mathematics Magazine*, 77(2), 140–145.

Schlatter, Mark (2005), How to create monolinear mirror curves, *Visual Mathematics*, 7(2).*

Sica, Giandomenico, ed. *What mathematics from Africa?* Monza: Polimetrica, 2005.

Vitturi, Mattia de Michieli & Favilli, Franco (2006), Sona drawings, mirror curves and pattern designs, *Proceedings of the 3rd International Congress on Ethnomathematics* (in press).

Zaslavsky, Claudia (1999), *Africa Counts: Number and Pattern in African Cultures*, Lawrence Hill, Westport [First edition 1973].

Finite Dynamical Systems: A Mathematical Framework for Computer Simulation

A.S. Jarrah and R. Laubenbacher

Summary. Dynamical systems over finite fields provide a natural mathematical framework for interaction-based computer simulation of complex systems. This paper provides an introduction to a theory of these systems. Motivating examples of agent-based simulations are given.

1 Introduction

Modeling and simulation are playing an increasingly important role in the analysis of highly complex natural and technological systems. A variety of modeling frameworks, each with its own advantages and disadvantages, are available. In particular, the analysis of social, socio-technical, and biological systems, such as social networks, road traffic networks, epidemiological networks, or the immune system can benefit from approaches other than differential-equations-based mathematical modeling, since their dynamics is generated by the local interactions of a large number of heterogeneous individuals. Interaction-based, or rule-based, simulations are being used successfully to simulate the dynamics of networks of this type, predominantly using cellular automata and Boolean network approaches. By an interaction-based simulation, we mean a collection of variables, each equipped with a function or a set of rules, that computes the state of each variable from the state of other variables that are interacting with it. An excellent survey of recent work in this direction can be found in the talks at the November 2003 "Hot Topics Workshop on Agent-based Modeling and Simulation" at the Institute for Mathematics and its Applications at the University of Minnesota [22].

One of the main advantages of such models is their ability to reflect individual differences in behavior rather than an average over a large number of individuals. An important disadvantage of interaction-based models, however, is that there are very few mathematical tools available for their design and subsequent analysis of their dynamics.

The basic approach to a mathematical specification of an interaction-based model is to represent it as a dynamical system of some sort. Typically, in such simulations, each variable is allowed to take on finitely many different states. One possible mathematical framework, therefore, is that of discrete dynamical systems

$$f = (f_1, \ldots, f_n) : R^n \longrightarrow R^n,$$

where R is a finite field or the commutative ring $\mathbb{Z}/r\mathbb{Z}$, for some integer r. (It is important to impose some kind of mathematical structure on the set R of states of the variables, since otherwise one studies set functions only. In practice, this is often very easy to do. See, e.g., [27].) This framework includes many cellular automata and Boolean networks (by choosing R to be the field with two elements), in addition to other multi-state models. We refer to such a system as a *finite dynamical system*. In the case where R is a finite field, it is a well-known fact that any such system f can be described by a collection of polynomial functions in n variables [31, p. 369]. Polynomial algebra has seen tremendous progress over the last 15 years, both in conceptual and in computational terms, all of which can now be brought to bear on the problem of a mathematical analysis of finite dynamical systems over finite fields (See [16] for a recent survey of algorithms.)

While many computer models of complex systems are stochastic we will focus here on deterministic models, for simplicity and because their theory is much better developed.

2 Examples of Interaction-Based Computer Simulations

We first describe in some detail two examples of interaction-based models for concreteness and to motivate the mathematical developments.

2.1 Simulation of Socio-technical Networks

First we describe a simulation method for road traffic networks called *TRAN-SIMS* [24], which is exemplary for similar approaches to other socio-technical networks, such as wireless communication systems, or power grids. *TRAN-SIMS*, developed at Los Alamos National Laboratory, represents a new, disaggregate approach to traffic demand modeling, and is designed to give traffic planners more accurate, complete information on factors impacting travel demand and traffic flow in urban areas. It is part of the Travel Model Improvement Program sponsored by the U.S. Department of Transportation, the U.S. Environmental Protection Agency, and the U.S. Department of Energy. *TRANSIMS* creates a virtual metropolitan region, with a complete representation of the population, at the level of individuals, their daily activities, and a faithful representation of the transportation infrastructure. Disaggregated demographic data and detailed activity surveys of the population are used to create activity schedules and travel choices for each individual. *TRANSIMS* then simulates the movement of travelers and vehicles across the transportation grid, using multiple modes, such as car, bus, bicycle, and foot, on a second-by-second basis. The interaction of individual vehicles and travelers produces realistic traffic dynamics, whose features can then be analyzed.

Simulation at this level of resolution requires a very complex software design on a parallel computation architecture. A single 24-hour simulation run can take several hours to complete and produce gigabytes of output. In light of such complexity, simulation design, implementation, software verification, validation, and analysis all become crucial issues in determining the validity and usefulness of large-scale simulations such as *TRANSIMS*. These issues are magnified when *TRANSIMS* is combined with other simulations, such as in *EPISIMS* [19], a simulation of the spread of infection by an airborne pathogen in an urban area. *EPISIMS* combines a model of the spread of a pathogen cloud over an urban area with an epidemiological model of infectivity, and movement of humans in relation to the pathogen cloud is generated by *TRANSIMS*.

In order to address these issues, the designers of *TRANSIMS* and *EPISIMS* initiated a research program to provide a mathematical foundation for interaction-based computing and simulation. The goal of the program is to create mathematical objects which are general enough to capture the key features of interaction-based simulations, but which have enough structure to allow a rich mathematical theory that provides tools to address the issues raised above. We will return to this topic in the next section.

2.2 Computational Immunology

Next we discuss an interaction-based simulation of certain aspects of the human immune system. Comprised of a large number of interacting cells whose motion is constrained by the body's anatomy, the immune system lends itself very well to simulation by interaction-based models. In particular, these models can take into account three-dimensional anatomical variation as well as small-scale variability in cell distributions. For instance, while the number of T-cells in the human body is astronomical, the number of antigen-specific T-cells, for a specific antigen, can be quite small, thereby creating many spatial inhomogeneities. Also, little is known about the global structure of the system to be modeled.

The first discrete model to incorporate a useful level of complexity was *ImmSim* [10,11], developed by Seiden and Celada, a stochastic cellular automaton simulation. It includes B cells, T cells, antigen presenting cells (APCs), antibodies, antigens, and antibody-antigen complexes. Receptors are represented by bit strings, and antibodies use bit strings to represent their epitopes and peptides. The bit string approach was initially introduced in [20]. Specificity and affinity are defined by using bit string similarity. The model is implemented on a regular two-dimensional grid, which can be thought of as a slice of a lymph node, for instance. It has been used to study various phenomena, including the optimal number of human leukocyte antigens in human beings [10], the autoimmunity and T lymphocyte selection in the thymus [35], antibody selection and hyper-mutation [12], and the dependence of the selection and maturation of the immune response on the antigen-to-receptor's affinity [7].

The computational limitations of the Seiden-Celada model have been overcome by a modified model, *CImmSim*, implemented on a parallel architecture. Its complexity is several orders of magnitude larger than its predecessor. It has been used to model hypersensitivity to chemotherapy [8], the selection of escape mutants from immune recognition during HIV infection [6], and mechanisms leading to persistence of the Epstein-Barr virus [9].

An important application for *CImmSim* is as a tool to explore different types of interventions that affect immune response to pathogens. With an appropriate mathematical specification for this model one could develop an appropriate control theory that might provide a mathematical basis for the study of interventions. Mathematical methods to control dynamical systems play an extremely important role in engineering, and there is a vast literature on the subject, in both engineering and in mathematics. There have been several promising applications in computational immunology, with the goal of discovering ways to enhance immune response to pathogen attacks. A well-developed control theory for immune system models would represent an invaluable tool. Working hand-in-hand with model building, it would allow the *in silico* exploration of known control mechanisms and would aid in the discovery of new ones.

In [39], an optimal control theory approach is applied to a model for the response of the innate immune system to infection and to therapy. The model, consisting of four nonlinear ordinary differential equations, is an enhancement of one in [1], with control variables added. The mathematical analysis can suggest single- and multi-agent therapies that enhance the innate response of the immune system. In [37], control theory applied to ODE models is used to study optimal choice of effectors during an immune response. A similar theme is studied in [38], again by an ODE approach. In [36] a model is studied that represents the coupling between the immune, nervous, and endocrine systems. The model is a hybrid ODE and logical model that is used to study optimal immune response to infections that are accompanied by immune toxicity.

For discrete computer models like *CImmSim*, no mathematical tools are available for rigorous approaches to the systematic study of immune response modification, since a mathematical specification is typically absent. In particular, no method exists to understand the structure of the enormous phase spaces of such systems. In [27] tools were developed that can be used in principle to construct a deterministic, finite dynamical system model of individual *CImmSim* components, for which a control theory framework has been developed [28, 32, 33]. In order to apply this method to build a comprehensive mathematical specification of a simulation as complex as *CImmSim*, it is necessary to develop mathematical methods for the decomposition of finite dynamical systems, while keeping control of the decomposition's effect on the dynamics. It is also necessary to develop tools for the design of finite dynamical systems from modular components and understand the resulting dynamics. Finally, one needs to have methods to compare finite dynamical systems in a meaningful way, such as through the use of transformations. We will address some of these issues below.

3 Background

Some of the mathematical problems raised here in the context of specific interaction-based computer simulations have also appeared in other contexts, which we review briefly, together with some solutions. The problem of understanding the relationship between the structure of a discrete model and the resulting dynamics goes back at least half a century. In [18], the author studies the question for Boolean networks, from the point of view of applications to the length of sequences produced by feedback shift registers. He also points out applications to radar and communication systems and automatic error correction circuits. He provides a method to compute the length of all *limit cycles* for *linear Boolean networks* (that is, networks whose Boolean functions are constructed using the logical operation XOR). The paper also contains a generalization to networks that take values in an arbitrary finite field with a prime number of elements, that is, in $\mathbb{Z}/p\mathbb{Z}$ where p is prime. For *affine* Boolean linear networks (that is, networks whose local functions are Boolean linear polynomials which might have constant terms), a method to analyze cycle length has been developed in [34]. After embedding the matrix of the transition function, which is of dimension $(n \times (n + 1))$, into a square matrix of dimension $n + 1$, the problem is reduced to the linear case.

The main objective in [40, 34] is to study the inverse problem of constructing Boolean networks with specified dynamics, in the context of logical neural networks (LNNs). The authors point to the lack of a theoretical understanding of the link between structure and dynamics as an important limiting factor in the theory of neural networks. In [15] a very elegant construction is given to reduce the problem for the nonlinear case to the study of a linear system, at least for a determination of the limit cycles. Unfortunately, if the system has dimension n, then the corresponding linear system has dimension 2^n, so that the approach is algorithmically impractical.

In the context of cellular automata some theoretical work on the problem of relating structure to dynamics has been done by Wolfram [41]. For instance, in [42], the authors prove (Theorem 4.3) that the phase space of an additive one-dimensional cellular automaton (CA) has the identical tree structure of transients at each node of each limit cycle. But such a CA can be viewed as a linear n-dimensional system over the field with two elements, where n is equal to the number of cells. The question of describing the phase space of a linear cellular automaton from the structure of the rule alone can now be answered completely, since it is a special case of a more general recent result about linear finite dynamical systems over an arbitrary finite field [21]. In particular the result in [42] mentioned above follows as a very special case from this paper. We briefly describe the result, as an example of how progress can be made by formulating the question within a rich mathematical framework.

Let k be an arbitrary finite field, and let $f : k^n \longrightarrow k^n$ be a linear finite dynamical system over k of dimension n. For instance, k could be the field with two elements, and f could be the transition rule of a 1-dimensional cellular automaton. Then, after choosing a basis for the vector space k^n, the function f can be described by an $(n \times n)$ square matrix M. An algorithm for the structure of the limit cycles of f had already been determined in [18]. Hernandez [21] shows that the exact number and length of each limit cycle as well as the structure of the transients can be determined from the factorization of the *elementary divisors* of M. It is shown that the structure of the *tree of transients* at each node of each limit cycle is the same, and can be completely determined from the elementary divisors of the form x^a. It is a fact that the system f is invertible if and only if f has no such elementary divisors, equivalently, if zero is not an eigenvalue of f.

Even the one-dimensional case of the general problem is very interesting and very challenging. Here, we are given a finite field k and a function $f : k \longrightarrow k$. It is well-known [31, p. 369] that f can be represented as a polynomial in one variable. The problem is to infer the phase space of f from the structure of this polynomial. The related question about when such a function is invertible has been studied extensively. Such polynomials are known as *permutation polynomials*. Many results have been obtained about special classes of permutation polynomials, notably monomials and binomials. For a survey of known results and related conjectures, see [31, Ch. 7] and [29, 30].

4 Definitions and Examples

This section contains the basic definitions and examples of finite dynamical systems. The main outline of the theory for sequentially updated systems is contained in the series of papers [2–5]. The mathematical concept at the core of the theory is that of a sequential dynamical system. For simplicity we will focus on systems over the field with two elements.

Definition 4.1. *Let $k = \{0, 1\}$ be the field with two elements. A* sequential dynamical system (SDS) $\mathfrak{F} = \mathfrak{F}(Y, \{f_i\}, \pi)$ *in n binary variables $x_1, \ldots, x_n$ is a function*

$$f : k^n \longrightarrow k^n,$$

constructed from the following data:

   1. a finite graph Y on the vertices $1, 2, \ldots, n$, called the dependency graph *of $\mathfrak{F}$;*

2. *a family of "local" update functions* $f_i : k^n \longrightarrow k^n$, $i = 1, \ldots, n$, *in the variables* $x_1, \ldots, x_n$, *which computes the binary state of the variable* x_i *and leaves the other coordinates unchanged. The* f_i *are assumed to be symmetric in their inputs. Furthermore,* f_i *depends only on* x_i *and those variables whose index is connected to* i *in the dependency graph* Y;

3. *an "update schedule"* π, *which specifies an order on the vertices of* Y, *represented by a permutation* $\pi \in S_n$.

The function f *is then constructed by composing the local update functions according to the update schedule* π, *that is,*

$$f = f_{\pi(n)} \circ \cdots \circ f_{\pi(1)} : k^n \longrightarrow k^n.$$

The dynamics of $\mathfrak{F}$ *is generated by iteration of the function* f.

The function f is then a finite dynamical system, obtained from the SDS $\mathfrak{F}$. The concept of an SDS incorporates the key features of an interaction-based simulation: a collection of entities (the x_i) which interact with each other, the interaction is "local" (given by the 1-neighborhood of the variables in the dependency graph), and the entities act according to a specified update schedule. In the language of agent-based simulation, the pairs (x_i, f_i) can be thought of as the agents that interact with each other. Thus, an SDS is a special type of time-discrete, sequentially updated dynamical system over the field with two elements.

Example 1. Consider the SDS $\mathfrak{F} = \mathfrak{F}(Y, \{f_i\}, \pi)$, where Y is the graph in Fig. 1, and the local functions f_i be the Nand function. For example,

$$f_2(x_1, x_2, x_3) = (x_1, \mathtt{Nand}(x_1, x_2, x_3), x_3), \quad \text{where } \mathtt{Nand}(x_1, x_2, x_3) = 1 + x_1 x_2 x_3.$$

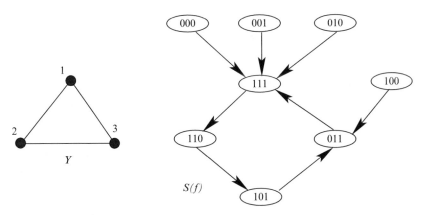

Fig. 1. The dependency graph Y and the phase space $\mathcal{S}(f)$ of the system in Example 1.

For the permutation $\pi = \begin{pmatrix} 1\ 2\ 3 \\ 3\ 2\ 1 \end{pmatrix}$, the FDS of $\mathfrak{F}(Y, \{f_i\}, \pi)$ is

$$f = f_1 \circ f_2 \circ f_3 : k^3 \longrightarrow k^3.$$

To understand the dynamics of a FDS f, one needs to analyze and understand the phase space of f.

Definition 4.2. *Let $f : R^n \longrightarrow R^n$ be a finite dynamical system. The phase space $\mathcal{S}(f)$ of f is the directed graph with the elements of R^n as vertices. (Observe that R^n has 2^n elements, in the case of an SDS.) There is a directed edge from an n-tuple u to another n-tuple v if and only if $f(u) = v$. Those vertices that are part of a directed cycle are called* periodic points, *and a directed cycle is called a* limit cycle. *The other vertices are* transients.

It is straightforward to see that the phase space of a system consists of several components, each of which consists of a single directed cycle, with "trees" feeding into the nodes of the cycle.

Example 2. The phase space of the system in Example 1 is given in Figure 1.

In order to see how an interaction-based simulation could be represented as a finite dynamical system we refer back to the two examples given earlier. In the case of *TRANSIMS* the variables are the cells making up the road network. The cell state contains information about whether or not the cell is occupied by a vehicle as well as the velocity of the vehicle. One may assume that each cell takes on states from the same set of possible states, which may be chosen to support the structure of a finite field.

The cells/variables interact with each other, but typically a cell only interacts with a small subset of other cells, its *neighbors*. Through such an interaction a cell changes its state based on the states (or other aspects) of the cells with which it interacts. We will refer to the process where a variable modifies its state through interaction as a *variable update*. The precise way in which a variable modifies its state is governed by the nature of the particular variable. In TRANSIMS the neighbors of a cell are the adjacent road network cells. From this adjacency relation one obtains a dependency graph of the variables. The local update function for a given variable can be obtained from the rules governing traffic flow between cells.

The updates of all the agents may be scheduled in different ways, e.g., synchronous, asynchronous or event-driven schemes. The choice will depend on system properties or particular considerations about the simulation implementation.

In the case of *CImmSim*, the situation is somewhat more complicated. Here the variables are also the spatial units of the system, each representing a small volume of lymph tissue. The total volume is represented as a 2-dimensional cellular automaton, in which every variable has 4 neighbors, so that the dependency graph is a regular 2-dimensional grid. The state of each variable is a collection of counts for the various immune cells and pathogens that are present in this particular unit of space (volume). Movement between spaces is implemented as diffusion. Immune cells can interact with each other and with pathogens while they reside in the same volume. Thus, the local update function for a given cell of the simulation is made up of the two components of movement between cells and interactions within a cell. For instance, a B cell could interact with the Epstein–Barr virus in a given volume and transition from uninfected to infected by the next time step. Interactions as well as movement are stochastic, resulting in a stochastic finite dynamical system. The update order is parallel.

Several results that link the structure of the local update functions f_i to the structure of the phase space $\mathcal{S}(f)$ are known. For instance, it is shown [2, Prop. 5] that the Nand system (logical Nand functions as local update functions) on a complete graph does not have any fixed points, see Figure 1. Other results pertain to the number of fixed points of special types of systems and classifications of certain families of invertible systems, that is, systems that do not have any transients. See [2–4] for details.

   One of the first research problems about SDS was the question of how a change of update schedule affects the dynamics of an SDS. This question is motivated by the sequential nature of $TRANSIMS$ and other large simulations like it. Moving the simulation from one platform to another can quite possibly change the update schedule. Precisely formulated, the question becomes that of how many different finite dynamical systems one obtains simply by varying the update schedule of an SDS, that is, how many different functions f : $k^n \longrightarrow k^n$ result from changing π in the data for $\mathfrak{F} = \mathfrak{F}(Y, \{f_i\}, \pi)$. The answer is given as a sharp upper bound in terms of the number of acyclic orientations of the dependency graph Y [2, Sect. 2.2]. (The number of acyclic orientations of a graph is closely related to its chromatic number.)

   In summary, SDS were created in order to provide a mathematical foundation for large-scale interaction-based computer simulations. The goal is to create mathematical specifications of simulations whose properties can guide the design of simulations, and can aid in the analysis of their dynamics. Results obtained focus on the effect of certain structural components of an SDS on its dynamics, mainly the update schedule. The complexity of their proofs make it clear that the effect of the update schedule of a sequentially updated simulation is very subtle and difficult to understand.

5 Mathematical Results

As we mentioned already, an important problem is the determination of the dynamics of a finite dynamical system $f : R^n \longrightarrow R^n$ from the structure of the transition function alone. Observe that f can be described in terms of its coordinate functions:

$$f = (f_1, \ldots, f_n),$$

with $f_i : k^n \longrightarrow k$. It is well-known [31] that each f_i can be represented as a multivariate polynomial function, that is, an element in the polynomial ring $k[x_1, \ldots, x_n]$. Furthermore, this polynomial can be chosen uniquely so that every variable appears to a power less than the number of elements in the field. In particular, if k is the field with two elements, then this result implies that every Boolean function can be described as a square-free polynomial function with binary coefficients. Thus, the set of all functions $f_i : k^n \longrightarrow k$ over the field k with two elements is in bijection with the quotient ring $k[x_1, \ldots, x_n]/\langle x_j^2 - x_j, j = 1, \ldots, n \rangle$. This allows the application of tools from computational commutative algebra, as demonstrated in [27].

Systems of polynomials generated by monomials have proven to play a very special role in the theory and are amenable to characterization through combinatorial invariants of their exponent vectors. It is natural therefore to study *monomial dynamical systems* over finite fields, that is, systems where the local functions f_i are monomials. An example of a monomial system is

$$f = (x_1 x_2, x_3, x_1 x_3) : k^3 \longrightarrow k^3.$$

One problem that has been studied in this context is the characterization of all monomial systems that have only fixed points as their limit cycles, so called *fixed-point systems*. The problem is motivated by the use of polynomial models to describe biochemical networks. The two papers [14, 13] contain necessary and sufficient conditions for a monomial system over a finite field to be a fixed point system. This result was proved first in [14] for *Boolean monomial systems* (here, the field has two elements, $k = \mathbb{F}_2$). This class includes, in particular, all systems whose functions are constructed using only the logical AND operator. To describe the main result of this paper, we introduce a slightly modified definition of the dependency graph of a finite dynamical system.

Definition 5.1. *Let*

$$f = (f_1, \ldots, f_n) : k^n \rightarrow k^n$$

be a dynamical system, described in terms of its coordinate functions. Then each f_i can be assumed to be in the polynomial ring $k[x_1, \ldots, x_n]$. The dependency graph of f is the directed graph on the vertex set $\{1, \ldots, n, \epsilon\}$. For $i, j \in \{1, \ldots, n\}$, there is a directed edge $i \rightarrow j$ if and only if x_j appears in f_i. There is a directed edge $i \rightarrow \epsilon$ if $f_i = 0$.

A directed graph G is *strongly connected* if there is a directed path from any vertex to any other vertex. For a strongly connected directed graph G we can define a numerical invariant that reflects its loop structure. Let $v \in G$ be a vertex, and consider the set of all directed loops at v. For any pair of loops we can form the absolute value of the difference of their lengths, that is the difference of the respective numbers of edges in the loops. The minimum of all nonzero differences is called the *loop number* of G. Alternatively, it can be defined as the greatest common divisor of the lengths of all directed loops based at v. It is straightforward to show that this number does not depend on the choice of v. One of the main results in [14] is the following.

Theorem 5.2. *Let G be the dependency graph of the Boolean monomial system $f : k^n \longrightarrow k^n$. Then the following are equivalent.*

1. *The system f is a fixed point system.*
2. *One of the following three conditions holds for each vertex $v \in G$:*
   a) the strongly connected component of G containing v has loop number 1;
   b) there is a walk in G from v to ϵ;
   c) there is no walk of length greater than or equal to 1 from v to v.

The paper contains a polynomial time algorithm to compute the strongly connected components of the dependency graph and their loop numbers.

The paper also introduces a "glueing operation" that joins monomial systems to form larger ones, and we show that if one glues two fixed point systems together, then one obtains again a fixed point system, see [14].

In the second paper [13] fixed point monomial systems over arbitrary finite fields are studied, that is, systems of the form $f = (f_1, \ldots, f_n) : \mathbb{F}_q^n \longrightarrow \mathbb{F}_q^n$, where $f_i = x_1^{\alpha_{i1}} \cdots x_n^{\alpha_{in}}$, and $0 \leq \alpha_{ij} < q$. One can define two new systems from f.

Definition 5.3. *Consider the system f above. Define the following two systems. Let*

- *$g = (g_1, \ldots, g_n) : \mathbb{F}_2^n \longrightarrow \mathbb{F}_2^n$, be the Boolean monomial system, where $g_i(a) = f_i(a) \bmod 2$, and*
- *$h = (h_1, \ldots, h_n) : \mathbb{Z}_{q-1}^n \longrightarrow \mathbb{Z}_{q-1}^n$, be the linear system, where $h_i = \alpha_{i1}x_1 + \cdots + \alpha_{in}x_n$.*

We proved the following theorem in [13].

Theorem 5.4. *The system f is a fixed point system if and only if g and h are fixed point systems.*

To see whether the Boolean system g is a fixed point system we use the theorem above from [14]. Notice that, since the system h is over the ring $\mathbb{Z}/(q-1)\mathbb{Z}$ which, in general, is not a field, we can NOT use the results from [18, 21] to find out if h is a fixed point system. In [13] this question is studied and necessary and sufficient conditions are given for h to be a fixed point system.

Using the Chinese Reminder Theorem, it follows that, if $r = p_1^{\gamma_1} \cdots p_t^{\gamma_t}$, then

$$\mathbb{Z}/r\mathbb{Z} \cong \mathbb{Z}/p_1^{\gamma_1}\mathbb{Z} \times \cdots \times \mathbb{Z}/p_t^{\gamma_t}\mathbb{Z}.$$

Now let $f = (f_1, \ldots, f_n) : (\mathbb{Z}/r\mathbb{Z})^n \longrightarrow (\mathbb{Z}/r\mathbb{Z})^n$ be a linear system over $\mathbb{Z}/r\mathbb{Z}$. For each $i = 1, \ldots, t$, define the linear system $h_{p_i} : (\mathbb{Z}/p_i^{\gamma_i}\mathbb{Z})^n \longrightarrow (\mathbb{Z}/p_i^{\gamma_i}\mathbb{Z})^n$ such that $h_{p_i}(a) = [f(a)]_{p_i^{\gamma_i}}$, where $[-]_\mu$ is the vector of remainders after dividing each coordinate by μ.

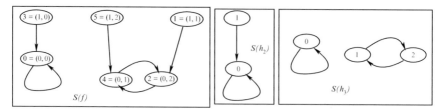

Fig. 2. The phase spaces of f, h_2 and h_3.

Theorem 5.5. *The state space of f is isomorphic to the product of the state spaces of $h_{p_1}, \ldots, h_{p_t}$. That is $\mathcal{S}(f) = \mathcal{S}(h_{p_1}) \otimes \cdots \otimes \mathcal{S}(h_{p_t})$.*

Example 3. Let $f : \mathbb{Z}/6\mathbb{Z} \longrightarrow \mathbb{Z}/6\mathbb{Z}$ be given by $f(x) = 2x$. Then $h_2 : \mathbb{Z}/2\mathbb{Z} \longrightarrow \mathbb{Z}/2\mathbb{Z}$ is given by $h_2(x) \equiv 0$, and $h_3 : \mathbb{Z}/3\mathbb{Z} \longrightarrow \mathbb{Z}/3\mathbb{Z}$ is given by $h_3(x) = 2x$. Figure 2 shows that $\mathcal{S}(f) = \mathcal{S}(h_2) \otimes \mathcal{S}(h_3)$. Therefore, it is sufficient to study systems only over the ring $\mathbb{Z}/p^\gamma\mathbb{Z}$.

In [13], the following theorem is proved.

Theorem 5.6. *Let $f : (\mathbb{Z}/p^\gamma\mathbb{Z})^n \longrightarrow (\mathbb{Z}/p^\gamma\mathbb{Z})^n$ be a linear map, and let g be the projection map of f on $\mathbb{Z}/p$. That is $g = (g_1, \ldots, g_n) : (\mathbb{Z}/p)^n \longrightarrow (\mathbb{Z}/p)^n$, where $g_i = f_i \mod p$. Then the phase space of g is isomorphic to a subgraph of the phase space of f.*

In particular, the theorem above implies that if g is not a fixed point system, then f is not a fixed point system.

As with other classes of mathematical objects it is natural to study transformations between them. For finite dynamical systems such transformations play a very important role. For instance, a transformation between models of a complex system could represent a dimensional reduction of models or the simulation of one model by another. In [25, 26] the notion of *sequential dynamical system* was generalized to allow for a richer theory, and the notion of a transformation, or morphism, of sequential dynamical systems was defined. The definition is quite complicated, largely due to the subtleties introduced by the presence of an update schedule (which was generalized to be a partially ordered set rather than a permutation). (The next section contains a description of transformations in the context of parallel-update polynomial systems, which is substantially simpler.) This definition is reasonable since the composition of transformations is again a transformation and, most importantly, that a transformation of SDS induces a transformation of the corresponding phase spaces.

6 An Open Problem

An important motivation for studying transformations of finite dynamical systems is to develop a mathematical process by which one can amalgamate systems along common subsystems, in a way that allows some control over the

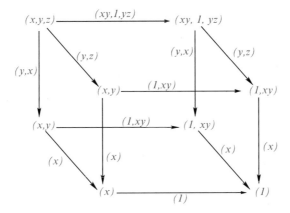

Fig. 3. The system f as a fiber product of g_1 and g_2 over h.

resulting dynamics. We give a simple example in terms of monomial systems over $k = \mathbb{F}_2$. Consider the system in three variables

$$f = (xy, 1, yz) : k^3 \longrightarrow k^3.$$

That is, $f(x, y, z) = (xy, 1, yz)$. We can construct this 3-dimensional system by amalgamating two copies of the 2-dimensional system $g(x, y) = (1, xy)$ over the 1-dimensional system $h(x) = (1)$, see Fig. 3. We denote this decomposition as

$$(xy, 1, yz) = (1, xy) \times_{(1)} (1, xy).$$

We can think of this construction as a fiber product of systems over a common subsystem. In fact, it is very tedious but straightforward to show that f satisfies the appropriate universal property that characterizes it as a fiber product.

The phase space of f is the *amalgamation* of the phase spaces of $g_1 : (\mathbb{Z}/2\mathbb{Z})^2 \longrightarrow (\mathbb{Z}/2\mathbb{Z})^2$ and $g_2 : (\mathbb{Z}/2\mathbb{Z})^2 \longrightarrow (\mathbb{Z}/2\mathbb{Z})^2$ over the common phase space of $h : \mathbb{Z}/2\mathbb{Z} \longrightarrow \mathbb{Z}/2\mathbb{Z}$, where $g_1(y, x) = (1, xy)$, $g_2(y, z) = (1, yz)$, and $h(y) = 1$.

It is clear that $\mathcal{S}(g_1) = \mathcal{S}(g_2)$, and the subgraph of $\mathcal{S}(g_1 \times g_2)$ on the vertex set (y, x, y, z) is isomorphic to the phase space of f, see Fig. 4.

It is an open question what happens in general. That is, when we carry out this fiber product construction, how can one identify the result with a finite dynamical system? Fiber products of this type play two important roles. On the one hand, they allow the decomposition of systems into systems of lower dimension, thus making their analysis easier. On the other hand, they can be used as an important design principle for systems by building them up from smaller, overlapping "local" systems.

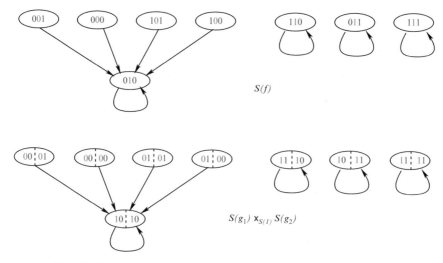

Fig. 4. Phase space of f is isomorphic to a subgraph of $\mathcal{S}(g_1 \times g_2)$.

7 Conclusion

In this paper we have introduced the class of dynamical systems over finite fields as a mathematical framework for interaction-based computer simulation. It is a natural framework that allows the development of a body of mathematical results important for applications. However, finite dynamical systems are of mathematical interest in their own right and deserve to be studied more extensively. A very useful tool for the exploration of a variety of questions about them is the simulation and visualization program *Discrete Visual Dynamics (DVD)* which can be used as a web-based tool or downloaded [23]. It uses the software package *GraphViz* [17] for the visualization part. *DVD* takes as input a ring $\mathbb{Z}/r\mathbb{Z}$ and a system of polynomial functions. For $r = 2$, the user can also input a system of Boolean functions, which the software translates into polynomial functions. It then computes the complete phase space and visualizes it, for small systems. For very large systems, the user can specify a particular initialization, and *DVD* computes the transient and limit cycle starting at that initialization. The user has the choice to update in parallel or sequentially.

Acknowledgments

The authors were supported partially by NSF Grant DMS-0511441. The second author was supported partially by NIH Grant R01 GM068947-01, a joint computational biology initiative between NIH and NSF. The authors thanks Bodo Pareigis for the example in Sect. 6.

References

1. A. Asachenkov, G. Marchuk, R. Mohler, and S. Zuev, *Disease dynamics*, Birkhäuser, Boston, 1994.
2. C. Barrett, H. Mortveit, and C. Reidys, *Elements of a theory of computer simulation ii: Sequential dynamical systems*, Appl. Math. Comput., 107 (2000), pp. 121–136.
3. C. Barrett, H. Mortveit, and C. Reidys, *Elements of a theory of computer simulation iii: Equivalence of sds*, Appl. Math. Comput., 122 (2001), pp. 325–340.
4. C. Barrett, H. Mortveit, and C. Reidys, *Elements of a theory of computer simulation iv. sequential dynamical systems: fixed points, invertibility and equivalence*, Appl. Math. Comput., 134 (2003), pp. 153–171.
5. C. Barrett and C. Reidys, *Elements of a theory of computer simulation i: Sequential ca over random graphs*, Appl. Math. Comput., 98 (1999), pp. 241–259.
6. M. Bernaschi and F. Castiglione, *Selection of escape mutants from immune recognition during HIV infection*, Immunol. Cell Biol., 80 (2002), pp. 307–313.
7. M. Bernaschi, S. Succi, and F. Castiglione, *Large-scale cellular automata simulations of the immune system response*, Phys. Rev. E, 61 (2000).
8. F. Castiglione and Z. Agur, *Analyzing hypersensitivity to chemotherapy in a cellular automata model of the immune system*, in Cancer modeling and simulation, L. Preziosi, ed., Chapman and Hall/CRC Press, London, 2003.
9. F. Castiglione, K. Duca, A. Jarrah, R. Laubenbacher, D. Hochberg, and D. Thorley-Lawson, *Simulating Epstein–Barr virus infection with C-ImmSim*, Bioinformatics, 23 (2007), pp. 1371–1377.
10. F. Celada and P. Seiden, *A computer model of cellular interactions in the immune syste*, Immunol. Today, 13 (1992), pp. 56–62.
11. F. Celada and P. Seiden, *A model for simulating cognate recognition and response in the immune system*, J. Theor. Biol., 158 (1992), pp. 235–270.
12. F. Celada and P. Seiden, *Affinity maturation and hypermutation in a simulation of the humoral immune response*, Eur. J. Immunol., 26 (1996), pp. 1350–1358.
13. O. Colón-Reyes, A. Jarrah, R. Laubenbacher, and B. Sturmfels, *Monomial dynamical systems over finite fields*, (2004). Preprint.
14. O. Colón-Reyes, R. Laubenbacher, and B. Pareigis, *Boolean Monomial Dynamical Systems*, Ann. Combinat., 8 (2004), pp. 425–439.
15. P. Cull, *Linear analysis of switching nets*, Kybernetik, 8 (1971), pp. 31–39.
16. D. Eisenbud, D. Grayson, M. Stillman, and B. Sturmfels, *Computations in Algebraic Geometry with Macaulay2*, Springer Verlag, New York, 2002.
17. J. Ellson and S. North, *Graphviz – graph visualization software*. World Wide Web. http://www.graphviz.org/.
18. B. Elspas, *The theory of autonomous linear sequential networks*, IRE Trans. Circuit Theor., (1959), pp. 45–60.
19. S. Eubank, *Scalable, efficient epidemiological simulation*, in Proc. 2002 ACM Symp. on Applied Computing, Madrid, Spain, 2002, ACM Press, pp. 139–145.
20. J. Farmer, N. Packard, and A. Perelson, *The immune system, adaptation, and machine learning*, Phys. D, 2 (1986), pp. 187–204.
21. A. Hernández-Toledo, *Linear finite dynamical systems*, Commun. Algebra, 33 (2005), pp. 2977–2989.

22. IMA, *Hot topics workshop: Agent based modeling and simulation*, University of Minnesota, November 2003, Institute for Mathematics and its Applications. http://www.ima.umn.edu/talks/workshops/11-3-6.2003.
23. A. Jarrah, R. Laubenbacher, and H. Vastani, *Dvd: Discrete visual dynamics*. World Wide Web. http://dvd.vbi.vt.edu.
24. LANL, *Transims: transportation analysis simulation system.* World Wide Web. http://transims.tsasa.lanl.gov/.
25. R. Laubenbacher and B. Pareigis, *Decomposition and simulation of sequential dynamical systems*, Adv. Appl. Math., 30 (2003), pp. 655–678.
26. R. Laubenbacher and B. Pareigis, *Decomposition and simulation of sequential dynamical systems*, Discrete Appl. Math., (2003).
27. R. Laubenbacher and B. Stigler, *A computational algebra approach to the reverse-engineering of gene regulatory networks*, J. Theor. Biol., 229 (2004), pp. 523–537.
28. M. LeBorgne, A. Benveniste, and P. LeGuernic, *Polynomial dynamical systems over finite fields*, in Algebraic Computing in Control.
29. L. Lidl and G. Mullen, *When does a polynomial over a finite field permute the elements of the field?* Am. Math. Monthly, 95 (1988), pp. 243–246.
30. L. Lidl and G. Mullen, *When does a polynomial over a finite field permute the elements of the field?*, Am. Math. Monthly, 100 (1993), pp. 71–74.
31. R. Lidl and H. Niederreiter, *Finite fields*, Cambridge University Press, New York, 1997.
32. H. Marchand and M. LeBorgne, *On the optimal control of polynomial dynamical systems over $\mathbb{Z}/p\mathbb{Z}$*, in Fourth Workshop on Discrete Event Systems, IEEE, Cagliari, Italy, 1998.
33. H. Marchand and M. LeBorgne, *Partial order control of discrete event systems modeled as polynomial dynamical systems*, in IEEE International Conference on Control Applications, Trieste, Italy, 1998.
34. D. Milligan and M. Wilson, *The behavior of affine boolean sequential networks*, Connect. Sci., 5 (1993), pp. 153–167.
35. D. Morpurgo, R. Serentha, P. Seiden, and F. Celada, *Modelling thymic functions in a cellular automaton*, Int. Immunol., 7 (1995), pp. 505–516.
36. E. Muraille, D. Thieffry, O. Leo, and M. Kaufman, *Toxicity and neuroendocrine regulation of the immune response: a model analysis*, J. Theor. Biol., 183 (1996), pp. 285–305.
37. L. Segel and R.L. Bar-Or, *On the role of feedback in promoting conflicting goals of the adaptive immune system*, J. Immunol., 163 (1999), pp. 1342–1349.
38. E. Shudo and Y. Iwasa, *Inducible defense against pathogens and parasites: optimal choice among multiple options*, J. Theor. Biol., 209 (2001), pp. 233–247.
39. R. Stengel, R. Ghigliazza, and N. Kulkarni, *Optimal enhancement of immune response*, Bioinformatics, 18 (2002), pp. 1227–1235.
40. M. Wilson and D. Milligan, *Cyclic behavior of autonomous synchronous boolean networks: Some theorems and conjectures*, Connect. Sci., 4 (1992), pp. 143–154.
41. S. Wolfram, *Cellular Automata and Complexity: collected papers*, Westview Press, Colorado, 1994.
42. S. Wolfram, O. Martin, and A. Odlyzko, *Algebraic properties of cellular automata*, Comm. Math. Phys., 93 (1984), pp. 219–258.

Part IV

e-Learning

New Pedagogical Models for Instruction in Mathematics

W. Greenberg and M. Williams

Summary. A computer emporium is a large ensemble of computers accessible to students and faculty, where courses and coursework can be addressed. A model of emporium instruction of mathematics, developed at Virginia Tech, will be described. The model was initially created to deal with instruction under the burden of increased class sizes and increasing demands on faculty time. It has turned out to be an effective pedagogical method with particular advantages for instruction in less developed nations. In this article, we will describe the emporium model: its structure, software development and impact on pedagogy.

1 Setting

With approximately 20,000 undergraduate students and 5,000 graduate students, Virginia Tech is the largest university in the state of Virginia. Because of its large College of Engineering, with more than 5,000 students, and because of the requirement that every undergraduate student take a mathematics courses, the number of students serviced by the Department of Mathematics is typically in excess of 10,000 in each semester.

The reduction of government support for higher education, which has occurred in Virginia over the past 15 years, has significantly increased faculty teaching-loads. Although the Mathematics faculty numbers about 60 professors and instructors, the burden of teaching so many students has motivated the Mathematics Department to build a Mathematics Emporium. This somewhat whimsical term extends the definition of emporium as an open market place: the Mathematics Emporium should be a place where the market of ideas would be freely exchanged among students, faculty and computers. Despite having initially viewed the project as a response to the need to teach very large numbers of students, we have found that teaching mathematics in an emporium style has a number of advantages for the students over traditional lecture courses.

The Emporium itself contains a large ensemble of computers, located in a now-defunct supermarket building adjacent to campus. A particular style of designing and presenting mathematics courses has been developed both to deliver expository information and to provide quizzes on material the student is expected to have mastered. It is this style of course development that is the topic of this presentation.

At the present time, the Mathematics Emporium is available to every student enrolled at the university in any disciple, and to the faculty of every department and institute. In fact, however, while there is utilization of the Emporium by nearly all departments of the university, the largest segment of its use is by the Mathematics Department. Indeed, at present, three mathematics courses with annual enrollment of approximately 4,500 students are taught entirely at the Emporium (i.e., no classroom component), and half a dozen additional courses handling more than 5,500 students each year have major segments of the course taught at the Emporium. A great number of the remaining courses in mathematics have occasional Emporium assignments.

In this article, we will describe two emporium models: their structure, software development and impact on pedagogy.

2 Program Structure

In the United States, the vast majority of university courses consist of a set of weekly lectures, with weekly or periodic homework. Homework, for example in mathematics courses, consists of a set of problems based on the current lectures, which the student is expected to work out independently on his own time and submit for grading. Then periodically there are written examinations based on a collection of homework sets, and generally a written final examination at the end of the semester, based on all homework sets. The students grade is determined by his performance on the homework sets, the periodic exams and the final exam. (In some courses, especially upper level courses, the only examinations may be a written mid-semester exam and the written final exam.) Oral examinations are much rarer in the United States than in many other countries. Office hours by the faculty member, where students can get help with their course, are scheduled by the faculty member at a frequency determined by him to be adequate.

The chief burden on the faculty member, in addition to the preparation and delivery of the lectures, is in grading: homework, exams and the final examination. With increased class size, many faculty members have considered the burden to be onerous.

The notion of an emporium was first considered as a response to these burdens [4]. However, our experience with several models of instruction indicate that emporium instruction not only relieves the faculty of much of this burden, but is in fact an improvement for the student over the conventional methods of pedagogy used for generations at American universities. We begin with a description of the two models we employ at the mathematics emporium.

A primary distinguishing feature of the independent model is that the students never meet in a classroom. Their introduction to the on-line courses is an orientation session, which exposes them to course resources that include tutoring labs, on-line course web pages, the testing system, on-line videos, etc. Thereafter, subject to unit deadlines, the students have complete autonomy over their schedules.

The course is made up of a sequence of weekly lesson-practice-quiz cycles with periodic exams, all accessible from the Emporium computers. The lessons are given by mildly interactive web page presentations. Individual lesson web pages consist of three overlays: explanation, examples and a challenge problem. The default schedule would require a student to master about ten of these units per week, followed by a quiz.

Additionally, there are five examinations and a final exam. A student may proceed at any pace he wishes as long as he does not fall behind the default schedule. Theoretically, the able and energetic student could finish the entire course in a few weeks, although that is rarely the case.

What distinguishes this model from the typical on-line course is that the software consists of modules related to each lesson and exam. These modules are entirely accessible to the individual professor, and can be modified or changed with great ease even by professors who have little or no prior experience with programming. A further distinction is the powerful test engine (in both models), which will be described below.

In the lecture model we continue to meet the students in a classroom setting several times a week for conventional lectures, covering the material to be mastered for the course. However, in lieu of homework, there is a weekly practice quiz/quiz-for-credit, which can be taken by the student from any Internet site anywhere in the world. In particular, he does not have to appear at the Emporium, as long as he has access to a computer and a reliable Internet connection. Since Virginia Tech requires each entering student to obtain a computer, most students are able to take the quizzes from their apartments or dormitories.

The weekly quiz consists of approximately eight mathematics problems, presented in the format of a multiple choice quiz. Each mathematics problem is designed to test the understanding of a specific concept covered in the lecture, and is presented with, in general, six to twelve possible answers, a number and choice of wrong answers intended to reduce the importance of guessing and to catch many of the more common mistakes. The actual problems are created by a test generator which will be described below. Because the problems are written in code which allows a variety of choices, each individual problem typically has 1,000 or more variants. Therefore, the student is encouraged to practice the quiz as often as he wishes (the practice quiz) before he chooses to take it one time for credit (the quiz-for-credit). Whether it is a practice quiz or a quiz-for-credit, the quiz is graded immediately upon its completion, the student is given his answers and the correct answers, and he may request further information on each problem he has answered incorrectly. This help information can be of a general nature about solving the problem, or can be programmed to refer specifically to the particular answer given by the student. Moreover, it is geared specifically to the individual variant of each problem generated on the specific quiz taken by the student.

One should compare this to homework in the traditional setting, where a single set of problems is assigned, the student in most cases will not know the correct answers until the homework is graded and returned, if the answer is wrong the student is unlikely to get help until he can meet the instructor in an office hour, and, in any case, he has no real opportunity to redo the homework (or, if such opportunity is available, it is very unlikely to be utilized).

It is essential, in understanding (both the models), to appreciate that the practice quizzes are, in our opinion, where nearly all of the "learning" takes place [1, 2, 3]. They replace the practicum, recitation, or homework employed in various nations as follow-ups to the lecture, and, we believe, they are actually a better, more effective product than these other alternatives for the reasons iterated above. Moreover, once the course machinery is set up, there is no effort or time required by the professor.

Approximately every five weeks, the student is required to take an examination which will cover problems from a number of previous quizzes, and at the end of the course a final exam. Unlike the quizzes, the examinations and final must be taken at the Mathematics Emporium, where they are subject to fail-safe proctoring. Although the quizzes (which are not proctored) in total may count only about 15% of the final grade, and the exams (which are proctored) up to 85%, because the student knows that the pool of problems for the quizzes and for the examinations is the same, there is every motivation for the student to practice the quizzes as often as necessary until he is confident he has mastered the material covered by each quiz problem.

While the student is allowed to take the quizzes outside the Emporium, there are considerable advantages for him to go to the Emporium even for the practice quizzes. An important feature of the Emporium is the availability of one-on-one help provided typically within one minute of the request by a support staff versed in all offered subjects [4]. The support staff varies from Full Professor to advanced undergraduates. Just-in-time help is naturally more effective than a general lecture and a much later office hour.

Whether the student chooses to take quizzes at the Emporium or at an outside location, he is able to do this at any time of the day or night (although one-on-one help will not be available at all times). This is because the Mathematics Emporium is open 24 h a day, 7 days a week (not quite 52 weeks a year, since there are short closures during important vacation and inter-session periods), and, of course, the servers which provide Internet access are available at all times. Surprisingly, we have found this at times to present a cultural difficulty in some nations. It appears that the notion of keeping a university facility open day and night seems sometimes to be a difficult adaptation, even in nations where labor costs are exceptionally low (since the emporium site needs to be manned with at least a minimal check-in staff, and in some locations a guard).

For both the independent model and the lecture model, a robust test engine is required. A test engine is a system set up to deliver assessments (practice quizzes, quizzes, exams) to an end user, a student, presenting the practice tests and examinations on demand, grading them, providing the student with access to the results and to correct answers, recording, saving, and sending spreadsheet data to the instructor, etc. Each time a practice quiz, quiz or examination is called up, the student sees a different set of problems. Consequently, the engine has to be capable of delivering large volumes. In fact, our test engine generates more than 10 million mathematics problems each year. We believe it to be the largest such engine in existence. Our typical courses have 1,500 students. For each quiz, each student takes on average about ten practice quizzes. The typical quiz has eight problems, with exams having from 15 to 30 problems. Since each mathematics problem must be individually generated for each student each time the problem is accessed, it is evident that the volume is quite massive.

In addition, these volumes must be delivered in a fashion which is insensitive to the users platform, and indifferent to the network connection – dial up, T1 link, Emporium connection. We are not aware of any other test engines which are currently capable of delivering these volumes.

Because of the very high level of usage, a much richer set of problems is required for this setup than is usually found in a test database. One needs to be sure that students see a different problem each time a problem is accessed. This dictates that the model of problem database that relies on static lists of problems is insufficient, and that problems must be generated from programs that introduce sufficient variety to distinct problems.

To be platform independent at the student end requires use of a common interface that is available ubiquitously. The obvious choice is the web browser.

On the server side, we constrained ourselves to open standards and modular design. In this way as course components are improved, and the delivery system upgraded, these changes can be easily implemented. A paramount consideration at all times was to ensure that the system was scalable to large transaction volumes.

3 Hardware Requirements

These simple design requirements imply that the server consist of a standard web server (we use Apache) and a page delivery service capable of very high volumes of transactions. Apache is public domain, which means it is standards-compliant and free to use. Apache has proven to be extremely robust. In order to retain sufficient programming control to attain these transaction volumes, the most practical environment is java server-pages (JSP). JSP permits a maximum of computational speed and processing in a rich environment.

With this architecture, a high-powered machine as server is not necessary. We utilize a pair of aging Sun 3000 servers with 4G of memory and RAID storage. Though they are more than 7 years old, and we have not found the need to upgrade. If we were to replace them at this time, we would use off-the-shelf LINUX systems. One of the machines does the web serving, while the other accommodates the database. Except when we do builds, we rarely see usage exceed 10% of capacity on any machine.

We use Oracle as the database management system. This relational system holds and links data on the students (ID numbers, major discipline, course registrations, email addresses, etc.), on the raw assessments and on the completed exams and quizzes. An instructors gradebook server regularly taps this database to update this information in gradebook for each student, and passes relevant information to the faculty.

New tests and exams are supplied to the main server through a separate engine that uses Mathematica to generate the individual tests on-demand. As a practical matter we use the test engine to establish caches of new exams in every category. This prevents slowdowns or gaps in service should the test engine server be unavailable, or should the demand for specific quizzes exceed available generation resources.

Each of these engines currently resides on one of the two Sun systems or one of the handful of Apple XServes we have available. Because of our modular design, there are no constraints, other than adequate communications links, on where the individual services are located. In this way we gain easy redundancy, guarding against downtime.

At the student or instructor end, as mentioned above, any personal computer, which can support a modern browser, independent of platform, can access the system.

4 Operations

In order to gain economies of scale from our program, we maximized the amount of sharing of resources. Therefore, we established the Math Emporium in a $7,000\,\mathrm{m}^2$ former department store, whose interior is undivided. It is populated with 550 computer workstations on hexagonal pods of six stations evenly spaced throughout this large open area. The students, independent of which course they are enrolled in, come to the Emporium to use the computer systems and to avail themselves of the mathematics instructional support. The instructional support staff is available approximately 14 h per day, although the Emporium itself is open 24 h per day.

Instructional support is provided by a cadre of senior undergraduate students, graduate students and some regular instructional faculty, who are available to roam throughout the floor and respond to any student in any course who has signaled for help.

This support is a key component of retaining human interactions with the students, especially in the independent model. The support consists of personalized just-in-time help for each individual student as he encounters an issue in solving a mathematics problem, and should be contrasted with the traditional lecture delivery system of one teacher many students, which can neither adapt to the different speeds of comprehension of each student, nor deal with individual problems each student encounters in trying to solve a specific problem [5].

At the present time, for reasons of security, quizzes and examinations for credit must be taken at the Mathematics Emporium itself. However, we reiterate that the practice quizzes and examinations themselves are accessible to any students enrolled in the corresponding courses at any time and at any place, through the Internet. The Emporium is of course available to any student who does not otherwise have access to a computer, or who chooses to study the practice quizzes in an environment where he or she has immediate access to instructional support.

5 Test Problems

For the types of courses we have developed, the practice quiz plays the most fundamental role in the learning process, and also serves as the pool from which all exam problems are chosen. This introduces a burden on the problem writer of generating questions of variety and repeatability.

A key feature of the test engine is that it has been made accessible to all of the faculty. We have created an environment where any faculty member can create his own course on-line, with virtually no prior experience in programming. Indeed, we have a number of faculty who are currently involved in placing their courses in the Emporium and whose prior use of computers was restricted to receiving and sending email!

Mathematica, due to its versatility and sophistication, has been chosen as the environment for creating these tests. We have created a series of utilities that simplify the task. The utilities provide a series of commands, functions and shortcuts to compile and display the quiz problems. The instructor needs only to insert the mathematical specifics for each quiz problem into pre-constructed quiz modules, placing, for example, the correct answer in its appropriate location, a series of wrong answers in a corresponding appropriate location, the statement of the problem likewise in its location, etc.

We must be careful not to oversimplify the burden on the instructor. The problems have to be created in a manner that is programmable and allows for distinct versions of the same problem. Moreover, there are many nuances to assuring that duplicate answers are not created, that a wrong answer might be correct for extraneous reasons, that patterns are not unconsciously established among the set of answers which will tip off students as to the correct choice, etc. Finally, there is the overriding pedagogical need to assure that the quiz problems teach and test the desired mathematical principles.

Because of the modular construction of the utilities, and their assistance in programming the problems, we have found by experience that mathematics faculty with no prior experience in computer programming can learn to effectively start making quiz problems after one afternoon of preparatory instruction and a selection of model problems.

6 An Example Problem

Here we present an example of a Mathematica program that generates a simple geometry problem.

```
ProblemInfo["Q2.09"]={1,"Tests understanding of collision points"};
makeproblem["Q2.09",str_]:=Module[{a,b,c,f,e,a0,a1,a2,b1,b2,b3,c1,
    c2,c3,p,q,r,px,qy,qy2,x1,y1,x2,y2, test1},
ClearAll[t,x,y,s];
test1=True;
While[test1,
{a,b}=ChooseRandom[Range[1,5],2];
{c}=ChooseRandom[Range[1,3],1];
{f}=ChooseRandom[Range[0,1],1];
{a0,a1,a2}=ChooseRandom[Range[1,5],3];
{b1,b2,b3}=ChooseRandom[Range[1,5],3];
{c1,c2,c3}=ChooseRandom[Range[1,5],3];
e=c+f;
p[t_]=a2 t^2+a1 t+a0;
q[t_]=-b3 t^3+b2 t^2+ b1 t;
r[t_]=c3 t^3-c2 t^2+ c1 t;
px[t_]=Simplify[(t-a) (t-b)+p[t]];
qy[t_]=Simplify[q[t]-q[a]+c q[a]];
qy2[t_]=Simplify[r[t]-r[a]+e q[a]];
{x1[t_],y1[t_]}={px[t],qy[t]};
{x2[s_],y2[s_]}={p[s],qy2[s]};
test1=Length[Union[{{x1[a],y1[a]}, {x1[b],Abs[y1[b]]}, {x1[a+1],
Abs[y1[b-1]]},
{x1[b+1],Abs[y1[a-1]]}, {x1[a-3],Abs[y1[b+3]]}, {x1[b-1], Abs[y1[a+1]]}}]]<6
||{{f==1}&&{qy(a)==qy2(a)}}||{{f==1}&&{qy(b)==qy2(b)}}||{{f==1}&
    &{q(a)==0}};];
ProblemText[str]={{"Problem",{StringJoin["Suppose a particle travels on
the path given by x = ", tF[ x1[t]],", y = ", tF[ y1[t]], " at time ",tF[t],", and
another particle travels on the path given by x = ",tF[ x2[s]],", y = ", tF[
y2[s]]"\ n at time ",tF[s],". Do they collide, and, if so, at what point?" ]}},
```

{"Right",{If[f<1,StringJoin["Collide at (",tF[x1[a]],", ",tF[y1[a]],") "],"Do not collide."] }},
{"Wrong",{If[f>0,StringJoin["Collide at (",tF[x1[a]],", ",tF[y1[a]],") "]," Do not collide."] }},
{"Wrong",{StringJoin["Collide at (",tF[x1[b]],", ",tF[Abs[y1[b]]],") "]}},
{"Wrong",{StringJoin["Collide at (",tF[x1[a+1]],", ",tF[Abs[y1[b-1]]],") "]}},
{"Wrong",{StringJoin["Collide at (",tF[x1[b+1]],", ",tF[Abs[y1[a-1]]],") "]}},
{"Wrong",{StringJoin["Collide at (",tF[x1[a-3]],", ",tF[Abs[y1[b+3]]],") "]}},
{"Wrong",{StringJoin["Collide at (",tF[x1[b-1]],", ",tF[Abs[y1[a+1]]],") "]}},
{"Comment",{ "Since collisions occur at the same time, change both parameters to t. Then set the xs and ys equal. Solve whichever is easier for t. The other equation will give either an identity or a contradiction." }} };];

One iteration of the compiled problem:

Suppose a particle travels on the path given by

$$x = 5t^2 - 4t + 14, \qquad y = -t^3 + 3t^2 + 4t + 24$$

at time t, and another particle travels on the path given by

$$x = 4s^2 + 3s + 2, \qquad y = 2s^3 - 3s^2 + 4s - 3$$

at time s. Do they collide, and, if so, at what point?

Collide at (47,36)
Do not collide
Collide at (78,24)
Collide at (78,36)
Collide at (119,36)
Collide at (14,144)
Collide at (47,24)

While the program may appear daunting to the novice, a template and a multitude of exemplars from current production programs substantially ease the difficulty of writing new problem programs. The experienced Mathematica user will notice many functions that are not part of the Mathematica package. These add-ons simplify the recurrent tasks. For example, makeproblem [ ] := Module[, and ProblemText[str] = "Problem", are part of the templates provided to instructional faculty wishing to author quizzes. Our own experience is that new faculty users with no prior programming experience can write effective problems with a half day of instruction, a number of example problems, and occasional assistance in debugging.

For this problem, the Comment (which the student receives only if he requests help from the quiz) is general to any variant of the problem and any answer. However, this help aid can also give different responses for different wrong answers, can refer to the particular values of parameters in the problem, and can give references to precise locations in course texts or in online course material.

It should be easy to see that changing an existing course into another language is, with the assistance of a translator whose mathematics knowledge need extend only to translating a few common mathematics words, a rather trivial task. Indeed, to change this problem into Bambara, say, one need only replace the words surrounded by quotation marks in the line beginning with ProblemText and the line beginning with Comment with their Bambara translation. In that manner, an entire course should be translatable in no more than a day.

7 Economics and Conclusions

The conversion of traditional classroom-based course offerings to the models described in the preceding has reduced costs by an average of 75%. This is due to the fact that the demand for one-on-one help (sometimes provided by advanced undergraduates) is dwarfed by the costs of more senior lecturers in so many classrooms. Furthermore, when the performance in later courses of students who have completed courses under the new models are compared to those who learned the material under the traditional model, we find that the former significantly out-perform the latter. We believe this is due to increased time-on-task and an increased independence.

More important, the Math Emporium has permitted us to make a discontinuous change to the methods of operation of our course offerings. We have used this unique asset to transform traditional classroom-based courses into more efficient technology-based learning programs. The structure of the program is very different from the conventional one, having a new set of expectations and motivations. To summarize, the results include:

Students are more self-reliant
Students budget their time more effectively and satisfyingly
We enjoy a substantial cost savings
Students demonstrate increased proficiency in down-stream course work
We have learned much about distance learning
We have a new fully automated testing system which broadens the impact of testing
We have a stream-lined process which takes source material into structured web resources effectively and quickly
We have shown the models we are using scales well (e.g., savings increase) to larger enrollment courses

We continue to aggressively pursue further course conversions.

References

1. F. Quinn and M. Williams, *Lessons from the Emporium 1: Goals and economics* (November 2003), preprint, at www.math.vt.edu/ people/quinn/education.
2. F. Quinn and M. Williams, *Lessons from the Emporium 2: Help for computer-based learning* (November 2003), preprint, at www.math.vt.edu/people/quinn/education.
3. F. Quinn and M. Williams, *Lessons from the Emporium 3: Testing and course design* (July 2004), preprint, at www.math.vt.edu/ people/quinn/education.
4. M. Williams, *The Math Emporium: The Changing Academy or Changing the Academy*, in Developing Faculty to Use Technology, D. Brown, ed., Anker Publishing, Boston, 2003, pp. 285–287.
5. W. Greenberg and M. Williams, *A Model for Computer-Assisted Lecture Courses in Mathematics*, in Third International Conference on Education and Information Systems Technologies and Applications, vol. 1, F. Malpica, F. Welsch, A. Tremante, J. Lawler, eds., Orlando, International Institute of Informatics and Systemics, 2005, pp. 26–31.

Printing: Krips bv, Meppel, The Netherlands
Binding: Stürtz, Würzburg, Germany